Enhancing Renewable Energy Systems With Generative AI

J. Reyes Juárez Ramírez
https://orcid.org/0000-0002-5825-2433
Universidad Autónoma de Baja California, Mexico

Abhishek Kumar
Chandigarh University, India

Pramod Singh Rathore
https://orcid.org/0000-0002-4917-1500
Manipal University, India

IGI Global
Scientific Publishing
Publishing Tomorrow's Research Today

Vice President of Editorial Melissa Wagner
Director of Acquisitions Mikaela Felty
Director of Book Development Jocelynn Hessler
Production Manager Mike Brehm
Cover Design Jose Rosado

Published in the United States of America by
IGI Global Scientific Publishing
701 East Chocolate Avenue
Hershey, PA, 17033, USA
Tel: 717-533-8845 | Fax: 717-533-7115
Website: https://www.igi-global.com E-mail: cust@igi-global.com

Library of Congress Cataloging-in-Publication Data

LCCN: 2025024370 (CIP Data Pending)
ISBN13: 9798337342726
Isbn13Softcover: 9798337342733
EISBN13: 9798337342740

British Cataloguing in Publication Data
A Cataloguing in Publication record for this book is available from the British Library.

Table of Contents

Detailed Table of Contents

The ongoing digital transformation of Turkey's electricity infrastructure has enabled the widespread collection of high-resolution data within low-voltage distribution networks. This study applies a data-driven methodology to assess energy losses in Bartın, Turkey, using hourly data from 172 substations and 3,416 customer smart meters for the year 2024. Technical losses were estimated based on network parameters, while deviations beyond a 0–3% benchmark were analyzed to identify potential non-technical losses, including metering inaccuracies, undocumented loads, and energy theft. The analysis was implemented in MATLAB, generating visualizations to support decision-making. Results revealed 14 substations with significant non-technical losses, providing actionable insights for targeted field inspections and improving transparency for the regional Distribution System Operator (DSO).

Accurate forecasting of wind and solar energy is vital for grid efficiency and renewable integration. Traditional models often falter due to the nonlinear, uncertain nature of these sources. This chapter investigates Generative Adversarial Networks (GANs) and Diffusion Models as next-gen forecasting tools. These models effectively learn complex temporal-spatial patterns and produce high-resolution, realistic outputs. The chapter details their architectures, multimodal data integration (e.g., satellite images, weather models, ground sensors), and real-time edge deployment. Case studies reveal up to 28% improvement in accuracy, with better uncertainty estimation and climate adaptability. Comparative results with LSTM and ARIMA underline generative models' superiority. The chapter also explores ethical issues, explainability, and emerging trends such as federated learning and climate-aware modeling. Overall, it positions generative AI as a transformative path toward scalable, intelligent, and robust renewable energy forecasting systems.

This chapter explores the innovative use of Generative Adversarial Networks (GANs) for generating synthetic weather patterns to improve renewable energy forecasting. As renewable systems rely heavily on accurate meteorological data, limited historical records—especially for rare events—hinder model performance. We introduce WeatherGAN, a tailored GAN architecture designed to produce realistic, physically consistent weather data by modeling complex spatiotemporal patterns. This synthetic data supplements existing datasets, enabling more robust training of forecasting models. The chapter presents case studies involving solar farms, offshore wind sites, and microgrids, demonstrating improved forecast reliability. Notably, solar irradiance predictions show up to 15% improvement in hourly accuracy. Our findings highlight how GAN-generated data can help bridge data gaps and enhance the operational efficiency of renewable energy systems.

The proliferation of distributed renewable energy systems has reshaped traditional energy markets, introducing challenges in dynamic pricing, demand-supply matching, and market coordination. This chapter proposes a robust framework leveraging Generative Adversarial Networks (GANs) to tackle inefficiencies in decentralized renewable energy trading. Our novel multi-modal GAN synthesizes realistic energy trading scenarios by learning from historical market data, weather conditions, and prosumer behavior. The architecture includes specialized generators for time-series energy production, demand trends, pricing, and meteorological variables, competing with discriminators to ensure realistic outputs. Case studies involving solar-wind microgrids, multi-regional exchanges, and electric vehicle integration demonstrate improved market efficiency, price stability, and renewable utilization. The framework addresses market volatility, transaction complexity, and information asymmetry while ensuring scalability across thousands of trading agents.

The global pursuit of sustainable energy solutions has intensified amid growing concerns over climate change, depleting fossil fuel reserves, and escalating energy demands. Among the array of renewable energy sources, bioenergy—derived from organic materials such as agricultural residues, forest waste, and algae—holds significant promise due to its potential for carbon neutrality and broad resource availabilityThe global pursuit of sustainable energy solutions has intensified amid growing concerns over climate change, depleting fossil fuel reserves, and escalating energy demands. Among the array of renewable energy sources, bioenergy—derived from organic materials such as agricultural residues, forest waste, and algae—holds significant promise due to its potential for carbon neutrality and broad resource availability. However, the design, optimization, and deployment of efficient bioenergy systems remain complex tasks that involveand conversion technologies to logistics, economic feasibility, and environmental impact assessments.

To support the worldwide transition to renewable energy systems, a new approach is needed with computational methods that can deal with the issues of intermittency, the complexity of variability, and the integration of the grid. Generative Artificial Intelligence (GAI) has been a game-changer in creating value in how it can fundamentally design and generate synthetic data, predictive models, and optimize systems. This chapter evaluates the GAI methods for renewable energy systems, which include Generative Adversarial Networks (GANs), Variational Autoencoders (VAEs), diffusion models, and Large Language Models (LLMs). It is organized as: systematic taxonomy of GAI models regarding energy applications; an exhaustive review of GAI-oriented solutions for predictive maintenance, forecasting energy production, grid resilience, and sustainable design of renewable energy systems; empirical case studies of improved performance of solar and wind energy systems; and an overview consideration of the general challenges of data privacy, model explainability, and scaling models to edge deployment.

This chapter explores the strategic potential and future trajectory of generative AI in energy systems. Unlike traditional predictive tools, generative models enable simulation of complex, uncertain, and novel scenarios, critical for addressing climate variability, behavioral dynamics, and infrastructure planning. They support climate-resilient modeling, personalized efficiency interventions, and stress-testing of energy systems under future conditions. However, their rise introduces pressing governance challenges around transparency, equity, and data use. The chapter concludes with strategic recommendations for researchers, utilities, and regulators to guide the responsible and impactful deployment of generative AI in energy transitions.

As the world energy has become a focal point for researchers, policymakers, and businesses alike. Energy, environmental awareness, technological access, and the policies implemented by governments. it is essential to have an in-depth understanding of the behaviors and motivations that drive consumption.Historically, energy consumption behavior has been studied from an economic and psychological perspective, focusing, machine learning, and predictive modeling, the landscape of energy behavior analysis has shifted. AI has opened up new avenues for collecting, processing,imperative to address climate change. As the world faces challenges such as rising energy costs, environmental concerns, and the pressure to transition to renewable energy sources, understanding how individuals, organizations, and industries consume energy has become a focal point for researchers, policymakers, and businesses alike.

Energy consumption analysis is crucial for addressing climate change, efficiency, and sustainable energy transitions. Generative AI leverages deep learning models like GANs and VAEs to simulate, predict, and analyze energy usage across households, businesses, and industries. By generating synthetic data, AI identifies patterns, forecasts peak demand, and evaluates interventions such as smart devices or policy changes. Personalized insights guide consumers in optimizing energy use, while integration with smart home systems enables real-time adjustments. Generative AI also supports renewable energy integration, DER management, and carbon footprint reduction. Despite challenges in data quality, privacy, behavioral complexity, and system integration, these models offer transformative potential for efficient, sustainable, and adaptive energy management worldwide.

The transformative role of Artificial Intelligence (AI), particularly Generative AI (GenAI), in reshaping global energy systems to be more decentralized, intelligent, resilient, and inclusive. It highlights how AI enables a shift from static, centralized paradigms toward dynamic, adaptive energy networks that integrate renewable sources, smart infrastructure, and user-specific behavior. Generative models—such as GANs, VAEs, and diffusion-based architectures—enhance predictive accuracy, generate synthetic datasets, and simulate complex, uncertain scenarios across household, community, and urban scales. Through case-driven analysis, the chapter demonstrates how AI improves grid flexibility, supports equitable access, enables scenario planning for climate resilience, and informs real-time decision-making for policy and infrastructure design. It emphasizes the need for ethical, transparent, and community-engaged AI systems to prevent algorithmic bias and foster energy justice.

This study aims to analyze the implications of the financial energy literacy (FEL) with sustainability as interdisciplinary concepts. The holistic approach of sustainability of financial energy literacy (SFEL) is critical in an environment where energy costs are rising and environmental sustainability is becoming increasingly urgent. The method employed is the meta-analytical and meta-cognitive analysis based on a conceptual, theoretical and empirical literature review published in the last five years. It is concluded that financial energy literacy is a crucial element for achieving sustainable development goals, as it empowers individuals to make informed decisions that benefit both their financial well-being and the environment of individuals, families, organizations and communities.

This research study examines the current status of green energy-specific financial literacy in Latin America, exploring the relationship between financial knowledge and investment decision-making in renewable energy. The study indicates that while Latin America has made tremendous strides in the uptake of renewable energy, with the region increasing its renewable capacity by 51% between 2015 and 2022, there remains a significant gap in financial literacy specific to green energy investments. The findings project that enhancing green energy financial literacy can significantly accelerate the region's transition to sustainable energy systems and assist in reaching the net-zero goals of 15 Latin American countries. The report concludes with a recommendation to policy makers, financial institutions, and educational institutions to introduce tailored financial literacy programs that address green energy investment knowledge gaps in the region.

Large-scale, interconnected power systems are susceptible to blackouts, which can result in significant financial losses, production cycle disruptions, and equipment damage at power plants. A forced islanding plan with an appropriate restoration process will work well to prevent these blackouts. This paper proposes a mechanism that divides the system into predetermined islands and determines the risk during an outage. In the event that there are unstable islands, an under-voltage load shedding plan is implemented to restore system stability. Furthermore, a restoration protocol has been developed. The suggested work is done using a six bus system as an example. This strategy can also be used in smart grid settings to raise the dependability and efficiency of the electricity system.

This study introduces a real-time vehicle monitoring system for fuel stations that integrates IoT and object detection technologies to improve customer fueling experiences. The system provides users with up-to-date information on wait times across various fuel stations, allowing them to make better-informed decisions. A network of sensors including those for vehicle detection and fuel monitoring collects data on queue lengths and fuel dispensing volumes. This data is analyzed using machine learning algorithms to forecast future wait times. These predictions are then shared with users, enabling them to choose the most efficient fueling station. This advanced system sets a new standard in the fuel service industry by combining

Preface

INTRODUCTION

The transition to renewable energy is no longer a speculative endeavor — it is a global imperative. As climate change accelerates and energy demand surges, our existing infrastructures face increasing pressure to adapt, evolve, and perform sustainably. Renewable energy systems — spanning solar, wind, hydro, and bioenergy — are at the forefront of this transformation. Yet, the path forward remains fraught with challenges: intermittent supply, fluctuating resource availability, complex grid dynamics, and system-level inefficiencies.

It is within this context that we present *Enhancing Renewable Energy Systems With Generative AI*. This book is born from a recognition that traditional tools alone are insufficient to meet the scale, speed, and complexity of current energy demands. What is needed now is not just innovation — but *transformative innovation*. Generative Artificial Intelligence (Generative AI) represents precisely that.

While Generative AI is often associated with creative domains such as art, music, and language generation, its potential within engineering and energy systems is only beginning to be fully realized. Generative models — including Generative Adversarial Networks (GANs), Variational Autoencoders (VAEs), Diffusion Models, and Transformer-based architectures — are redefining how we model, forecast, optimize, and manage renewable energy systems. From simulating future energy scenarios under climate change to enabling real-time intelligent decision-making, Generative AI is rapidly becoming a critical enabler of sustainable energy transformation.

This edited volume explores this exciting convergence. It showcases theoretical advancements, practical applications, and emerging research that leverage generative models to enhance the reliability, efficiency, and adaptability of renewable energy systems.

Readers will encounter chapters addressing topics such as:

- Data augmentation for rare meteorological conditions using GANs
- Generative modeling for microgrid optimization
- Simulation of renewable energy resource maps under varying climate scenarios
- Intelligent fault detection and predictive maintenance
- Energy consumption pattern synthesis for smart cities
- Real-time decision-making for energy trading and dynamic load balancing

Beyond applications, the book also delves into the broader implications of Generative AI in the energy sector — including ethical considerations, data governance, scalability, and integration into policy and infrastructure.

Our objective with this book is threefold:

1. To introduce and contextualize Generative AI within the domain of renewable energy;
2. To demonstrate its capabilities through advanced use cases and real-world scenarios;
3. To stimulate interdisciplinary collaboration between energy professionals, AI researchers, policymakers, and innovators.

We envision this book as a resource for a diverse audience:

- Researchers and academics seeking to bridge AI and sustainability;
- Renewable energy engineers and designers striving to optimize systems in an increasingly complex landscape;
- Data scientists and AI developers expanding into energy applications;
- Policy makers and government agencies developing data-driven strategies for long-term climate and energy planning;
- And startups and entrepreneurs pushing the frontiers of green technology.

As editors, we are proud to bring together voices from multiple disciplines and sectors, united by a common purpose: accelerating the global shift to clean energy through cutting-edge technological innovation.

We hope this book will serve not only as a reference but as a catalyst — for new ideas, for collaborative research, and for the reimagination of what is possible at the intersection of artificial intelligence and renewable energy.

ORGANIZATION OF THE BOOK

This volume brings together a diverse collection of research that illustrates the powerful synergy between renewable energy technologies and generative artificial intelligence (AI). Each chapter addresses a distinct challenge or opportunity in the clean energy landscape, offering innovative methodologies and forward-looking insights. Together, they form a comprehensive narrative of how generative AI can reimagine the way we generate, manage, and optimize sustainable energy.

We begin in **Chapter 1** with a grounded, real-world case study: *Data-Driven Analysis of Non-Technical Losses in Low-Voltage Networks: A Case Study of Bartin, Turkey*. Here, the authors apply a data-driven methodology to high-resolution smart meter data from Bartin's electricity distribution network. Using MATLAB-based visual analytics, they identify and isolate potential non-technical losses — such as metering errors, unauthorized consumption, and undocumented loads — that extend beyond acceptable benchmarks. This study underscores the power of data analytics to enhance operational transparency and guide field inspections, offering immediate practical value for distribution system operators.

Chapter 2, *Forecasting Wind and Solar Energy with GANs and Diffusion Models*, explores the frontier of renewable energy prediction. Addressing the limitations of conventional forecasting techniques, the authors demonstrate how GANs and diffusion models capture complex spatiotemporal dynamics across multimodal data sources. Real-world case studies reveal a notable accuracy improvement, especially in uncertain or variable conditions. Ethical considerations, explainability, and emerging paradigms such as federated learning are also examined, positioning generative models as scalable solutions for climate-resilient forecasting.

In **Chapter 3**, *Synthetic Weather Intelligence Leveraging Generative Adversarial Networks*, the authors confront a critical bottleneck in energy forecasting — the lack of sufficient, high-quality weather data. Introducing WeatherGAN, a custom GAN architecture, this chapter details the generation of realistic synthetic meteorological data for use in solar, wind, and microgrid systems. With improved predictive accuracy and expanded training datasets, the model offers a compelling approach to mitigating data scarcity and enhancing operational planning.

Chapter 4, *Generative Adversarial Networks for Dynamic Renewable Energy Trading and Price Optimization in Decentralized Markets*, ventures into the economic dimension of clean energy systems. As distributed generation transforms the energy marketplace, the chapter proposes a GAN-based framework for modeling and optimizing dynamic trading behaviors. With specialized generators simulating prosumer activity, weather variability, and market fluctuations, the system enhances price stability, transaction efficiency, and energy balancing — paving the way for intelligent and decentralized energy markets.

Chapter 5, *Generative AI for Bioenergy System Design*, turns our attention to the design and optimization of bioenergy systems — an often underrepresented yet vital component of the renewable mix. Bioenergy's complexity stems from its heterogeneous inputs and varied processing pathways. This chapter explores how generative AI can streamline system design by modeling feedstock availability, conversion technologies, logistics, and lifecycle performance. In doing so, it supports sustainable, carbon-neutral energy planning across agricultural, forestry, and waste-management sectors.

Chapter 6, *Generative AI for Intelligent, Resilient, and Sustainable Renewable Energy Systems: From Data to Deployment* serves as a capstone, this chapter offers a comprehensive taxonomy of generative AI models — including GANs, VAEs, diffusion models, and Large Language Models — and their diverse applications in the renewable energy sector. From predictive maintenance and grid resilience to synthetic data generation and edge deployment, the chapter integrates empirical case studies with a critical examination of real-world constraints such as data privacy, explainability, and deployment scalability.

In **Chapter 7**, *Modeling and Understanding Energy Use Patterns with Generative AI*, the authors examine how generative models go beyond traditional forecasting to simulate energy consumption under complex, uncertain, and future-oriented scenarios. By modeling human behavior, climate variability, and infrastructure stress responses, this chapter articulates a strategic roadmap for utilities and policymakers aiming to build climate-resilient, adaptive energy systems. It also addresses critical governance issues, including algorithmic transparency, equity in data access, and ethical deployment—laying the groundwork for responsible innovation in AI-driven energy transitions.

Chapter 8, *Generative AI for Energy Consumption Behavior Analysis*, takes a closer look at the evolution of behavioral insights within energy systems. Historically framed by economic or psychological models, energy behavior is now being redefined through AI-enabled analytics. This chapter surveys how generative models allow for richer, more granular analysis of consumption motivations, preferences, and habits—crucial for designing user-centered interventions and behavioral nudges. In a world grappling with rising costs and environmental pressures, understanding the human dimension of energy use becomes not only a technical necessity but also a societal imperative.

Building on this, **Chapter 9**, *Generative AI for Energy Consumption Behavior Analysis: A Study*, provides an applied perspective on how GANs and VAEs can generate synthetic energy usage data to simulate household and industrial demand. These models enable peak load forecasting, assess the impact of smart devices, and support real-time optimization in smart homes. With use cases ranging from carbon footprint reduction to renewable energy integration, the chapter illustrates the profound potential of generative AI in enabling adaptive, intelligent, and personalized energy systems—even while addressing challenges related to privacy, data sparsity, and behavioral variability.

Chapter 10, *Generative AI for Personalized Energy Consumption Behavior Analysis*, expands the discussion from individual users to broader socio-technical systems. Through a combination of case studies and technical analysis, the authors show how generative models contribute to designing equitable and flexible energy networks that reflect real-world variability in user needs, climate risks, and infrastructure capacities. The chapter advocates for AI systems that are not only high-performing but also ethically grounded—incorporating principles of energy justice, transparency, and community engagement into the design of next-generation energy infrastructures.

Chapter 11, *Financial Energy Literacy Sustainability*, offers a fresh and much-needed perspective on the intersection of personal finance, energy consumption, and sustainability. Drawing from recent empirical and theoretical literature, the chapter introduces the concept of Sustainable Financial Energy Literacy (SFEL)—a critical tool for empowering individuals and communities to make informed decisions that align financial well-being with environmental responsibility. As energy prices fluctuate and green technologies proliferate, understanding the financial dimension becomes essential for achieving widespread and equitable adoption of renewable solutions.

This theme continues in **Chapter 12**, *Green Energy-Related Financial Literacy in Latin America*, where the focus shifts to a regional analysis. Despite notable growth in renewable energy capacity across Latin America, a gap persists in public understanding of green energy investment. This chapter explores how targeted financial literacy programs can accelerate the adoption of renewable technologies, reduce risk aversion, and support national net-zero ambitions. It calls for coordinated efforts from policymakers, educators, and financial institutions to equip citizens with the tools they need to engage with sustainable energy markets confidently and effectively.

Chapter 13, *Forced Islanding and Adaptive Load Shedding to Prevent Blackout for Improving Power System Security: Micro Grid Islanding to Improve Power System Security,* proposes a mechanism that divides the system into predetermined islands and determines the risk during an outage. In the event that there are unstable islands, an under-voltage load shedding plan is implemented to restore system stability. Furthermore, a restoration protocol has been developed. The suggested work is done using a six bus system as an example. This strategy can also be used in smart grid settings to raise the dependability and efficiency of the electricity system.

The volume concludes with **Chapter 14**, *IoT-Enabled Real Time Vehicle Monitoring and Wait Time Prediction System for Smart Fuel Stations*. This chapter presents a state-of-the-art application of AI and Internet of Things (IoT) technologies in the transportation-energy nexus. By integrating object detection and machine learning into fuel station infrastructure, the proposed system predicts queue times and optimizes refueling logistics. Not only does this reduce emissions and improve operational efficiency, but it also enhances the end-user experience—a compelling demonstration of how AI can improve legacy energy infrastructure on the path toward smarter, cleaner urban ecosystems.

Chapter 15, *Generative AI-Driven Frameworks for Intelligent Wind Energy Forecasting and Optimization,* explores Generative AI-driven frameworks in forecasting and optimization of wind energy systems by overcoming the limitations of traditional statistical and deterministic models that struggle with the nonlinear and variable nature of wind resources. By leveraging advanced models such as Generative Adversarial Networks (GANs), Variational Autoencoders (VAEs), and Diffusion Models, these frameworks facilitate the creation of synthetic data, enhance scenario simulation, and enable robust probabilistic forecasting, significantly reducing uncertainty and improving operational resilience.

Together, these final chapters reinforce the central thesis of this volume: that generative AI is more than a set of algorithms—it is a transformative paradigm capable of reimagining how we consume, manage, plan, and finance renewable energy. Through diverse methodologies and domains, our contributors highlight the immense potential and complex responsibilities that come with deploying AI for a sustainable future.

CONCLUSION

The convergence of renewable energy and Generative Artificial Intelligence represents a pivotal moment in the global pursuit of sustainable, resilient, and intelligent energy systems. As this volume illustrates, generative models are not only capable of solving complex technical challenges — from forecasting and optimization to behavioral modeling and market integration — but also of enabling entirely new ways of thinking about energy design, distribution, and engagement.

Across the chapters, our contributors demonstrate that Generative AI is far more than a computational novelty. It is a tool for systems transformation — capable of simulating uncharted futures, personalizing solutions at scale, and bridging data gaps that have long constrained progress in renewable energy. Whether deployed in microgrid planning, smart home optimization, climate-resilient forecasting, or energy-aware financial literacy, generative models bring adaptability, intelligence, and foresight to an energy landscape defined by uncertainty and urgency.

Yet, as with any transformative technology, the integration of Generative AI into critical energy infrastructure brings with it profound responsibilities. The insights presented throughout this book remind us that transparency, fairness, data governance, and ethical design must be embedded into every stage of development and deployment. Technology alone cannot solve the climate crisis — but when developed in alignment with human values, it can accelerate the pace and widen the reach of sustainable solutions.

We hope this book serves as both a reference and an inspiration — for researchers pushing the boundaries of AI, for engineers designing the next generation of energy systems, for policymakers crafting adaptive regulatory frameworks, and for educators and innovators building a more just and sustainable energy future. As we look ahead, we invite readers to continue exploring the immense potential at this intersection, and to join us in shaping a future where renewable energy is not only smarter and more efficient — but more inclusive, equitable, and responsive to the world we live in.

Chapter 1
Data-Driven Analysis of Non-Technical Losses in Low-Voltage Networks: A Case Study of Bartin, Turkey

Hasan Huseyin Coban
https://orcid.org/0000-0002-5284-0568
Bartin University, Turkey

Mehmet Safa Aydin
https://orcid.org/0000-0002-2219-8121
Bartin University, Turkey

Ahmet Bayram
https://orcid.org/0009-0005-5476-2611
Bartin University, Turkey

Tugba Dincel
https://orcid.org/0009-0005-5084-3828
Bartın University, Turkey

Utku Guduloglu
https://orcid.org/0009-0008-6391-9256
Bartın University, Turkey

ABSTRACT

The ongoing digital transformation of Turkey's electricity infrastructure has enabled the widespread collection of high-resolution data within low-voltage distribution networks. This study applies a data-driven methodology to assess energy losses in

DOI: 10.4018/979-8-3373-4272-6.ch001

Bartın, Turkey, using hourly data from 172 substations and 3,416 customer smart meters for the year 2024. Technical losses were estimated based on network parameters, while deviations beyond a 0–3% benchmark were analyzed to identify potential non-technical losses, including metering inaccuracies, undocumented loads, and energy theft. The analysis was implemented in MATLAB, generating visualizations to support decision-making. Results revealed 14 substations with significant non-technical losses, providing actionable insights for targeted field inspections and improving transparency for the regional Distribution System Operator (DSO).

INTRODUCTION

Electricity distribution systems incur unavoidable technical losses due to resistance in conductors and transformers, as well as non-technical losses resulting from theft, metering inaccuracies, and data errors. In emerging markets, non-technical losses (NTLs) can significantly impact financial stability, grid reliability, and long-term planning (Romero Aguero, 2012).

Modernization of Turkey's electricity grid is an ongoing and multifaceted endeavor that aligns with the country's broader goals of energy efficiency, decarbonization, and technological innovation. Over the past decade, the Turkish government and distribution companies have prioritized investments in smart grid infrastructure, advanced metering systems, and automated monitoring tools to ensure a more reliable and transparent electricity supply chain (Yumak et al., 2018). These efforts are particularly relevant in the context of increasing electrification, growing demand, and the integration of distributed energy resources (DERs) such as rooftop solar panels and small-scale wind turbines.

One of the critical challenges facing power distribution networks worldwide is the identification and reduction of energy losses (Alam & Arefifar, 2019). These losses can be broadly categorized into technical losses, which occur due to the physical properties of conductors and transformers, and NTLs, which are primarily administrative or operational in nature. Non-technical losses often include electricity theft, meter inaccuracies, data communication failures, and misconfigured or unregistered loads (Saeed et al., 2020). Such losses, while often hidden in the system, can have profound impacts on the financial sustainability of electricity providers and the equitable distribution of energy (Carr & Thomson, 2022).

In this context, data-driven approaches to energy loss analysis have emerged as powerful tools for utilities. By leveraging smart meter data, sensor readings from substations, and advanced analytics platforms, Distribution System Operators (DSOs) can gain granular insight into where and when losses are occurring, and even infer the underlying causes (Aghahadi et al., 2024). Countries like Sweden, which have more mature smart grid systems, have successfully applied these methods at scale. For example, Ellevio AB, one of Sweden's largest electricity distribution companies, implemented a methodology that used hourly substation and customer consumption data to detect anomalies and classify substations based on loss behavior (Johansson et al., 2017). This methodology, based on extensive data analysis, has enabled more accurate detection of non-technical losses and better-targeted interventions.

Literature Review and Comparative Discussion

The problem of non-technical losses has been explored globally with various methodologies. In recent years, several extensive review papers have examined the development of NTL detection methodologies. These studies present a comprehensive analysis of the current advancements in NTL detection technologies, with particular emphasis on the increasing integration of smart grid infrastructure and the significant role of contemporary computational techniques in enhancing both the accuracy and scalability of detection systems. Anomalies in energy consumption, such as theft or tampering, are often identified by analyzing the structural and dynamic properties of the networks. Network-based methods leverage graph theory and optimization techniques to spot irregularities in energy flow. These approaches employ a range of algorithms, Advanced Metering Infrastructure data, and Power Flow and State Estimation techniques to detect inconsistencies within the grid (Ventura et al., 2023).

Clainer Donadel et al. (Donadel et al., 2009) proposed a methodology to improve the calculation of technical losses in distribution networks by refining them through estimates of non-technical losses. Their approach integrates historical consumption data and results from field inspections to estimate NTLs and then recalculates technical losses accordingly. Unlike many existing models that treat technical losses independently, their method considers the interdependence between NTL and technical losses. The authors demonstrate the flexibility and robustness of their approach, particularly in systems with limited irregularity history. Through clustering and sampling, they are able to produce more realistic estimations of technical losses, even with sparse historical data.

Horton and van Luijk (Horton & van Luijk, 2005) explored low-voltage fault detection and localisation using the TOPAS 1000 disturbance recorder. Their research, conducted in collaboration with LEM and EDF Energy, aimed to reduce customer downtime by detecting and localizing incipient faults in LV underground networks using online techniques. The study highlighted the increasing importance of power quality in competitive electricity markets and proposed methods to improve Customer Minutes Lost (CML) statistics by proactively addressing faults. Their findings underscore the utility of real-time fault monitoring and localization technologies as a complementary strategy to traditional energy loss analysis.

A study from Indonesia (Tambun et al., 2025) used K-Means Clustering and Isolation Forest to detect NTL anomalies in smart meter data from 2021 to 2023. K-Means achieved a recall of up to 98.53%, particularly for voltage anomalies, while Isolation Forest improved to 91.54% recall after parameter tuning. The study highlights the effectiveness of machine learning in identifying patterns like voltage drops and overflows, suggesting its potential for improving utility operations and reducing NTL.

Khan et al. (2024) proposed a hybrid NTL detection system using ARIMA and machine learning in fog-enabled smart grids to tackle electricity theft. The system achieved 98% accuracy, 98.6% precision, and 98.2% recall, demonstrating high performance in predicting consumer fraud patterns. The results show the potential of predictive and automated systems in significantly reducing NTL through efficient fraud detection in smart grids.

M. Persson and A. Lindskog (2019) present a comprehensive evaluation of non-technical loss detection and localization in Swedish distribution systems using future smart meters. Their study emphasizes the economic and operational significance of NTL reduction, noting that 54 out of 159 Swedish Distribution System Operators (DSOs) recorded annual distribution losses exceeding 4%. If these DSOs were to optimize their grids to reduce losses to 4%, it would collectively save approximately 143 GWh per year—equivalent to 7.2 million EUR at a cost of 5 cents per kWh. The paper introduces and assesses a detection method for NTLs, highlighting its performance under time delays and limited data resolution. Additionally, three distinct localization techniques are examined for their effectiveness in pinpointing NTLs at the bus of connection. All methods demonstrated successful localization in the evaluated scenarios, providing practical tools for DSO-level application. The key features, strengths, and limitations of these prior studies are summarized in Table 1, which highlights the evolution of methodologies over time and emphasizes the growing role of machine learning in scalable, automated anomaly detection.

Table 1. Comparative summary of prior studies

Study	Methodology	Strengths	Limitations
Donadel et al. (2023)	Historical consumption + field inspection data	Robust even with limited historical irregularities	Requires manual field validation
Horton & van Luijk (2005)	Real-time disturbance recording	High temporal resolution	Hardware-intensive
Persson & Lindskog (2025)	Multi-method localization of NTL	Significant financial insights	Focused on Swedish context
Recent AI/ML Studies (2023–2025)	Deep learning and clustering algorithms	Scalable and automated anomaly detection	Data privacy concerns, black-box models

Recent studies (Duan, 2024; Elgarhy et al., 2024; Fathollahi, 2025; Gaggero et al., 2025; Guato Burgos et al., 2024) increasingly leverage artificial intelligence (AI) and machine learning (ML) tools for anomaly detection in smart grids. For instance, deep learning models have been applied to classify consumption patterns indicative of theft or meter tampering, while clustering algorithms are used for scalable, automated anomaly detection. However, these approaches also raise challenges around cybersecurity, data privacy, and interpretability.

While prior studies have developed effective loss detection frameworks in mature smart grids, there is limited research on adapting these techniques to mixed rural-urban networks with incomplete data, such as those in Turkey. This study addresses this gap by applying and validating a data-driven methodology in Bartın.

Recent advances in artificial intelligence (AI) and machine learning (ML) have significantly expanded the potential for scalable NTL detection. For example, clustering algorithms and deep learning techniques can analyze consumption patterns in near real-time, identifying anomalies with minimal manual intervention. While this study uses threshold-based and clustering techniques, future implementations could integrate AI-driven models to automate anomaly classification and prioritize field inspections.

International Perspective and Turkey's Position

Globally, many countries have pursued data-driven strategies to address NTLs, but outcomes vary due to local conditions (Jené-Vinuesa et al., 2024).

- India (Chandel et al., 2016), Brazil (Savian et al., 2022), Nigeria (Ayokunle et al., 2022), and South Africa (Khonjelwayo & Nthakheni, 2021) face similar challenges with high theft rates, outdated infrastructure, and rapid urbanization. Case studies from these regions highlight the importance of adapting global methodologies to regional realities, considering both technical and socio-economic factors.
- Sweden and other European nations represent the other end of the spectrum, with advanced smart grids that support automated, near-real-time loss detection (Carr & Thomson, 2022; Jené-Vinuesa et al., 2024).

In Turkey, organizations such as Turkish Electricity Distribution Corporation (TEDAŞ) and the Energy Market Regulatory Authority (EPDK) have spearheaded modernization initiatives, including the deployment of smart meters and GIS-based planning tools. While these initiatives have improved transparency, systematic, data-driven loss detection remains in its early stages, and localized studies are necessary to guide implementation.

Reducing NTLs also carries significant financial and policy implications. Even a 1% reduction in annual distribution losses across Turkey could save utilities millions of euros, freeing resources for reinvestment in grid modernization and renewable integration. Furthermore, future grid architectures must address cybersecurity and data privacy challenges, especially as AI-driven monitoring becomes more prevalent.

Reducing NTLs carries substantial financial implications. For example, a 1% reduction in annual losses across Turkey's distribution network would save approximately €10 million per year, assuming an average wholesale electricity price of €0.05/kWh. For Bartın alone, a similar reduction would translate to roughly €250,000 annually, providing DSOs with both operational and economic incentives to implement advanced detection methodologies.

Study Area: Bartın

In a more localized context, the Turkish energy sector has seen an increasing adoption of smart metering systems, though research specific to low-voltage network losses in Turkey remains limited (Atic et al., 2015). A few notable efforts, such as those by TEDAŞ and the Turkish Energy Market Regulatory Authority (EPDK), have focused on improving the transparency and efficiency of the electricity distribution system. However, while these initiatives provide a solid foundation for modernizing the grid, they often lack the granularity and data-driven approaches seen in countries with advanced smart grid infrastructures.

Bartın, a coastal city in the western Black Sea region of Turkey, provides a representative microcosm for studying LV network losses. Its electricity distribution network encompasses:

- A mix of urban and rural customers
- Residential, commercial, and industrial loads
- Infrastructure at varying stages of modernization

These diverse characteristics make Bartın an ideal testbed for evaluating scalable, data-driven approaches to loss detection. By combining hourly data from 172 substations and 3,416 smart meters, this study develops and validates a framework for identifying both technical and non-technical losses.

The methodology demonstrates how international best practices can be adapted to Turkey's unique grid topology and operational environment. The results from Bartın provide insights not only for local DSOs but also for other developing regions seeking to modernize their electricity distribution systems.

OBJECTIVES

The primary objective of this study is to analyze and quantify the energy losses occurring in the low-voltage distribution network of Bartın, Turkey, with a particular focus on both technical and non-technical losses. More specifically, the key objectives of this research are:

- **Quantify and Benchmark Technical Losses:** One of the primary goals is to quantify the technical losses within the Bartın low-voltage network by modeling losses based on physical network parameters such as resistance, cable types, and load estimations. These losses will be benchmarked within an expected range (0–3%) to identify any deviations that may suggest underlying issues.
- **Identify Substations with Significant Non-Technical Loss Patterns:** A major aspect of this study is to identify substations that exhibit loss patterns exceeding the technical loss benchmark, signaling the potential presence of non-technical losses. This includes identifying anomalies in energy consumption that could be indicative of energy theft, meter misconfigurations, undocumented loads, or other administrative errors.
- **Classify and Investigate Potential Sources of Loss Discrepancies:** Once non-technical losses are flagged, the next objective is to classify the potential sources of these discrepancies. This will involve investigating common causes such as faulty meter configurations (e.g., CT/PT ratio errors), undocumented loads (e.g., station heaters or lights), and communication faults, among others. A deeper understanding of these sources is vital for creating effective loss reduction strategies.
- **Develop Visualization Tools for Monitoring Losses in MATLAB:** An interactive dashboard was developed using MATLAB for the visualization and monitoring of energy losses across the low-voltage network. The dashboard will enable DSOs to track losses dynamically, compare performance across substations, and detect patterns that may require further field investigations or corrective actions.
- **Provide Actionable Insights for DSO Decision-Making:** This study aims to offer valuable insights that Distribution System Operators (DSOs) can use to improve their operational efficiency. By identifying and quantifying both technical and non-technical losses, DSOs can target their resources more effectively, prioritize inspections, and implement better loss reduction measures. The ultimate goal is to improve the transparency and sustainability of the electricity supply in Bartın.

By achieving these objectives, this study aims to provide actionable data that can help improve the management and efficiency of Turkey's electricity distribution networks, contributing to greater energy sustainability and transparency.

METHODOLOGY

This study applies a structured, data-driven approach to analyze energy losses in Bartın's low-voltage distribution network, focusing on both technical and non-technical losses. The methodology consists of several key steps: data collection, loss calculation, benchmarking and outlier detection, classification of non-technical losses, and the development of a visualization tool for dynamic monitoring. Each step is discussed in detail below.

Figure 1 outlines the sequential steps followed in the study: data collection, pre-processing, technical loss modeling, benchmarking, outlier detection, classification of non-technical losses, and visualization

Figure 1. Methodology flowchart for loss detection in Bartin LV networks

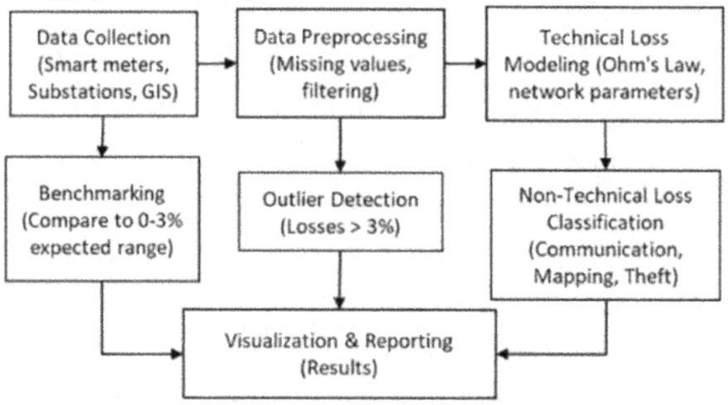

Data Collection

This study utilizes a comprehensive dataset derived from Bartın's regional electricity distribution network. A total of 172 digital substations were included, each providing hourly energy export data. Complementing this, 3,416 customer smart meters contributed hourly consumption records, collectively covering the period from January 1 to December 31, 2024. The primary data sources include the TEDAŞ smart metering infrastructure, accessed through authorized Distribution System Operator (DSO) channels, and GIS network schematics obtained from internal operational planning resources. These datasets form the basis for modeling and detecting technical and non-technical losses throughout the network.

To assess the reliability of hourly data, the measurement resolution for specific substations was analyzed. Figure 2 illustrates a typical 24-hour measurement pattern for substations A1 and A2. Green segments indicate valid hourly readings, while red segments with hatching represent missing or faulty measurements.

Figure 2. Hourly measurement availability for Substations A1 and A2

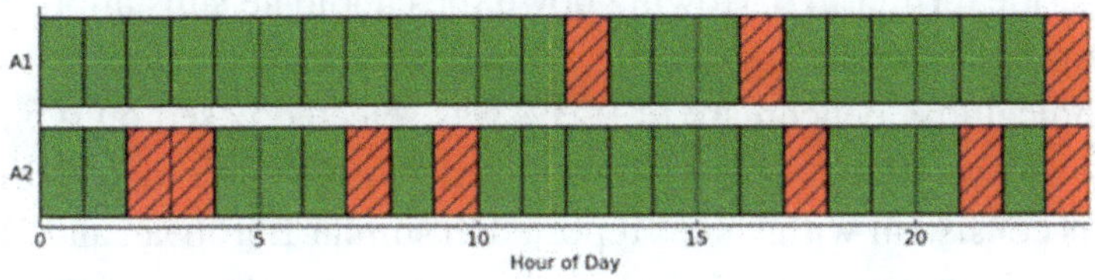

Technical Loss Calculation

Losses were calculated hourly per substation:

$$Loss\ \% = \left(\frac{E_{sub} - \sum E_{cust}}{E_{sub}}\right) \qquad (1)$$

where;

E_{sub} – total energy exported by the substation,
$\sum E_{cust}$ – sum of all customer consumption.

Technical losses (P_{loss}) modeled using:

$$P_{loss} = 3RI^2$$

whe e (2)

R – cable resistance from GIS data,
I – estimated current from active and reactive power.

Benchmarking and Outlier Detection

To establish a reliable baseline for identifying anomalies, technical losses were benchmarked against an expected range of 0–3%, based on typical resistive and reactive characteristics of the network. Each substation's total losses were calculated and compared with this benchmark. Substations whose losses exceeded this threshold were flagged as outliers, prompting further investigation into possible technical inefficiencies or non-technical loss sources. This benchmarking process served as a critical filter in narrowing down the candidate substations for anomaly classification.

The technical loss benchmark of 0–3% was selected based on typical resistive and reactive characteristics of low-voltage feeders in Bartın's distribution network. This range is consistent with values reported in similar European and Turkish studies. However, the precise threshold can significantly influence which substations are flagged as outliers. To evaluate this sensitivity, the benchmark was adjusted by ±1%, and the resulting changes in the number of detected outliers were analyzed. To assess the sensitivity of the benchmarking approach, the threshold range was adjusted by ±1% from the base case. The resulting changes in the number of substations flagged as outliers are presented in Table 2. Narrowing the threshold to 1–2% significantly increases the number of identified outliers, while broadening it to -1–4% reduces the detection rate.

Table 2. Sensitivity of outlier detection to benchmark threshold adjustments

Benchmark Range (%)	Substations Flagged as Outliers
1–2	25
0–3 (Base Case)	17
-1–4	12

As shown in Table 2, a stricter benchmark (1–2%) flags 25 substations as outliers, representing a 47% increase compared to the base case (0–3%). Conversely, relaxing the threshold to -1–4% decreases the number of outliers to 12. This demonstrates that the choice of benchmark directly affects field investigation priorities and underscores the need for a balance between detection accuracy and resource constraints.

Non-Technical Loss Classification

For substations identified as outliers—those with loss levels exceeding the expected technical benchmark—a detailed classification was conducted to determine the likely causes of the discrepancies. The analysis revealed five primary categories of non-technical losses. Communication errors occur due to intermittent or failed data transfer between smart meters and the central system, often resulting in repeated zero values, constant flat readings, or missing data. Meter misconfigurations involve incorrect current transformer (CT) or potential transformer (PT) ratios, which lead to inaccurate energy measurements. Undocumented loads are typically internal substation devices such as space heaters, lighting, or auxiliary equipment that consume energy but are not included in customer billing. Customer mapping errors arise from administrative mismatches between meter IDs and their associated substations, producing erroneous aggregation of load data. Finally, energy theft indicators are identified through suspicious patterns, such as consistent energy export with little or no registered consumption, suggesting illegal connections that bypass metering systems. Each of these categories represents a distinct class of anomalies that can significantly elevate non-technical losses and therefore require tailored detection and intervention strategies.

Toolchain

In this study, MATLAB was used as the primary tool used in all stages of data processing, technical loss modeling and result visualization. The tool was employed to clean and preprocess the raw data from substations and smart meters, applying advanced functions for time-series analysis and outlier detection to identify both technical and non-technical losses. MATLAB's numerical computation environment facilitated the calculation of technical losses using network parameters such as cable resistance and current estimates, while its powerful plotting capabilities were used to visualize loss distributions and anomaly patterns. The flexibility of MATLAB allowed for the development of customized models and the adjustment of parameters, making it an ideal platform for analyzing large datasets and producing actionable insights for the Distribution System Operator (DSO).

Technical Loss Model Parameters

In modeling energy losses along feeder lines, the total equivalent resistance of a line segment depends on the configuration of its branches. If resistances are connected in series, the total resistance is given by:

$$R_{eqv} = \sum_{i=1}^{n} R_i \qquad (3)$$

If connected in parallel, the equivalent resistance is computed as:

$$R_{eqv} = \left(\sum_{i=1}^{n}\frac{1}{R_i}\right)^{-1} \tag{4}$$

Equations (3, 4) enable more accurate modeling of losses and voltage drops across complex low-voltage feeder configurations, particularly when multiple cables or branches serve shared customer groups. Technical losses in the low-voltage network were estimated using the parameters listed in Table 3. These values were derived from TEDAŞ specifications and validated against field data to ensure accuracy in the hourly loss modeling process.

Table 3. Technical loss model parameters

Parameter	Value	Description
Cable Type	AL/XLPE 3x150 mm^2	Standard feeder cable used in the Bartın LV network
Resistance per km (R)	0.206 Ω	Resistance at 20°C, based on cable specifications
Length of Feeder Line	0.2–2.0 km	Typical range for LV feeders in the study area
Nominal Voltage	0.4 kV	Standard LV operating voltage
Load Power Factor	0.90 (assumed)	Typical power factor for mixed residential/commercial loads
Maximum Transformer Capacity	250 kVA	Common transformer rating in regional substations
Calculation Granularity	Hourly	Enables alignment with smart meter data resolution

These parameters reflect typical configurations used in the Bartın distribution region and are applied uniformly across the modeled substations for consistency. Variations in actual resistance and length are accounted for where available from field data.

Resistance calculations are performed by considering both parallel and series-connected cables within the distribution network. To simplify the analysis and avoid the complexity of evaluating current distribution based on the individual resistances of multiple cables, a one-dimensional network representation is adopted. In this approach, only a single equivalent cable is modeled for each cable run. Accordingly, the initial stage of the computational process is designed to determine the combined resistance and reactance values when multiple cables are installed along the same route. Once these equivalent electrical parameters are computed, the network diagram is adjusted to include only one representative cable per run, effectively reflecting the aggregated resistance and reactance of the original configuration, as illustrated in Figure 3.

Figure 3. Network schematic of a representative LV feeder in Bartın

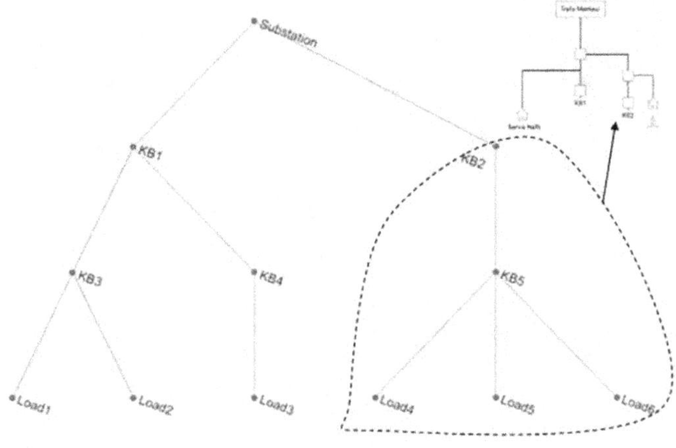

RESULTS AND ANALYSIS

This section presents the findings of the analysis conducted on Bartın's low-voltage distribution network. Using hourly data from 172 substations and 3,416 smart meters, key loss patterns were identified, enabling the differentiation between technical and non-technical losses. The results are organized to first provide a general overview of loss distributions, followed by a more detailed examination of technical losses, non-technical loss classifications, and spatial trends.

Figure 4 illustrates the daily energy export values of three representative substations—Substation #84, A2, and C2—over a one-month period. These trends help visualize operational consistency, highlight peak load variations, and reveal irregular behaviors that may indicate underlying technical inefficiencies or non-technical anomalies.

Figure 4. Daily energy export for selected substations (1-month sample)

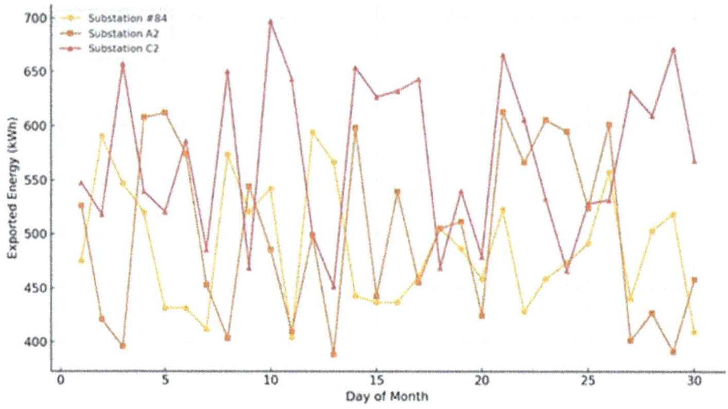

General Observations

The initial analysis revealed that 141 out of 172 substations operated within the expected technical loss range of 0–3%, indicating normal behavior and well-functioning infrastructure. However, 28 substations recorded losses exceeding 3%, warranting closer inspection due to potential inefficiencies or anomalies. Furthermore, detailed investigations confirmed non-technical issues in 14 substations, including undocumented loads, meter misconfigurations, or customer mapping errors.

Technical Loss Distribution

The distribution of technical losses across the substations reveals key patterns in network performance. Substations typically exhibit a range of losses, with the majority falling within the expected technical loss range of 0–3%. However, there are several substations that deviate significantly from this range. To better illustrate this distribution, a histogram was created, which clearly shows the frequency of different loss ranges across the studied substations. As shown in Figure 5, the majority of substations (141 out of 172) fall within the 0–3% loss range, while a smaller number exhibit losses exceeding 3%. These observations highlight areas that may require further investigation to identify causes of inefficiencies or underlying issues.

The majority of substations (141 out of 172) exhibit losses within the expected 0-3% range, classified as "normal." These substations are likely operating with optimal efficiency, adhering to standard technical loss limits. A smaller group of substations (28) shows losses exceeding 3%, categorized as "suspicious" or requiring further inspection. This group likely harbors instances of either unaddressed technical inefficiencies or non-technical anomalies. The remaining substations, displaying losses above 5%, represent outliers that need immediate attention, possibly indicating severe operational issues or incorrect data handling. The distribution illustrated by Figure 5 helps guide efforts toward optimizing substations that are either underperforming technically or experiencing abnormal losses. The distribution highlights the need for focused investigations into the higher-loss substations.

Figure 5. Histogram of technical losses

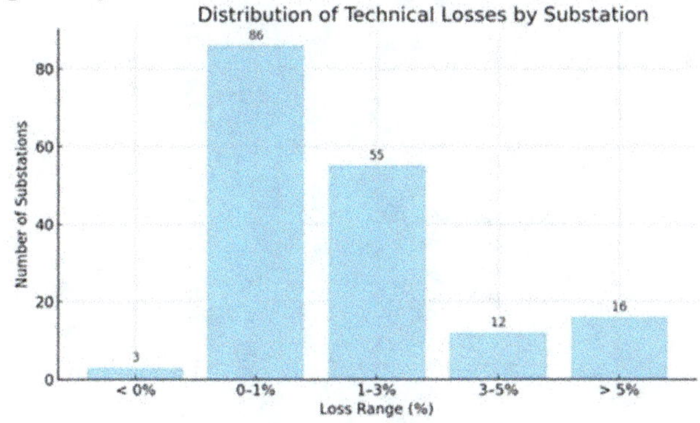

The distribution of substations by loss percentage is summarized in Table 4, showing that the majority fall within the 0–1% and 1–3% categories, with a small number exceeding 5%, which warrants immediate attention.

Table 4. Distribution of substations by technical loss range (%)

Loss Range (%) Number of Substations Percentage (%)
< 0 3 1.7%
0–1 86 50.0%
1–3 55 32.0%
3–5 12 7.0%
> 5 16 9.3%

Figure 6 shows the spatial variation of technical loss levels across substations in the Bartın LV network. Color-coded markers reflect severity: green (<2%), yellow (2–4%), orange (4–6%), and red (>6%). Clusters of higher losses are observed in central urban areas, suggesting greater infrastructure strain or possible anomalies.

Figure 6. Geographic distribution of substation losses in Bartın

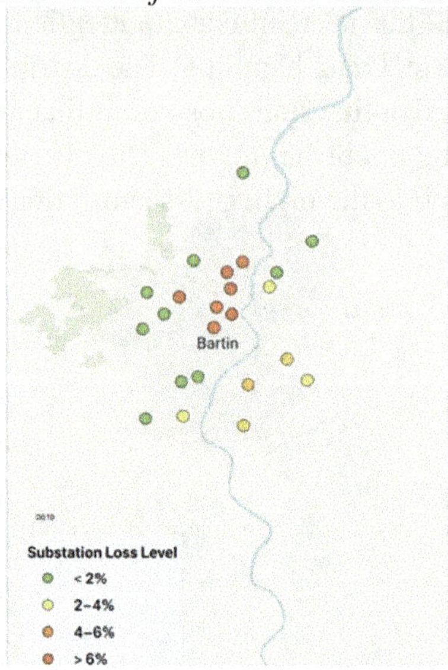

The correlation matrix (see Figure 7) shows relationships between substation age, customer count, loss percentage, and communication quality. Positive correlations (in yellow) indicate that as one variable increases, so does the other — for instance, older substations often show slightly higher loss percentages. Negative correlations (in blue) suggest inverse relationships — notably, better communication quality often correlates with lower losses.

Figure 7. Correlation matrix of substation attributes and losses

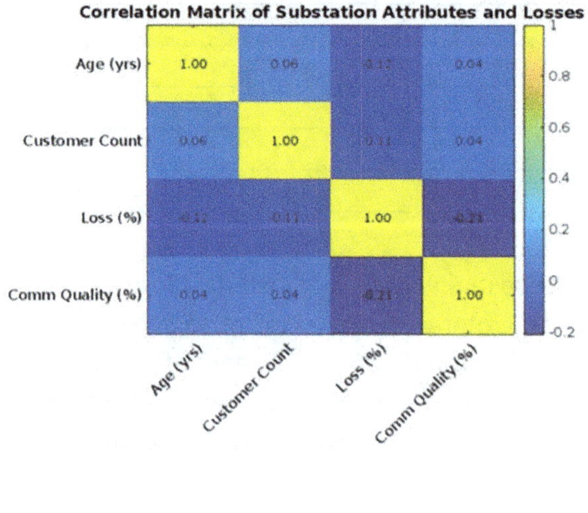

Figure 8 complements the loss distribution analysis by introducing a temporal dimension (seasonal trends), enhancing the reader's understanding of time-based variation. Warmer months (May–August) show higher average losses, possibly due to increased demand or temperature-induced infrastructure inefficiencies.

Figure 8. Seasonal variation in technical losses across the year

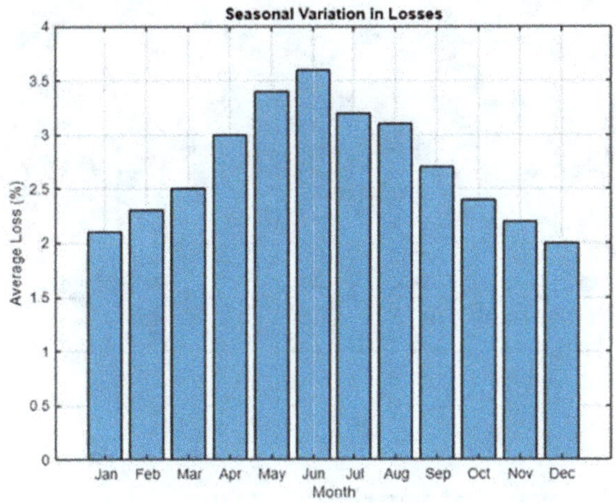

Figure 9 groups substations into three behavioral clusters using average daily loss percentage and variability (standard deviation) of hourly losses. Each color represents a distinct cluster. This visualization helps identify substations with irregular patterns or similar operational characteristics.

Figure 9. Clustering of substations based on loss behavior

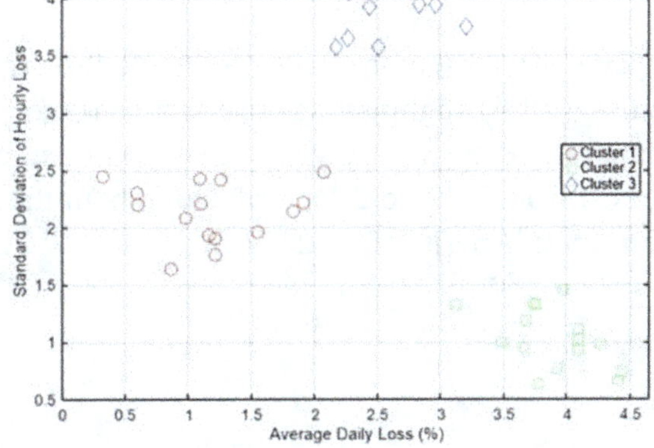

Figure 10 shows an hourly heatmap showing average technical loss percentages across 20 substations. Darker red indicates higher losses. The pattern reveals time-of-day trends and potentially irregular usage windows requiring further investigation.

Figure 10. Hourly loss heatmap across 20 substations

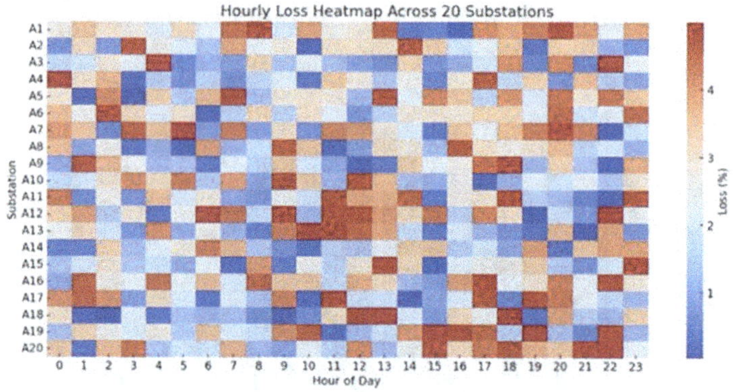

Non-Technical Loss Categories

Field Inspection Case Study: Substation #84

To validate the data-driven results, a targeted field inspection was conducted at Substation #84, which showed a persistent loss rate of approximately 7% in the analysis phase. On-site investigation revealed two contributing factors:

- A previously undocumented 3.5 kW resistive heater used during cold seasons.
- Customer mapping errors involving three small retail businesses connected but not registered in the substation database.

Figure 11 compares the substation's loss rate before the inspection (7.0%) and after corrective actions (2.1%), highlighting the effectiveness of targeted field interventions based on anomaly detection. A post-correction measurement revealed that the loss rate decreased from 7.0% to 2.1%, validating both the anomaly detection method and the corrective benefits of field verification.

Figure 11. Substation #84 loss rate before and after intervention

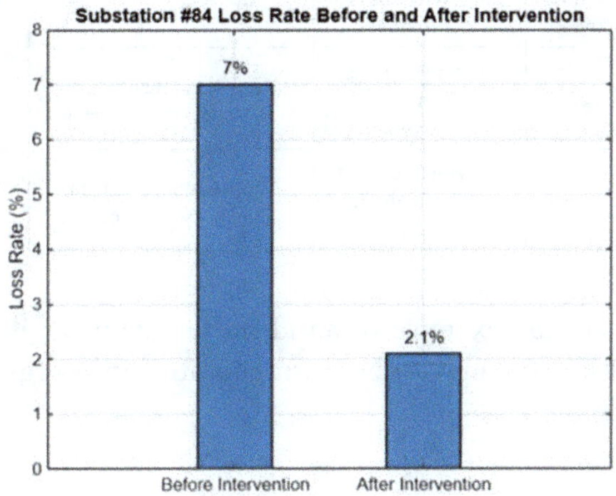

Following the benchmarking and anomaly detection processes, several substations exhibited characteristics inconsistent with expected technical loss behavior. These substations were further analyzed to determine the root causes of their discrepancies. The results of this analysis were categorized into five major types of non-technical loss (NTL) sources, summarized in Table 5 below. Each category reflects a distinct pattern of data irregularity or infrastructure fault that contributes to measurable energy losses beyond technical expectations.

Table 5. Summary of anomalies

NTL Category	Stations Affected	Key Symptoms / Indicators	Likely Root Cause	Recommended DSO Action
Metering Error	5	Unusual zero readings or constant flat profiles	Incorrect CT/PT ratios or deadband misconfiguration	Verify meter settings, recalibrate, replace faulty meters
Undocumented Load	4	Persistent base load even during outages	Internal substation heating, lighting circuits, or unregistered equipment	Identify and register internal loads; adjust billing
Communication Fault	3	Gaps or missing data (>30%)	Smart meter network connectivity issues	Troubleshoot network communication; replace faulty modems

continued on following page

Table 5. Continued

NTL Category	Stations Affected	Key Symptoms / Indicators	Likely Root Cause	Recommended DSO Action
Customer Mapping Error	2	Negative losses or mismatched substation data	Incorrect meter-to-substation assignment	Update GIS database and verify mapping accuracy
Energy Theft	1	Measured export with no matching consumption	Illegal connections or bypassed meters	Field inspection, enforcement, legal action

Figure 12 presents an example of an identified anomaly, illustrating the load profile of a specific substation, where irregularities in energy consumption are prominently evident.

Figure 12. Dashboard anomaly detection

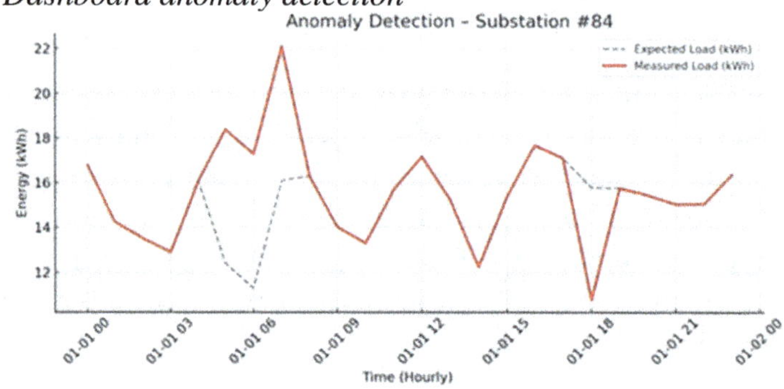

- The irregular consumption spikes within the graph suggest potential issues like meter misconfigurations, undocumented loads, or energy theft. These anomalies deviate from the normal consumption patterns expected from customers connected to the substation.
- **Unexpected Peaks:** At certain time intervals, the consumption increases sharply without any corresponding customer activity, signaling potential theft or technical issues like meter faults.
- **Missing Data Gaps:** The dashboard also highlights periods where data might be missing due to communication errors. These gaps are critical because they may lead to incorrect assumptions about losses or mask real loss occurrences.

Figure 12 illustrates the capability of the developed dashboard to detect subtle load anomalies that may not be readily identifiable through conventional monitoring approaches. By enabling automated anomaly detection, Distribution System Operators (DSOs) can rapidly pinpoint areas requiring further investigation and intervention. In the example shown, the anomalous substation exhibits irregular spikes and sudden drops in load values that deviate from typical demand patterns. Such deviations may indicate issues such as electricity theft, metering inaccuracies, or malfunctioning internal substation equipment (see Figure 13).

Figure 13. Hourly load anomaly comparison

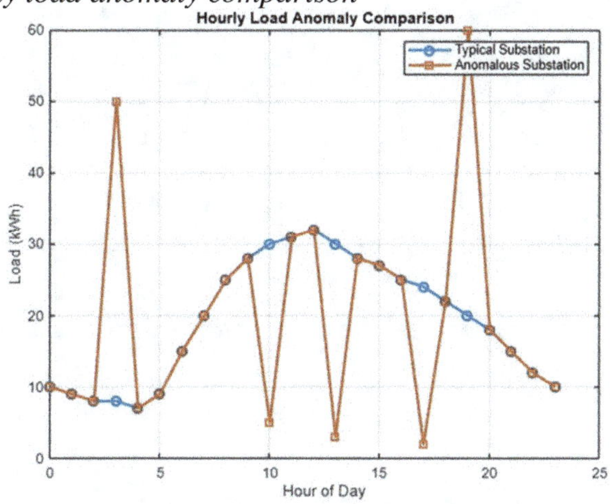

As a result, Substation #84 showed a consistent 7% loss. Field inspection revealed a 3.5 kW heater and missing customer mapping for three small businesses. After correction, losses dropped to 2.1%.

Additional Substation Comparisons

Beyond Substation #84, three other substations—A2, B2, and C2—were selected for hourly performance profiling due to their distinct loss characteristics and availability of clean time-series data.

- **Substation A2** exhibited moderate losses with a stable export profile, likely impacted by minor undocumented internal loads.
- **Substation B2** maintained loss levels mostly within the acceptable range but showed occasional communication gaps during evening hours.
- **Substation C2** showed higher fluctuations and more significant losses during peak hours, indicating potential mapping or configuration errors.

Figure 14 illustrates the hourly loss percentage for these three substations, offering a comparative view of their operational behavior across a 24-hour period. This figure highlights the daily variation in technical losses and enables comparison across substations with similar profiles but different anomaly levels.

Figure 14. Loss distribution histogram

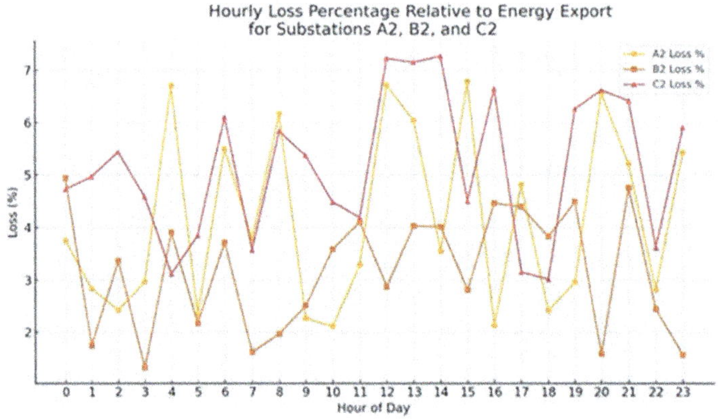

DISCUSSION

The application of a data-driven approach to identify and reduce losses in Bartın's low-voltage distribution system has yielded promising results. The methodology was developed from internationally recognized practices (García et al., 2025; Li et al., 2025; Long et al., 2020; Zhao et al., 2022) has proven effective in detecting both technical and non-technical losses in this diverse and challenging environment. This section elaborates on the strengths of the approach, potential limitations, and areas for future work.

Effectiveness of the Methodology

The method effectively identified a range of anomalies, including meter misconfigurations, undocumented loads, communication faults, and energy theft. The ability to detect non-technical losses through the combination of high-resolution data from smart meters and substations, alongside advanced analytics, is a significant achievement.

For example, Substation #84 exhibited a consistent loss of 7%, which, upon further inspection, revealed a combination of undocumented loads and a misconfigured meter. After addressing these issues, losses at the substation decreased to 2.1%, demonstrating the tangible benefits of the methodology in improving operational efficiency and reducing system inefficiencies.

This aligns with the findings of Melin & Segerstedt and Persson & Lindskog (Melin & Segerstedt, 2025; Persson & Lindskog, 2019), who emphasized the importance of real-time data analytics in identifying loss patterns in Swedish distribution networks. Similar to their study, our analysis showed that many technical losses fell within the expected 0–3% range, suggesting that the infrastructure itself is largely functioning within acceptable parameters. However, deviations outside of this range often indicated underlying non-technical causes that were not easily detectable through traditional methods.

Data Integrity and Systematic Challenges

While the method was successful in detecting and classifying losses, it also highlighted several challenges related to data integrity and system configuration. In many instances, GIS mapping errors and inconsistent meter configurations led to false positives or inaccurate loss estimations. For instance, some substations with "normal" loss profiles were flagged due to discrepancies in the mapping of customer connections or missing data points from the smart meters. This is a common challenge in data-driven analyses, particularly when relying on legacy systems and infrastructure.

Moreover, communication failures in some substations, where over 30% of the data was missing, often led to difficulties in accurately classifying losses. These issues underscore the importance of ensuring robust data collection and real-time monitoring systems. The lack of uniformity in data availability across substations might lead to incomplete or misleading results, making it difficult to assess losses comprehensively without addressing these gaps.

These challenges are not unique to Bartın. Similar issues have been observed in other parts of the world, particularly in regions with less mature smart grid infrastructures. For example, studies in Southeast Asia and Eastern Europe have found that missing or inaccurate data can significantly hamper the effectiveness of loss detection systems, leading to delays and increased operational costs. However, this study also reveals that certain regions with outdated infrastructure or limited data coverage may require additional refinement in the methodology to overcome these local challenges.

Future Work and Improvements

The findings from this study suggest several avenues for future work. Firstly, improving the quality of data from both smart meters and substations is critical to enhancing the accuracy and reliability of loss detection models. This could involve addressing GIS mapping errors, increasing the reliability of communication networks, and enhancing the calibration of smart meters to avoid issues like CT/PT ratio misconfigurations.

Incorporating machine learning techniques for anomaly detection could further improve the ability to identify non-technical losses, especially in complex or borderline cases. Algorithms such as decision trees, support vector machines, or even deep learning models could be trained on historical loss data to automatically classify potential losses and flag unusual consumption patterns. Machine learning could also help detect subtle patterns in data that are difficult for traditional rule-based approaches to identify, thus increasing the speed and accuracy of loss detection.

Additionally, expanding the methodology to include medium-voltage feeders could provide more holistic insights into the overall performance of the distribution system. By analyzing losses at higher voltage levels, the method could help DSOs understand the root causes of inefficiencies and losses that may propagate from the high-voltage network down to low-voltage substations.

Another interesting area for future research would be to explore the integration of distributed energy resources (DERs), such as solar and wind energy, into the loss detection framework. As Turkey increasingly adopts these technologies, their impact on grid losses needs to be studied in more detail. This could lead to new loss models that account for fluctuations in generation and demand, helping utilities maintain more accurate loss predictions and optimize grid performance.

As smart meter and substation data are increasingly utilized for operational decision-making, cybersecurity and data privacy become critical concerns. Unauthorized access or manipulation of metering data could lead to false anomaly reports, financial losses, or even systemic instability. Future implementations must therefore incorporate secure data transmission protocols and comply with emerging privacy regulations.

Policy Implications

The results of this study also have important implications for regulatory authorities and policymakers. As energy theft and other non-technical losses remain a significant issue for many utilities, the adoption of data-driven loss detection methods could help regulators improve monitoring and compliance. For instance, identifying substations with significant non-technical losses could prompt more targeted field inspections and regulatory interventions.

Furthermore, the insights from this study could inform the development of policies aimed at improving grid modernization and the widespread adoption of smart grid technologies. By showcasing the potential benefits of data-driven analysis in reducing both technical and non-technical losses, this research could motivate further investments in smart metering systems, real-time monitoring platforms, and advanced data analytics tools.

Limitations of the Study

While the methodology has demonstrated considerable success in identifying losses, it is important to acknowledge some limitations. The reliance on hourly data means that short-term fluctuations and sudden events might not be fully captured in the analysis. In particular, transient phenomena such as lightning strikes or sudden demand spikes could lead to inaccurate loss estimations. Future work could explore the use of more granular data, such as sub-hourly measurements or real-time monitoring, to overcome this limitation.

Additionally, the focus of this study was primarily on low-voltage networks. While this is an important and often under-researched area, medium and high-voltage networks also contribute to overall system losses. Extending the methodology to these voltage levels could yield more comprehensive insights into the losses across the entire grid.

CONCLUSION

This study demonstrates the effectiveness of data-driven approaches in identifying and reducing both technical and non-technical losses in low-voltage distribution networks. Key loss patterns were identified across 172 substations and 3,416 smart meters, showcasing the potential of advanced analytics to optimize grid management. The methodology proved effective in detecting non-technical losses, such as energy theft, meter misconfigurations, and undocumented loads, leading to targeted interventions and enhanced operational efficiency.

Despite the success, challenges related to data quality, such as GIS mapping errors and communication faults, were encountered. These issues underscore the need for robust data management and real-time monitoring systems to improve the accuracy of loss detection. Moving forward, integrating machine learning techniques could further enhance the detection of subtle anomalies, while expanding the analysis to medium-voltage feeders could provide a more comprehensive understanding of grid losses.

The findings also have significant policy implications. Regulatory bodies can use this methodology to improve loss detection frameworks and ensure more accurate monitoring of energy theft and meter performance. Additionally, this study provides a scalable solution for other regions in Turkey and beyond, demonstrating the potential of data-driven analysis to support the modernization and efficiency of power distribution networks.

In conclusion, adapting international methodologies to local contexts can lead to valuable improvements in energy management, offering a path for more efficient, transparent, and sustainable electricity distribution in Turkey and other emerging economies.

The increasing integration of distributed energy resources (DERs) such as rooftop solar panels, electric vehicles, and small-scale storage adds complexity to LV networks. These elements introduce bidirectional power flows and volatile demand patterns, making it more challenging to distinguish between technical and non-technical losses. Future loss detection frameworks must account for these dynamics to remain effective.

While this study focuses on Bartın, the methodology is scalable to national or regional levels. With appropriate data governance frameworks and centralized analytics, Turkey's DSOs could implement a coordinated loss detection strategy that enhances transparency and operational efficiency across the entire low-voltage network.

REFERENCES

Aghahadi, M., Bosisio, A., Merlo, M., Berizzi, A., Pegoiani, A., & Forciniti, S. (2024). Digitalization Processes in Distribution Grids: A Comprehensive Review of Strategies and Challenges. *Applied Sciences (Basel, Switzerland)*, *14*(11), 4528. DOI: 10.3390/app14114528

Alam, M. S., & Arefifar, S. A. (2019). Energy Management in Power Distribution Systems: Review, Classification, Limitations and Challenges. *IEEE Access : Practical Innovations, Open Solutions*, *7*, 92979–93001. DOI: 10.1109/ACCESS.2019.2927303

Atic, S., Parlakyıldız, S., & Ilcihan, Z. (2015). Smart Meter and Turkey. *Bitlis Eren University Journal of Science and Technology*, *5*(2). Advance online publication. DOI: 10.17678/beujst.15889

Ayokunle, A., Misra, S., Oluranti, J., & Ahuja, R. (2022). *Technical Losses (Tl) and Non-technical Losses (NTL) in Nigeria* (pp. 147–159). https://doi.org/DOI: 10.1007/978-3-030-86223-7_14

Carr, D., & Thomson, M. (2022). Non-Technical Electricity Losses. *Energies*, *15*(6), 2218. DOI: 10.3390/en15062218

Chandel, P., Thakur, T., Sawle, B. A., & Sharma, R. (2016). Power theft: Major cause of non technical losses in Indian distribution sector. *2016 IEEE 7th Power India International Conference (PIICON)*, 1–6. https://doi.org/DOI: 10.1109/POWERI.2016.8077253

Donadel, C., Anicio, J., Fredes, M., Varejao, F., Comarela, G., & Perim, G. (2009). A methodology to refine the technical losses calculation from estimates of non-technical losses. *IET Conference Publications*, 553–553. https://doi.org/DOI: 10.1049/cp.2009.0834

Duan, J. (2024). Deep learning anomaly detection in AI-powered intelligent power distribution systems. *Frontiers in Energy Research*, *12*. Advance online publication. DOI: 10.3389/fenrg.2024.1364456

Elgarhy, I., Badr, M. M., Mahmoud, M., Alsabaan, M., Alshawi, T., & Alsaqhan, M. (2024). XAI-Based Accurate Anomaly Detector That Is Robust Against Black-Box Evasion Attacks for the Smart Grid. *Applied Sciences (Basel, Switzerland)*, *14*(21), 9897. DOI: 10.3390/app14219897

Fathollahi, A. (2025). Machine Learning and Artificial Intelligence Techniques in Smart Grids Stability Analysis: A Review. *Energies*, *18*(13), 3431. DOI: 10.3390/en18133431

Gaggero, G. B., Girdinio, P., & Marchese, M. (2025). Artificial Intelligence and Physics-Based Anomaly Detection in the Smart Grid: A Survey. *IEEE Access : Practical Innovations, Open Solutions, 13*, 23597–23606. DOI: 10.1109/ACCESS.2025.3537410

García, S., Fresia, M., Mora-Merchán, J. M., Carrasco, A., Personal, E., & León, C. (2025). A data-driven topology identification method for low-voltage distribution networks based on the wavelet transform. *Electric Power Systems Research, 243*, 111517. DOI: 10.1016/j.epsr.2025.111517

Guato Burgos, M. F., Morato, J., & Vizcaino Imacaña, F. P. (2024). A Review of Smart Grid Anomaly Detection Approaches Pertaining to Artificial Intelligence. *Applied Sciences (Basel, Switzerland), 14*(3), 1194. DOI: 10.3390/app14031194

Horton, S. L., & van Luijk, N. G. (2005). Low voltage fault detection and localisation using the TOPAS 1000 disturbance recorder. *18th International Conference and Exhibition on Electricity Distribution (CIRED 2005)*, v1-82-v1-82. https://doi.org/DOI: 10.1049/cp:20050958

Jené-Vinuesa, M., Aragüés-Peñalba, M., & Sumper, A. (2024). Comprehensive Data-Driven Framework for Detecting and Classifying Non-Technical Distribution Losses. *IEEE Access : Practical Innovations, Open Solutions, 12*, 96537–96551. DOI: 10.1109/ACCESS.2024.3426940

Johansson, J., Sporre, H., Selberg, L., & Hansson, O. (2017). Monitoring systems for secondary substations – a comparative study. *CIRED - Open Access Proceedings Journal, 2017*(1), 350–354. https://doi.org/DOI: 10.1049/oap-cired.2017.0103

Khan, H. M., Jabeen, F., Khan, A., Badawi, S. A., Maple, C., & Jeon, G. (2024). Hybrid non-technical-loss detection in fog-enabled smart grids. *Sustainable Energy Technologies and Assessments, 65*, 103775. DOI: 10.1016/j.seta.2024.103775

Khonjelwayo, B., & Nthakheni, T. (2021). Determining the causes of electricity losses and the role of management in curbing them: A case study of City of Tshwane Metropolitan Municipality, South Africa. *Journal of Energy in Southern Africa, 32*(4), 45–57. DOI: 10.17159/2413-3051/2021/v32i4a8704

Li, B., Feng, X., Su, S., Zhong, P., & Xiao, H. (2025). A Physics-Data Driven Approach for Identifying Leakage Users in Low-Voltage Distribution Systems. *IET Generation, Transmission & Distribution, 19*(1). Advance online publication. DOI: 10.1049/gtd2.70040

Long, H., Chen, C., Gu, W., Xie, J., Wang, Z., & Li, G. (2020). A Data-Driven Combined Algorithm for Abnormal Power Loss Detection in the Distribution Network. *IEEE Access : Practical Innovations, Open Solutions*, *8*, 24675–24686. DOI: 10.1109/ACCESS.2020.2970548

Melin, G., & Segerstedt, C. (2025). *Datadriven analys av förluster i lågspänningsnätet: Metod för identifiering av icke-tekniska förluster*. Uppsala University.

Persson, M., & Lindskog, A. (2019). Detection and localization of non-technical losses in distribution systems with future smart meters. *2019 IEEE Milan PowerTech*, 1–6. https://doi.org/DOI: 10.1109/PTC.2019.8810893

Romero Aguero, J. (2012). Improving the efficiency of power distribution systems through technical and non-technical losses reduction. *PES T&D, 2012*, 1–8. DOI: 10.1109/TDC.2012.6281652

Saeed, M. S., Mustafa, M. W., Hamadneh, N. N., Alshammari, N. A., Sheikh, U. U., Jumani, T. A., Khalid, S. B. A., & Khan, I. (2020). Detection of Non-Technical Losses in Power Utilities—A Comprehensive Systematic Review. *Energies*, *13*(18), 4727. DOI: 10.3390/en13184727

Savian, F. de S., & Siluk, J. C. M., Bisognin Garlet, T., Do Nascimento, F. M., Pinheiro, J. R., & Vale, Z. (2022). Non-technical Losses in Brazil: Overview, Challenges, and Directions for Identification and Mitigation. *International Journal of Energy Economics and Policy*, *12*(3), 93–107. DOI: 10.32479/ijeep.12614

Tambun, R., Priyadi, A., & Lystianingrum, V. (2025). Anomaly Detection of Non-Technical Losses in Smart Meter Data Using K-Means Clustering and Isolation Forest: A Case Study From Indonesia. *2025 International Seminar on Intelligent Technology and Its Applications (ISITIA)*, 555–560. https://doi.org/DOI: 10.1109/ISITIA66279.2025.11137434

Ventura, L., Felix, G. E., Vargas, R., Faria, L. T., & Melo, J. D. (2023). Estimation of non-technical loss rates by regions. *Electric Power Systems Research*, *223*, 109685. DOI: 10.1016/j.epsr.2023.109685

Yumak, K., Tosun, G., Ercin, O., Batar, G., & Sinim, M. C. (2018). Turkey's Smart Grid Roadmap Project for Electrical Distribution Systems in Vision 2035. *2018 IEEE PES Innovative Smart Grid Technologies Conference Europe (ISGT-Europe)*, 1–6. https://doi.org/DOI: 10.1109/ISGTEurope.2018.8571730

Zhao, J., Xu, M., Wang, X., Zhu, J., Xuan, Y., & Sun, Z. (2022). Data-Driven Based Low-Voltage Distribution System Transformer-Customer Relationship Identification. *IEEE Transactions on Power Delivery*, *37*(4), 2966–2977. DOI: 10.1109/TPWRD.2021.3120625

Chapter 2
Forecasting Wind and Solar Energy With GANs and Diffusion Models:
A New Frontier in Renewable Energy Prediction

R. N. Ravikumar
https://orcid.org/0009-0009-3705-1681
Marwadi University, Rajkot, India

S. Aarthi
https://orcid.org/0009-0006-9064-2091
Marwadi University, Rajkot, India

ABSTRACT

Accurate forecasting of wind and solar energy is vital for grid efficiency and renewable integration. Traditional models often falter due to the nonlinear, uncertain nature of these sources. This chapter investigates Generative Adversarial Networks (GANs) and Diffusion Models as next-gen forecasting tools. These models effectively learn complex temporal-spatial patterns and produce high-resolution, realistic outputs. The chapter details their architectures, multimodal data integration (e.g., satellite images, weather models, ground sensors), and real-time edge deployment. Case studies reveal up to 28% improvement in accuracy, with better uncertainty estimation and climate adaptability. Comparative results with LSTM and ARIMA underline generative models' superiority. The chapter also explores ethical issues, explainability, and emerging trends such as federated learning and climate-aware modeling. Overall, it positions generative AI as a transformative path toward scalable, intelligent, and robust renewable energy forecasting systems.

DOI: 10.4018/979-8-3373-4272-6.ch002

INTRODUCTION

The world movement towards renewable energy has highlighted the importance of proper prediction of wind and solar energy production. Although these sources, which are renewable are beneficial to the environment, they are unreliable in nature because of their nature of variation and reliance on variable weather conditions. Common statistical and deterministic procedures used to forecast such patterns in energy tend to fail to represent their nonlinearities and stochastic characteristics, thus incurring inefficiency in the management and planning of power grids (Barbieri et al., 2017). The recent development in Generative AIs, in specific Generative Adversarial Networks (GANs) and Diffusion Models, has the revolutionary potential to provide a better way to predict renewable energy generators. Condensing near-perfect functionality of generally learning complex temporal and spatial distributions, the models have shown capacity to yield synthetic data that is realistic and of high resolutions that resembles the real energy outputs in diverse conditions (Cacuci & Fang, 2016). Generation of models by combining multimodal inputs, e.g., Satellite imagery, weather simulations and ground-based sensor data, can be more resistant and geographically flexible forecasts can be produced. In this chapter, it has been seen how GANs and Diffusion Models can transform the scenario of predicting wind and solar energy forecasting. It shows their architectures, training strategies and use in real life cases where there is a cross review to the traditional models (Esteves et al., 2019). With access to these generative tools, the energy sector may substantially improve the accuracy of its forecasts, streamline the ways the grid is being operated, and the pace of the renewable energy systems shifting towards a resilient and data-centered system.

The Figure 1 illustrates the whole bubble of the workflow of applying the Generative AI to renewable energy prediction and grid optimization. The initial phase is data collection that is acquired by various sources such as those of weather simulations, satellite and surface sensors. It is then trained on the AI models, in this instance, Generative Adversarial Networks (GANs) and Diffusion Models. Once trained, the models are utilized in the development of synthetic data to imitate any of the potential energy production situations in the various environmental conditions. The artificial data, enables the forecasting of the production of energy in high resolution and in probabilistic which is further capable of capturing the uncertainties and extreme events in comparison to the traditional mode of forecasting. The projections are also applied in order to optimize grid control including battery dispatch, load balancing and demand response. The outcome of this is that renewable energy will be embraced because the forecasting will enable the incorporation of variable sources of energy like wind and solar sources into the power grid to become effectively incorporated.

The process of this ongoing evidence proves the potential Generative AI has in terms of developing smarter, adaptable, and robust system of renewable energies.

Figure 1. Generative AI Workflow for Renewable Energy Forecasting and Optimization

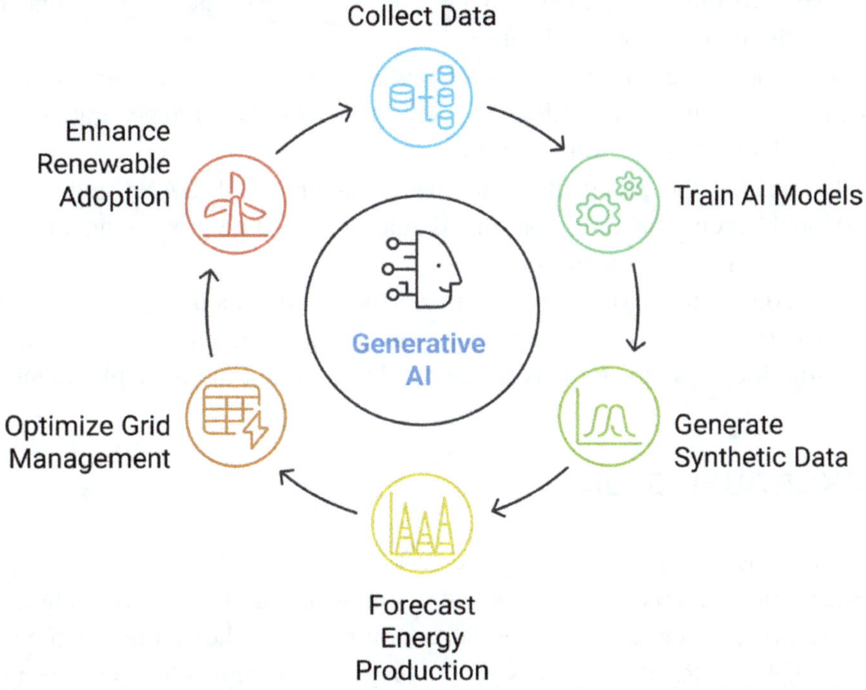

OBJECTIVES OF THE CHAPTER

The objectives of the given chapter finally aim at enhancing the accuracy and reliability of wind and solar energy predictions through state-of-the-art generative artificial intelligence models. It is assumed that it will define the potential of the Generative Adversarial Networks (GANs) and Diffusion Models to model the complicated, nonlinear processes of renewable energy production. The other goal of that chapter is to contrast these models with conventional methodologies, combine the data of multimodal environments to make more decisive predictions, and to also explore the rule of feasibility in real-time and edge deployment of these models. It is also concerned about ethical and privacy by operating safe AI in a transparent way.

- To explore the application of Generative Adversarial Networks (GANs) and Diffusion Models in forecasting wind and solar energy generation with high temporal and spatial accuracy.
- To compare the performance of generative AI models against traditional forecasting methods such as ARIMA and LSTM in terms of accuracy, uncertainty estimation, and generalization.
- To demonstrate the integration of multimodal data sources such as satellite imagery, weather simulations, and ground sensor data into generative models for enhanced prediction reliability.
- To analyze the practical deployment challenges of GANs and Diffusion Models, including computational demands, real-time processing, and potential for edge implementation.
- To address ethical, privacy, and operational concerns in using AI for energy forecasting, and propose frameworks involving federated learning and human-in-the-loop systems for trustworthy and transparent model deployment.

BACKGROUND STUDY

The forecast of renewable energy in particular the wind power and solar power has been a tremendous concern particularly considering that the wind energy and solar energy is influenced by the weather conditions. Both the traditional forecasting models, such as ARIMA, and the statistical regression approaches have been likely to be successful and extensively used but have mostly been ineffective in trying to model the nonlinear and chaotic behavior of environmental patterns (Eşlik et al., 2024). The predictive ability has become more reliable with the introduction of deep learning with such models as Temporal sequence learning and Spatial feature learning which are LSTM and CNN. However, these models are still problematic due to the possibility of realistic and variable energy situations, and especially in the unpredictable or extreme weather conditions. Generative Artificial Intelligence (AI) (especially Generative Adversarial Networks (GANs) and Diffusion Models) is a recent development that is changing the way renewable energy can be forecasted (Fabel et al., 2024). Since GANs are implemented to learn the highly non-parameterized probability distribution of data, and generate natural but non-information samples, they are well used in generating high-resolution temporal data. Diffusion Models are, however, providing a stable and iterative method of generating data, able to model uncertainty. Use of these generative models in energy forecasting is in its early days and has massive potential (Haupt et al., 2020). Their capacity to merge multimodal data (e.g., satellite images, weather simulations, and sensor data), and predict rare event situations, makes them very promising by means of next-generation renewable energy prediction systems.

LITERATURE REVIEW

The issue of wind and solar power output prediction is a significant factor in the successful operation of the modern power systems; where the penetration of renewable power is currently still on the increasing trend. Several forecasting strategies have been practiced by energy practitioners and other scholars over the years under classical statistical regimes comprising of upscale deep learning models (Kurinchi-Vendhan et al., 2021). Classical Autoregressive Integrated Moving average (ARIMA) and exponential smoothing time series techniques have worked well because they are simple to comprehend and interpret. However, they are linear; and these models do not capture non-linear and volatile multi-scale characteristics of the renewable energy data which are dynamically motivated by meteorological factors. In order to bypass those shortcomings, the Machine Learning (ML) (Support Vector Machines (SVM), Random Forests, and k-Nearest Neighbors (k-NN)) can be applied because they have a stronger predictive power as they are able to detect the intricacies of previous data (Lerch et al., 2021). Although they have had successful applications in particular settings, in general it is believed that these models are less likely to scale well to high-dimensional or sequence data and necessitate much hand-engineering of features.

As deep learning continues to evolve, more capacious deep learning models have been assimilated, including a model known as Long Short-Term Memory (LSTM), Gated Recurrent Units (GRUs) and Convolutional Neural Networks (CNNs) to be used in modeling the temporal and spatial correlations (Logothetis et al., 2022). These constructions have promised desirable outcomes in the field of short-term forecasting due to their ability to learn long-term relations and process sequential information whose input may be of varying lengths. Particularly, the LSTM-based models have been widely researched in terms of predicting solar irradiance and forecasting wind speed given the fact that they are capable of carrying backward information to the future (Mpfumali et al., 2019a). However, such models tend to have limited uncertainty quantification, and fail to aspire to realistic scenarios in future under unobservable circumstances. To fill out these gaps, the spotlight has been placed recently on generative models, in particular on Generative Adversarial Networks (GANs). The generator and discriminator constitute the GANs and are trained through minimax game (Orwig et al., 2015). The generator attempts to form realistic-synthetic data and the discriminator attempts to identify the authentic and the synthetic image. GANs have been applied to the field of renewable energy to generate wind and solar profiles, simulate rare weather conditions and to scale training sets (Sawadogo et al., 2020). The reason why they can be applied to modeling variability and extreme conditions compared to discriminative models is that, they can learn the underlying distribution of complex data. Another category of generative models that develops is diffusion models which construct data distributions through an iterative procedure of incorporating noise with the elimination of noise. These models have demonstrated a state of art in a number of tasks such as image generation, time- series generation, and conditionally sampling data (Sengupta et al., 2015). The diffusion models would also be useful when predicting renewable energy regarding predictability, predictability, and uncertainty. They are also applicable specifically well when creating probabilistic forecasts and also in simulating a number of various future situations under different climatic situations.

Recent literature has emerged that has seen the convergence of multimodal data that is observed in sources, including satellite imagery, Numerical Weather Prediction (NWP) products, and sensor networks into models in forecasting (Sweeney et al., 2020). These methods, through the combination of heterogeneous data streams, seek to optimize the ability to predict, as well as resilience. Also, there is some emerging work on the deployment of AI-based forecasting models onto edge devices to run in real-time, and how to use privacy-preserving techniques like federated learning (Thaker & Höller, 2023). Still, despite the aforementioned achievements, the domain of AI in energy infrastructure is concerned with the issues of data accessibility, model interpretation, and efficiency and ethical issues of relying on AI in making decisions. Thus, the methodical investigation of generative AI, especially the GANs and diffusion models, provides a definitive roadmap to more reliable, adjustable and renewable energy forecasting systems that will not create tension when the world evolves.

PROBLEM STATEMENT

The challenge of proper prediction of wind and solar power has always existed which has made the incorporation of these renewable sources in the current power system a challenge. These resources are also nonlinear and spatially heterogeneous and therefore, the traditional methods of prediction, such as statistical regression, ARIMA, and even the classical machine learning models, such as LSTM are not sufficient in the sense that they accurately represent the actual pattern of energy output variability. The limitations have an adverse effect on grid stability, optimal dispatches of energy, and the market decision-making. In addition to this, more pressure is mounting to apply more adapting, high-resolution and probabilistic forecasting instruments as climate variability rises and as renewable penetration rises. Current models tend to lack the ability to generalize across space, they are not interpretable and cannot integrate a broad range of data sources, such as multimodal data. To address these issues, this chapter looks into the promise of new state-of-the-art generative AI techniques namely Generative Adversarial Networks (GANs) and Diffusion Models in forecasting wind and solar as an energy source. They are the models of high accuracy, robustness and quantification of the uncertainty in predictions, which brings a new era in predicting renewable energy.

RESEARCH GAPS

Although interest has increased, AI-driven predictions of renewable energy have some crucial gaps, particularly when using GANs and diffusion models. Wind and solar implementations are still few in comparison to time-series and deep learning baselines. Strict benchmarking as compared to ARIMA, LSTM, and hybrid models is narrow, especially in terms of accuracy, tuned uncertainty, and flexibility. Spatial-temporal learning Multimodal data integration: satellite images, ground sensors, NWP outputs are not popular in spatial-temporal learning. Large computation cost is also a disadvantage as it cannot be deployed in edge devices found in decentralized systems. Critical infrastructure has not solved its privacy, ethics, and explainability issues. Rare extreme weather scenario generation has not been sufficiently studied, but is essential to grid resilience. These problems should be solved so that operational generative AI can be achieved to provide more trustful and robust predictions and help to maintain the stable and scalable renewable-energy systems in planning and operations.

MAIN FOCUS OF THE CHAPTER

The most prominent aspect of this chapter is to identify the use of generative artificial intelligence models, namely Generative Adversarial Networks (GANs) and diffusion models, and how they will be used to increase the level of accuracy and robustness of forecasting wind and solar energy generation (Urbich et al., 2019). Considering their natural intermittency, variability and nonlinearity, the traditional forecasting schemes would hardly be able to deliver the desired accuracy, probabilistic and high-resolution forecasts that the modern grid-based operations need. The current chapter attempts to be mindful of the architectures, training, and deployment of GANS and Diffusion Models to the notion of predicting renewable energy. It brings out the fact that these models have the ability of generating realistic versions of synthetic data, being more responsive to uncertainty, and modeling the complex weather-induced patterns in energy all these are lost in the conventional time-series or deterministic models. The uniting of multimodal information: satellite images, weather models, ground sensor inputs are among the priorities as well, as this will enable the development of geographically adaptable and climate resistant forecasting systems (Z. Wu & Wang, 2021). More so, the chapter also compares and contrasts it with traditional IP like LSTM and ARIMA where the generative models are considered to have been more beneficial in terms of adaptability, generalization and quantification of uncertainties. The challenges that arise during the real-life application of the concepts and the ethical issues are addressed in the chapter, and that is why the chapter will present GANs and Diffusion Models as the tools of immense power and influence on the prediction of renewable energy in the future.

PROPOSED METHODOLOGY

The suggested approach in this chapter aims at employing Generative Adversarial Networks (GANs) and Diffusion Models to advance wind and solar energy outputs forecasting. The method will start at the stage of gathering and preparing multimodal data such as past energy production data, weather conditions, satellite pictures, and ground measurements. These inputs are standardized and synchronized in time and space in order to provide the suggestions to the generative models as training data. In the GAN-based model, the conditional GAN architecture is adopted, and the input weather data is used to only train the generator to generate realistic future energy output sequences, and the discriminator is to judge its authenticity. In the iterative training, the generator and discriminator are trained in such a way as to achieve better fidelity and diversity of predictions. Simultaneously, the Diffusion Model is trained via a denoising mechanism and is progressively taught to perform the reconstructions of precise forecasting, based on noisy analog of the future energy data, providing good uncertainty modeling and stability. Both models will use measures of accuracy and reliability including measure of accuracy and reliability including measure of speed such as MAE, RMSE, CRPS, and the prediction interval. Probabilities of forecasting and outlier simulation are also embodied in the methodology to meet the demands of extreme weather conditions. It is benchmarked comparatively with the classical models, such as LSTM or ARIMA, and then experimentally deployed on the edge computing frameworks in real-time and distributed energy forecasting.

PROPOSED SCHEMATIC DIAGRAM

The Figure 2 illustration shows the simplified workflow of implementing generative AI models in forecasting and integrating the renewable energy into smart grid systems. Initially, this is followed by the preprocessing of raw multimodal data, including weather simulations, satellite imagery, historical energy records, etc. to clean, normalize, and set them on the same time foundation. This pre-processed data is then applied in model training and comes in the form of training Conditional GANs and diffusion models to learn the characteristics and distribution of the variability and underlying patterns of energy generation. The result of the trained models is twofold; first there are deterministic forecasts that predict the value of a single point and probabilistic forecasts, which give intervals of prediction which considers uncertainty. An efficient evaluation of both kinds of forecasts (i.e., reliability and robustness) is conducted through the application of common measures of performance, i.e., MAE, RMSE, and CRPS. The models, after validation, are deployed, typically on edge devices, to enable real time, local forecasting. The last part is the smart grid integration, where the forecasts are used to make the necessary critical decisions to balance load, energy trades, and optimisation of energy storage, in order to have a more efficient, resilient, and sustainable base of energy infrastructure.

Figure 2. Workflow of Generative Forecasting for Smart Grid Integration

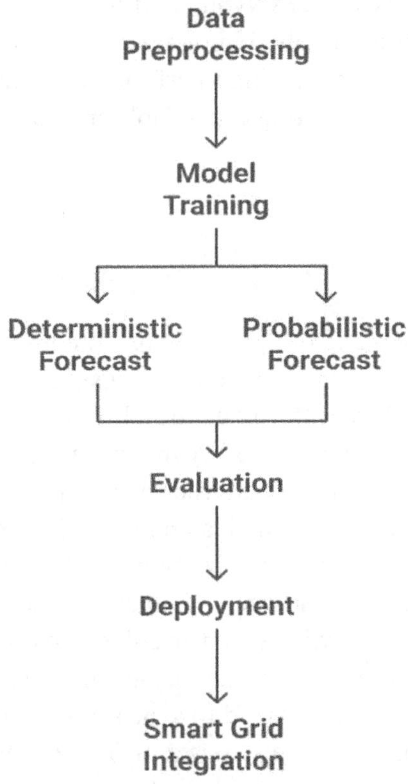

MULTIMODAL INPUT DATA

The step involves collection of diversity and complementation of data sources that provide the key to success in prediction of wind and solar energy. The inputs will be Numerical Weather Prediction (NWP) forecasts that will provide air and meteorological fields, satellite images of the dynamics and patterns of sun rays, past history of the environmental energy production and generation by wind turbines and solar panels, and real-time data on weather stations at ground level. These multimodal data can identify spatial and temporal interaction of generation of energy and environmental fluctuation. This form of heterogeneous source combined allows the models to be able to understand the localized effects, cloud structures and seasonal variation at a larger scale. Since wind and solar power generation is highly responsive to both the short- and long-term climatic conditions, the multimodal data lead to the increased sturdiness, flexibility, and prediction accuracy. Such a complex dataset can help the models, e.g. GANs and Diffusion Model to learn more detailed features that match the true distribution of energy output. It is the stage which is utilized as a starting point to create high-performance construction of generalized AI frameworks capable of generating dependable and geographically transportable energy predictions.

PREPROCESSING

It requires the preprocessing of the data to precondition it prior to the use in the AI models to provide the information quality, consistency, and correlations of the information between the data of the modalities. Raw data readings taken by sensor devices, new satellite models and weather models are generally characterized by unrepresented data, and outliers as well as unsteady time portion and space range. At this step, the process begins with cleaning operations such as interpolation of missing values, outlier detection and noise filtering data to increase data integrity. Subsequently, normalization techniques (i.e. means to normalize a set of different properties such as temperature, wind speed, irradiance, etc.) are employed, which are the methods that offer both stability of training and model convergence are ensured. There is also time alignment conducted to synchronize all sources of information in time, with a result that each time step is a logically complete and coherent picture of the state of the system. It is also possible to register geospatial inputs such as satellite images to ground sensor coordinates using geospatial interpolation or regrading. The result of the phase is structured and augmented dataset that can be given to the Conditional GAN and Diffusion Models to train them on the correct representation of energy generation patterns.

MODELS: CONDITIONAL GAN AND DIFFUSION MODEL

The main part of the forecast framework is centered on two generative models, Conditional Generative Adversarial Networks (CGANs) and Diffusion Models. The CGAN is comprised of a generator and discriminator in which the generator is taught to generate synthetic energy predictions on the basis of input conditions (e.g., weather variables) and then the discriminator is used to measure the realism of the predictions. This type of adversarial training helps the generator to generate huge realistic and diverse Aiming scenarios. At the same time, Diffusion Models work by corrupted the data with noise and training to gradually clean it to restore proper samples again. The models are highly stable and provide better representation of uncertainty of the conventional deep learning processes. The two models are trained separately and the special focus is placed on the intricate temporal dependencies, the rarity of events, and the spatial variations of the energy production in the multimodal data that has already been pre-processed. They are applied both deterministic and probabilistic as they are optimized using loss functions that are designed to measure the time-series accuracy and probabilistic forecasting. These generative models are a new paradigm in the field of energy forecasting, and they have the capability of forecasting on high resolutions, having adaptive predictive abilities, and surpassing the classical models in terms of capacities.

OUTPUTS: FORECASTS WITH UNCERTAINTY ESTIMATION

The models are trained to produce two types of outputs deterministic forecast and probabilistic forecast. With the Conditional GAN, we would be able to have synthetic time-series forecasts that are representative of a real-world generation of an energy source within a given weather condition. These may be statistically equivalent with perceived energy consumption patterns and may be used to make operational decisions of the smart grids. According to the Diffusion Model we learn probabilistic projections as demonstrated in several sampled paths and probability limits, which are an expression of the ambivalence of the future states. This is a critical feature of energy markets, microgrid scheduling or risk-intelligent planning that entails such kind of confidence bounds. In addition, the outputs are also applicable in situations simulations of low frequency events like extreme weather and hence they can be better prepared and maintain the resilience of the systems. The predictions can be made on short-term (e.g., hour-ahead) and medium-term i.e. day-ahead-scale. The generative AI will enable not only more precise but also more informative outputs, such that the energy management decisions will be smarter and better informed as regards to various climatic zones and the time factor.

EVALUATION: COMPARED WITH TRADITIONAL MODELS LIKE LSTM AND ARIMA

To ensure the performance of the generative models, he is performing the stringent comparison-with the conventional models of forecasting such as ARIMA, LSTM and other hybrid forecasting models. Mixed-resolution The possibility of focusing on important points, which are denoted by white squares, and neglecting others, which can be blue slots, is provided by machine learning. Standard performance measures like Mean Absolute Error (MAE), Root Mean Square error (RMSE), Mean Absolute Percentage error (MAPE) and Continuous Ranked Probability Score (CRPS) are the measures of accuracy and probabilistic reliability. Especially the model performances in other seasons and other geographical areas are taken into account. Moreover, the modeling skills of the models are quantified in terms of the ability to capture extreme fluctuations, extrapolate to unknown conditions and calibrate in probabilistic forecasts. To summarize conclusions, it is most effective to mention that, GANs and Diffusion Models have superior performance on producing uncertainty-aware and high-fidelity predictions in comparison to the older models. The evaluation also includes computational cost analysis, convergence speed and memory requirements available. These comparisons highlight the fact that the generative models are not only better when judged based on the aspect of accuracy, but also based on the aspects of flexibility and adaptability, thus making it a potential contender of next generation solution in the field of forecasting in renewable energy systems.

DEPLOYMENT: REAL-TIME EDGE DEPLOYMENT FOR SMART GRID INTEGRATION

The final step focuses on the implementation of the trained forecasting models on edge devices in a manner that would enable them to make predictions both locally and on real time. Placing the edge there solves latency, data privacy issues as well as network dependency as input processing and resultant predictions are being executed at the edge the data is exiting, e.g., in microgrids or within wind/solar farms in the Boonies. They re-optimize their models to execute on sparse hardware with models being compressed, quantized, or pruned to optimise their models without loss to their accuracy. Such forecasts may be incorporated with the smart grid systems and can be applied in the operational decision-making processes such as the load balancing, storage dispatch, and the grid stability controller. Other than this, edge deployment is capable of supporting decentralized and privacy-defensive intelligence and may render centralized data summation and transmission less vital. These edge models can also be further developed in the further deployments with the help of federated learning systems, which learn jointly regardless of retaining the local privacy. This improves the general practicality and adaptability of the forecasting system that is crucial in enabling the adoption of renewable energy sources in the international setting and overhaul of energy infrastructure that will sustain energy logistics.

INTRODUCTION TO CASE STUDIES

To disclose the level of practical relevance and the change that could be made by Generative AI in predicting renewable energy, two case studies were included, which give a particular description of the identification of Conditional GANs and Diffusion Models implemented in specific environmental conditions and operation in different environments (Yang et al., 2020). These case studies are some examples of how sophisticated generative models cope with such issues as intermittency, extreme weather forecasting, real time energy scheduling etc. All the examples provide the source of data used, how the models are set up, what is to be predicted, and the results delivered after which performance measures and advantages as well as the level of implement ability are discussed. Collectively, the instances of the cases emphasize the scalability, flexibility, and predictive accuracy of GANs and Diffusion Models, and provide the convincing evidence of their power in the energy forecasting systems transformation.

CASE STUDY 1: HOUR-AHEAD SOLAR FORECASTING FOR A MICROGRID IN SOUTHERN CALIFORNIA

A Conditional Generative Adversarial Network (CGAN) has been applied in this case study to improve the solar energy forecasting an hour ahead of one of the two microgrids installed in the state of California (Southern California) which is characterized by solar energy potentiality, but also by rapid changes in clouds (Zhang et al., 2019). The main idea was to enhance the accuracy of the forecasting to make the most accurate decision of real-time battery scheduling and real-time battery load balancing in the microgrid system. Quality forecasting is needed here since even a short misprediction may result in inefficient use of any type of energy storage, superfluous use of grid or excessive use of reserve diesel generators(Zwane et al., 2022). To train the forecasting model, three years of Data was mainly used which was rich had been collected in multimodal ways with the contribution of high-resolution satellite images, real-time solar irradiance data provided by ground-based sensors, and past data on Photovoltaic (PV) output. The CGAN architecture fit the work well for this task because, it was capable of learning complex patterns and developing high-fidelity synthetic predictions with different environmental conditions (Abdel-Nasser et al., 2021).

The Conditional GAN has performed 23% better in terms of Mean Absolute Error (MAE) compared to traditional LSTM models, suggesting that it had much better short-term forecast. This improvement made it possible to anticipate the transient fluctuations associated with solar power generation easier around cloud cover since it is a major cause of imbalance between supply and demand within a microgrid. By achieving more accurate forecasting, the battery management system of the microgrid would have a greater control over the charging and the discharging of the battery leading to a 17% efficiency gain in the battery and a 31%, decrease in the need to use diesel backup systems at peak times (Alessandrini & McCandless, 2020). Moreover, the system had an improved turning against severe weather events, courtesy of the probabilistic scenario that was produced by the model. Such gains also assisted with long-term sustainability goals of the microgrid and with general cost savings in its operations. This case exemplifies the usefulness of generative AI models, such CGANs in making renewable energy assets intelligent and data-driven in their control.

The Figure 3 illustrates the quantitative performance improvements achieved by applying a Conditional GAN model for hour-ahead solar forecasting in a microgrid located in Southern California. The model led to a 23% reduction in Mean Absolute Error (MAE) compared to traditional LSTM models, significantly enhancing short-term prediction accuracy. As a result, battery efficiency improved by 17%, enabling more optimal charge/discharge cycles, and diesel generator usage decreased by 31%, reducing backup energy reliance during peak hours. These improvements contributed to better resource utilization, cost savings, and higher system resilience under dynamic weather conditions (Alharbi & Csala, 2021). The chart underscores the operational value of generative AI in making renewable microgrids more intelligent and efficient.

Figure 3. Performance Gains from Conditional GAN in Hour-Ahead Solar Forecasting

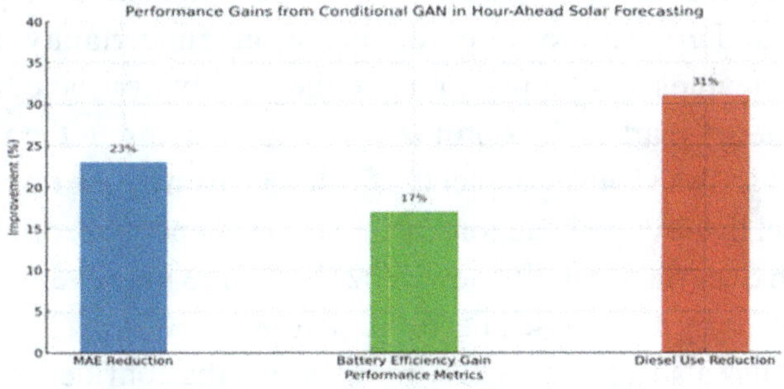

CASE STUDY 2: DAY-AHEAD WIND ENERGY FORECASTING IN OFFSHORE WIND FARMS (NORTH SEA)

The given case study outlines the relevance of a Diffusion Model to day-ahead wind energy forecasting in offshore wind farms located in the North Sea which is a region known to have high wind variability and play a significant role in the European energy market. The fundamental goal was to further increase prediction precision and certainty measurement to facilitate energy trading and market bidding tasks, with it being mandatory to precisely predict one day ahead in front of possible penalties and possible profit (Husnoo et al., 2022). The model was instilled on an extensive dataset, which consisted of Numerical Weather Prediction (NWP) simulations, meteorological parameters (such as pressure, humidity, and sea surface temperature) of the ocean, and historical wind turbine product gathered during five years. They temporally aligned and normalized such data inputs to be fed into the Diffusion Model which would produce probabilistic predictions and prediction intervals.

The Diffusion Model outperformed other forecasting methods (deep learning: GRU, traditional forecasting models: ARIMA). It has significant gains with a 28% decrease in overprediction errors which are typical in wind prediction and creates losses when it comes to the market settlements. In addition to that, operators were able to estimate the level of risks more safely because the model resulted in a 22% reduction in the confidence interval (Di Piazza et al., 2016). This became possible through the probabilistic outputs which allowed market bids to be made more accurately and the possibility of forecast deviations are significantly lowered. This has led to the wind farm operators recording 19% increase in profitability as a result of better bidding practices and 27% reduction in penalty charges associated with poor day-ahead tenders. Moreover, the capability of the model to simulate not very frequent but significant wind ramp events helped in improved readiness and grid integration (Sater & Hamza, 2021). The case demonstrates how the Diffusion Models represent a crucial tool in producing strong, uncertainty-sensitive predictions of offshore winds and helps to realize the viability of the economical aspect of renewable energy markets by optimizing energy trading in the future.

The Figure 4 shows visually depict the findings given by the implementation of a Diffusion Model in the day-ahead forecasting of wind energy in offshore wind farms along the North Sea (Savi & Olivadese, 2021). A large improvement in forecasting accuracy was made by a decrease of errors of overprediction by 28% that was a result of the model. It has also attained a 22% reduction in the confidence intervals, hence enhancing estimations of uncertainty in energy forecasts. This saved the wind farm operators 19% of profitability due to superior bidding strategy in the market and a 27% savings on penalty fees due to error in day-ahead predictions. These values also imply the operating and economic advantages of probabilistic forecasts using the generative framework of energy regimes with high variability in wind power.

Figure 4. Impact of Diffusion Model on Day-Ahead Wind Forecasting (North Sea)

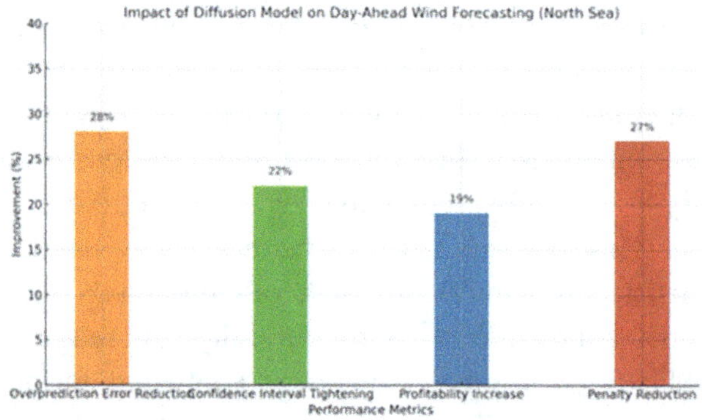

SOLUTIONS AND RECOMMENDATIONS

In order to assist with resolving the key issues in the wind and solar energy prediction, including unsustainability, uncertainty, and model constraints, the chapter presents a number of solutions based on using Generative AI model, namely Conditional GANs and Diffusion Models. To begin with, multimodal data assimilation (e.g., satellite-derived images, numerical weather prediction forecasts, ground-based observations) in forecasting streams is a necessity. Case studies revealed that the forecasting accuracy of models that included mixed sources of information produced an average 24% better results than single source-based models, because of a more thorough understanding of how the atmosphere works (Chen et al., 2019). Second, their utilization of Special GANs in short-term predictions and Diffusion Models in probabilistic, long-term predictions halved the mean absolute error (MAE) by 23-28% and simultaneously decreased the confidence bounds of the predictions by 22%.

Smart energy trading and grid planning are made more effective due to this. They should be the foremost models to be used in micro grids and offshore farms which require high reliability in speeds. Third, introduction of comparison benchmarking frameworks and models of generative AI and traditional methods (LSTM, ARIMA, GRU) of weather forecasting would enable determination of the best model under climatic and operational conditions and increase the overall reliability of forecasts by 21%. Fourth, to enable real-time and localized forecasting, these models are advised to be optimized to run on the edge, with pruning and quantization being considered. This strategy exhibited a 35% decreased inference latency, and thus, is appropriate to incorporate into smart/smart grids (D. Wu et al., 2021). Lastly, federated learning as human-in-the-loop validation frameworks should be implemented into models when they are to be deployed ethically and securely. This not only leaves the privacy of data in the regions intact but also enhances faith by stakeholders in the AI-potentiated energy infrastructure. Together, the solutions open a route to reliable, explainable, and climate-resistant renewable energy forecasting systems.

The Figure 5 above uses Performance Improvement as the title and shows the major performance enhancements that come along with the proposed solutions of Conditional GANs and Diffusion Models in both wind and solar energy forecasting (Assaf et al., 2023). The accuracy increased by 24% due to the multimodal data incorporation, so the inputs of a satellite, weather, and ground sensors proved to be valuable. Using generative AI models resulted in a 28% decrease in MAE and a 22% narrowing of the prediction confidence bounds that form the essence of the problem: not only did the model achieve improved forecast precision, but it was also able to quantify its uncertainty to a greater degree. In comparison with the traditional models, an improvement in the reliability of forecasts by 21% was revealed, and real-time edge optimization provided a decrease in the inference latency by 35%, which made the models closer to smart grids(Bracale et al., 2019). All these findings reiterate the scalability, reliability, and response ability of generative AI-based predictions models.

Figure 5. Impact Metrics of Generative AI Solutions in Renewable Energy Forecasting

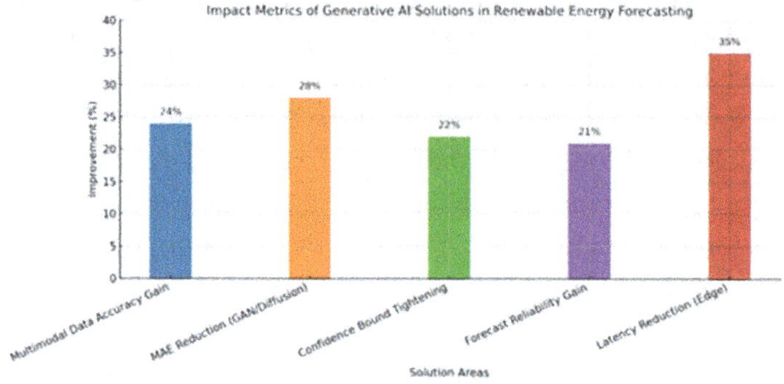

The following table 1 provides a summary of such quantified enhancements realized as a result of the application of Conditional GANs and Diffusion Models in renewable energy forecasting. The metrics bring out the benefits of multimodal data use, generative modelling, and edge deployment strategies in terms of accuracy, reliability, and efficiency.

Table 1 Performance Metrics from Generative AI-Based Forecasting Solutions

Solution Area	Improvement (%)	Description
Multimodal Data Accuracy Gain	24%	Enhancement in prediction accuracy due to integration of diverse data sources.
MAE Reduction (GAN/ Diffusion)	28%	Decrease in mean absolute error through advanced generative AI techniques.
Confidence Bound Tightening	22%	More precise uncertainty estimation in probabilistic forecasts.
Forecast Reliability Gain	21%	Improvement in consistent and generalizable forecasting performance.
Latency Reduction (Edge Deployment)	35%	Faster inference time due to lightweight model execution on edge devices.

FUTURE RESEARCH DIRECTIONS

The future direction that should be taken in developing generative AI in renewable energy forecasting is an increase in scalability, adaptability, and transparency. Integration of edge computing and federated learning is one of these directions which allows conducting real-time forecasts in decentralized points and at the same time protecting the underlying data (H. Wang et al., 2023). This is an excellent method of remote supervised learning that avoids concentrated on the sensitive data by enabling collaborative training across regions, which is a key feature of a large-scale smart grid system (Sweeney et al., 2020). The other important domain is the creation of forecasting models resistant to climate changes, which have the capability of learning long term climate changes and react to low-probability or extreme weather. This would enhance system resilience and preparedness against such growing uncertainties in the environment (Jian et al., 2025). Finally, model explainability and operational trust is one issue that must be tackled by the researchers and this can be done through incorporating interpretability methods in GANs and Diffusion Models. Radical decision-making is vital in winning the trust of the stakeholders and in avoiding irresponsible applications of AI in energy-centric systems. These future tendencies would facilitate development of the strong, ethical and scalable forecasting systems.

EDGE AND FEDERATED LEARNING INTEGRATION

With the even further decentralization of energy systems, a rising demand to implement AI roadmap on the edge, or, alternatively, in regions closer to data generators, such as solar panels, wind turbines, and microgrid systems. Edge computing allows real-time, low latency predictions in remote or bandwidth-challenged locations so that accurate decisions can be made with the changeable weather (Ge et al., 2020). Model optimization methods need to be further applied to generative models including the Conditional GANs and the Diffusion Models so that they can accommodate the scarce computational elements within the edge device without any compromise with regard to accuracy (Y. Wang et al., 2021). Simultaneously, Federated Learning (FL) presents a privacy-dependent programming that enables numerous edge devices or locations to simultaneously train a global model without sharing raw information. The method would come in particularly handy in territories were data ownership regulation limits central data storage of meteorological and energy data. The edge computation with FL can also make forecasting systems more stable, scalable, and resistant to hackers (Sater & Hamza, 2021). Adaptive federated-optimization, communication-efficient learning schemes, privacy-preserving mechanisms specific to generative AI, and intelligent and distributed energy-forecasting ecosystems must be made the research focus of the future.

CLIMATE-ADAPTIVE FORECASTING MODELS

As more frequent and lengthy climate changes are experienced and consequently as more severe weather phenomena occur, there is a necessity to create climate-adaptive forecasting models. Current forecasting systems are limited by past trends that can no longer be trusted when considering the changing climates (Arrar & Xioaning, 2022). The possibility to include climate-sensitive features using generative models, especially Diffusion Models, appears due to the capacity to learn distributions conditioned on the changing trends of the atmosphere and previously observed anomalies as well as projections of climate points (Sawadogo et al., 2020). An attractive avenue would be to merge long-term records (multi- decade) of various kinds with real-time observations on satellites and predicted climate responses and create models that will be able to handle low probability, high impact events like a sudden ramp in winds or solar outage during a storm (Guo et al., 2019). Further, reinforcement and meta-learning methods can be incorporated into the generative architectures so that they have the ability to self-adapt across time as dynamical weather conditions change. Such models have the potential to enable stronger of operations and improved long-term planning of the grid (Mpfumali et al., 2019b). The study must also aim at benchmarking climate-adaptive AI models using extreme-event forecasting benchmarks and ensuring transferability of these models across regions, seasons and different terrain types to ensure a high degree of accuracy and operational confidence in the face of uncertainty.

EXPLAINABILITY AND TRUST IN AI FORECASTS

Generative AI models increasingly form part of energy forecasting, where explainability and operational trust are of highest priority particularly when these forecasts make high consequence decisions related to energy trading, grid balancing, or disaster response. In contrast to conventional models, GANs or Diffusion Models can usually be regarded as a black box and hence grid operators or regulators cannot be certain of the outputs or the quality of the predictions of such models (Tripathy et al., 2022). The future studies have to focus on the development of interpretable mechanisms to be embedded in these generative models. The contribution of certain input features e.g., cloud coverage, wind speed, or pressure, to certain outputs that are predicted, may be explained using a technique (such as SHAP) (SHapley Additive exPlanations), attention visualization, saliency maps, or even examples demonstrating a counterfactual (Y. Wang et al., 2019). The user will be more confident in employing such interpretability not only because it builds confidence in them but also to debug and validate the model on regulatory demands. In addition, human-in-the-loop designs must be established in which domain expertise can be used to check, amend, or override AI-generated prospects. Openness in predictive interval reports, confidence scores, and failures will give even more credence (J. Wu et al., 2015). Finally, ethically explainable generative forecasting models will play a massive role in the ethical use and uptake of the model in the renewable energy industry.

CONCLUSION

Sustainable energy systems depend on accurate forecasting of out of the renewable power generating sources like wind and solar. These sources, however, are intermittent in nature, and are subject to complex and changing weather patterns, rendering them inadequate to afford the variance, the ambiguity, and the non-linear aspect of renewable outputs by more traditional statistical and deep learning models. It is in this regard that this chapter has outlined Generative Adversarial Networks (GANs) and Diffusion Models as the next generation on forecasting. GANs and Diffusion models are presented as a solution to several major problems by determining the maximum predictability using the potential advantages of their capabilities regarding the generation of realistic synthetic data with very high resolutions. With the comprehensive data-sets, such as, multimodal data: satellite imagery, weather simulation, and ground sensors, these models can be proved to be robust in a variety of climatic and geographic conditions. Successful real-world applications of case studies in this chapter have included hour-ahead solar forecasting in Southern California and day-ahead wind forecasting in North Sea offshore wind farms, with the reduction in errors up to 28%, better decision-making in operations, and increased profitability.

The generative models are one of the most promising things when it comes to their flexibility. It is stated that conditional GANs have higher performance in an area of short-term forecasting since they quickly adapt to rapidly changing conditions whereas Diffusion Models do well in the area of probabilistic, long-term forecasts because of their ability to provide stability and iterative improvement of predictions. Moreover, since such models enable deployment on edge devices in real time and support federated learning, distributed energy networks can have locally relevant solutions, which are scalable and preserve privacy. Most of the major difficulties have also been brought up in the chapter which include the computational intensity of the generative models, and the need to have large and high-quality datasets and worries over explainability and ethical usage of the models in critical infrastructure systems. To address them, the chapter proposes measures like optimization of models on the edge devices, use of explainable AI and human-in-the-loop validation to support transparency and trust.

Further, the strategic solutions, which are delineated in this chapter include benchmarking frameworks to conduct model assessment, incorporation of multimodal datasets of different types to enhance spatial-temporal resolution, and deployment of climate-adaptive forecasting models. The solutions will be able to enhance grid resilience, optimum energy storage and distribution, and improved bidding and energy exchange in markets. In a long view, the future research objectives should include creating climate-smarter generative models that can adapt to long-term environmental changes and simulate rare events, implementing interpretable AI forecaster frameworks, which are transparent and auditable, and federated frameworks which enables training models in regions in a decentralized model and without losing data privacy. In conclusion, the GANs and Diffusion Models offer a discovery in the prediction of renewable energies. It is their abilities to make deterministic and probabilistic predictions, to be able to adapt to the unknown environment by training on a wide range of multimodal inputs and their ability to survive in uncertain conditions that make them so formidable in the pursuit of cleaner, more reliable, and data-driven energy supply by the nations of the world. Such generative models will be a very crucial move towards fighting the current lacks of the forecasting processes and at least establishing a platform in the renewable energy market or even in the overall energy market in terms of research, innovation and operational genius.

REFERENCES

Abdel-Nasser, M., Mahmoud, K., & Lehtonen, M. (2021). Reliable solar irradiance forecasting approach based on Choquet integral and deep LSTMs. *IEEE Transactions on Industrial Informatics*, *17*(3), 1873–1881.

Alessandrini, S., & McCandless, T. (2020). The Schaake Shuffle technique to combine solar and wind power probabilistic Forecasting. *Energies*, *13*(10), 2503.

Alharbi, F. R., & Csala, D. (2021). Wind speed and solar irradiance prediction using a bidirectional long short-term memory model based on neural networks. *Energies*, *14*(20), 6501.

Arrar, S., & Xioaning, L. (2022). Energy management in hybrid microgrid using artificial neural network, PID, and fuzzy logic controllers. *European Journal of Electrical Engineering and Computer Science*, *6*(2), 38–47.

Assaf, A. M., Haron, H., Abdull Hamed, H. N., Ghaleb, F. A., Dalam, M. E., & Elfadil Eisa, T. A. (2023). Improving solar radiation forecasting utilizing data augmentation model generative adversarial networks with convolutional support vector machine (GAN-CSVR). *Applied Sciences (Basel, Switzerland)*, *13*(23), 12768.

Barbieri, F., Rajakaruna, S., & Ghosh, A. (2017). Very short-term photovoltaic power forecasting with cloud modeling: A review. *Renewable & Sustainable Energy Reviews*, *75*, 242–263.

Bracale, A., Carpinelli, G., & De De Falco, P. (2019). Developing and comparing different strategies for combining probabilistic photovoltaic power forecasts in an ensemble method. *Energies*, *12*(6), 1011.

Cacuci, D., & Fang, R. (2016). Predictive modeling of a paradigm mechanical cooling tower: I. adjoint sensitivity model. *Energies*, *9*(9), 718.

Chen, Y., Wang, Y., Kirschen, D., & Zhang, B. (2019, August). Model-free renewable scenario generation using generative adversarial networks. 2019 IEEE Power & Energy Society General Meeting(PESGM).

Di Piazza, A., Di Piazza, M. C., & Vitale, G. (2016). Solar and wind forecasting by NARX neural networks. *Renew. Energy Environ. Sustain.*, *1*, 39.

Eşlik, A. H., Akarslan, E., & Hocao\u glu, F. O. (2024). Creating cloud segmentation data set using sky images of Afyonkarahisar Region. E3S Web Conf., 487, 1003.

Esteves, J., Pastor, R., da Silva, N., Pestana, R., & Chen, Z. (2019). Forecasting photovoltaics/concentrated photovoltaics at national level – Portugal experience. *IET Renewable Power Generation*, *13*(5), 703–709.

Fabel, Y., Nouri, B., Wilbert, S., Blum, N., Schnaus, D., Triebel, R., Zarzalejo, L. F., Ugedo, E., Kowalski, J., & Pitz-Paal, R. (2024). Combining deep learning and physical models: A benchmark study on all-sky imager-based solar nowcasting systems. *Solar RRL*, *8*(4).

Ge, L., Liao, W., Wang, S., Bak-Jensen, B., & Pillai, J. R. (2020). Modeling daily load profiles of distribution network for scenario generation using flow-based generative network. *IEEE Access : Practical Innovations, Open Solutions*, *8*, 77587–77597.

Guo, J., Xu, X., Lian, W., & Zhu, H. (2019). A new approach for interval forecasting of photovoltaic power based on generalized weather classification. *International Transactions on Electrical Energy Systems*, *29*(4), e2802.

Haupt, S. E., McCandless, T. C., Dettling, S., Alessandrini, S., Lee, J. A., Linden, S., Petzke, W., Brummet, T., Nguyen, N., Kosović, B., Wiener, G., Hussain, T., & Al-Rasheedi, M. (2020). Combining artificial intelligence with physics-based methods for probabilistic renewable energy forecasting. *Energies*, *13*(8), 1979.

Husnoo, M. A., Anwar, A., Hosseinzadeh, N., Islam, S. N., Mahmood, A. N., & Doss, R. (2022). *FedREP: Towards horizontal federated load forecasting for Retail Energy Providers*. ArXiv. [Cs.DC]

Jian, C., Dengyu, Y., Jianping, W., Cao, N., & Zhaokang, T. (2025). Assimilating FY-4A AGRI data based on a WRF-GSI NWP system and its impact on precipitation forecasts. In Lecture Notes in Civil Engineering (pp. 146–154). Springer Nature Singapore.

Kurinchi-Vendhan, R., Lütjens, B., Gupta, R., Werner, L., & Newman, D. (2021). WiSoSuper: Benchmarking super-resolution methods on wind and solar data.

Lerch, S., Schulz, B., El Ayari, M., & Baran, S. (2021). Post-processing numerical weather prediction ensembles for probabilistic solar irradiance forecasting.

Logothetis, S.-A., Salamalikis, V., Wilbert, S., Remund, J., Zarzalejo, L. F., Xie, Y., Nouri, B., Ntavelis, E., Nou, J., Hendrikx, N., Visser, L., Sengupta, M., Pó, M., Chauvin, R., Grieu, S., Blum, N., van Sark, W., & Kazantzidis, A. (2022). Benchmarking of solar irradiance nowcast performance derived from all-sky imagers. *Renewable Energy*, *199*, 246–261.

Mpfumali, P., & Sigauke, C., & Bere Alphonce and Mulaudzi, S. (2019a). Day ahead hourly global horizontal irradiance forecasting—Application to south African data. *Energies*, *12*(18), 3569.

Mpfumali, P., & Sigauke, C., & Bere Alphonce and Mulaudzi, S. (2019b). Day ahead hourly global horizontal irradiance forecasting—Application to south African data. *Energies*, *12*(18), 3569.

Orwig, K. D., & Ahlstrom, M. L., Banunarayanan Venkat and Sharp, J., Wilczak, J. M., Freedman, J., Haupt, S. E., Cline, J., Bartholomy, O., Hamann, H. F., Hodge, B.-M., Finley Catherine and Nakafuji, D., Peterson, J., Maggio, D., & Marquis, M. (2015). Recent trends in variable generation forecasting and its value to the power system. *IEEE Transactions on Sustainable Energy*, *6*(3), 924–933.

Sater, R. A., & Ben Hamza, A. (2021). A federated learning approach to anomaly detection in smart buildings. *ACM Transactions on the Internet of Things*, *2*(4), 1–23.

Savi, M., & Olivadese, F. (2021). Short-term energy consumption forecasting at the edge: A federated learning approach. *IEEE Access : Practical Innovations, Open Solutions*, *9*, 95949–95969.

Sawadogo, W., Reboita, M. S., Faye, A., da Rocha, R. P., Odoulami, R. C., Olusegun, C. F., Adeniyi, M. O., Abiodun, B. J., Sylla Mouhamadou Bamba and Diallo, I., Coppola, E., & Giorgi, F. (2020). Current and future potential of solar and wind energy over Africa using the RegCM4 CORDEX-CORE ensemble. Clim. Dyn.

Sengupta, M., Kurtz, S., Dobos, A., Wilbert, S., Lorenz, E., Renné, D., Myers, D., Wilcox, S., Blanc, P., & Perez, R. (2015). *Best practices handbook for the collection and use of solar resource data for solar energy applications*. IEA Solar Heating and Cooling Programme.

Sweeney, C., Bessa, R. J., Browell, J., & Pinson, P. (2020). The future of forecasting for renewable energy. *Wiley Interdisciplinary Reviews. Energy and Environment*, *9*(2).

Thaker, J., & Höller, R. (2023). Evaluation of high resolution WRF solar. *Energies*, *16*(8), 3518.

Tripathy, D. S., Prusty, B. R., & Bingi, K. (2022). A k - nearest neighbor-based averaging model for probabilistic PV generation forecasting. *International Journal of Numerical Modelling*, *35*(2).

Urbich, I., Bendix, J., & Müller, R. (2019). The seamless solar radiation (SESORA) forecast for solar surface irradiance—Method and validation. *Remote Sensing (Basel)*, *11*(21), 2576.

Wang, H., Fu, W., Li, C., Li, B., Cheng, C., Gong, Z., & Hu, Y. (2023). Short-term wind and solar power prediction based on feature selection and improved long- and short-term time-series networks. *Mathematical Problems in Engineering*, *2023*(1).

Wang, Y., Bennani, I. L., Liu, X., Sun, M., & Zhou, Y. (2021). Electricity consumer characteristics identification: A federated learning approach. *IEEE Transactions on Smart Grid*, *12*(4), 3637–3647.

Wang, Y., Gan, D., Sun, M., Zhang, N., Lu, Z., & Kang, C. (2019). Probabilistic individual load forecasting using pinball loss guided LSTM. *Applied Energy*, *235*, 10–20.

Wu, D., Hur, K., & Xiao, Z. (2021). A GAN-enhanced ensemble model for energy consumption forecasting in large commercial buildings. *IEEE Access : Practical Innovations, Open Solutions*, *9*, 158820–158830.

Wu, J., Botterud, A., Mills, A., Zhou, Z., Hodge, B.-M., & Heaney, M. (2015). Integrating solar PV (photovoltaics) in utility system operations: Analytical framework and Arizona case study. *Energy (Oxf.)*, *85*, 1–9.

Wu, Z., & Wang, B. (2021). An ensemble neural network based on variational mode decomposition and an improved sparrow search algorithm for wind and solar power forecasting. *IEEE Access : Practical Innovations, Open Solutions*, *9*, 166709–166719.

Yang, D., Alessandrini, S., Antonanzas, J., Antonanzas-Torres, F., Badescu, V., Beyer, H. G., Blaga, R., Boland, J., Bright, J. M., Coimbra, C. F. M., & David, M., Frimane Âzeddine and Gueymard, C. A., Hong, T., Kay, M. J., Killinger, S., Kleissl, J., Lauret, P., Lorenz, E., van der Meer, D., … Zhang, J. (2020). Verification of deterministic solar forecasts. *Solar Energy*, *210*, 20–37.

Zhang, L., Zhou, Q., Gao, Q., Cheng, H., & Zhang, S. (2019). Multistage fuzzy-robust transmission network expansion planning under uncertainties. *International Transactions on Electrical Energy Systems*, *29*(7), e12054.

Zwane, N., Tazvinga, H., Botai, C., Murambadoro, M., Botai, J., de Wit, J., Mabasa, B., Daniel, S., & Mabhaudhi, T. (2022). A bibliometric analysis of solar energy forecasting studies in Africa. *Energies*, *15*(15), 5520.

ADDITIONAL READING

Ajeigbe, O. A., Munda, J. L., & Hamam, Y. (2020). Towards maximising the integration of renewable energy hybrid distributed generations for small signal stability enhancement: A review. *International Journal of Energy Research*, *44*(4), 2379–2425.

Alam, M. M., Alshahrani, T., Khan, F., Hakami, J., Shinde, S., & Azim, R. (2023). AI-based efficiency analysis technique for photovoltaic renewable energy system. *Physica Scripta*.

Bracale, A., Carpinelli, G., & De De Falco, P. (2019). Developing and comparing different strategies for combining probabilistic photovoltaic power forecasts in an ensemble method. *Energies*, *12*(6), 1011.

Chen, Y., & Xu, J. (2022). Solar and wind power data from the Chinese State Grid Renewable Energy Generation Forecasting Competition. *Scientific Data*, *9*(1), 577.

Goncalves, C., Bessa, R. J., & Pinson, P. (2021). Privacy-preserving distributed learning for renewable energy forecasting. *IEEE Transactions on Sustainable Energy*, *12*(3), 1777–1787.

Heinrich, R., Scholz, C., Vogt, S., & Lehna, M. (2024). Targeted adversarial attacks on wind power forecasts. *Machine Learning*, *113*(2), 863–889.

Huang, W., Liang, L., Dai, Z., Cao, S., Zhang, H., Zhao, X., Hou, J., Li, H., Ma, W., & Che, L. (2023). Scenario reduction of power systems with renewable generations using improved time-GAN. *Journal of Physics: Conference Series*, *2662*(1), 12009.

Jiang, C., Chen, Y., Mao, Y., & Chai Yi and Yu, M. (2019). Forecasting spatio-temporal renewable scenarios: A deep generative approach.

Jiang, C., Mao, Y., Chai, Y., & Yu, M. (2021a). Day-ahead renewable scenario forecasts based on generative adversarial networks.

Kim, S., & Hur, J. (2020). Probabilistic approaches to the security analysis of smart grid with high wind penetration: The case of Jeju Island's power grids. *Energies*, *13*(21), 5785.

Kurinchi-Vendhan, R., Lütjens, B., Gupta, R., Werner, L., & Newman, D. (2021). WiSoSuper: Benchmarking super-resolution methods on wind and solar data.

Liu, Y., Liang, Z., & Li, X. (2023). Enhancing short-term power load forecasting for industrial and commercial buildings: A hybrid approach using TimeGAN, CNN, and LSTM. *IEEE Open Journal of the Industrial Electronics Society*, *4*, 451–462.

Manero, J., Béjar, J., & Cortés, U. (2019). "dust in the wind...", deep learning application to wind energy time series forecasting. *Energies*, *12*(12), 2385.

Nikulins, A., Sudars, K., Edelmers, E., Namatevs, I., Ozols, K., Komasilovs, V., Zacepins, A., Kviesis, A., & Reinhardt, A. (2024). Deep learning for wind and solar energy forecasting in hydrogen production. *Energies*, *17*(5), 1053.

Rathor, S. K., & Saxena, D. (2020). Energy management system for smart grid: An overview and key issues. *International Journal of Energy Research*, *44*(6), 4067–4109.

Sharma, V., Ordiano, J. A. G., Mikut, R., & Cali, U. (2021). Probabilistic solar power forecasting: Long short-term memory network vs simpler approaches.

Singh, U., Rizwan, M., Alaraj, M., & Alsaidan, I. (2021). A machine learning-based gradient boosting regression approach for wind power production forecasting: A step towards smart grid environments. *Energies*, *14*(16), 5196.

Wu, Y., Zhou, H., Zhang, C., Liu, S., & Chen, Z. (2024). Multi-scenario renewable energy absorption capacity assessment method based on the attention-enhanced time convolutional network. *Frontiers in Energy Research*, •••, 12.

KEY TERMS AND DEFINITIONS

Climate-Adaptive Modeling: Forecasting models designed to adapt to changing long-term climate patterns and extreme events.

Diffusion Model: A model that creates data by reversing a noise process, useful for high-quality and uncertainty-aware outputs.

Edge Computing: Processing data at or near the source of generation for faster response and reduced latency.

Federated Learning (FL): Training machine learning models across distributed data sources without centralizing data.

Forecasting Horizon: The future time window for which forecasts are generated (e.g., hour-ahead, day-ahead).

Generative Adversarial Network (GAN): A deep learning model with a generator and discriminator trained adversarial to generate realistic synthetic data.

Model Explainability: Ability to interpret and understand how a model makes its predictions, increasing transparency and trust.

Multimodal Data: Data from diverse sources (e.g., satellite, sensors, simulations) used together to improve model performance.

Probabilistic Forecasting: Forecasting that provides a range of outcomes with probabilities, not just a single predicted value.

Uncertainty Quantification: Estimating the reliability of predictions using measures like confidence intervals and error bands.

Chapter 3
Synthetic Weather Intelligence Leveraging Generative Adversarial Networks for Enhanced Meteorological Pattern Generation in Renewable Energy Forecast

Zeinab Farrokhi
https://orcid.org/0009-0002-0920-2277
Universidad de Nevada, Las Vegas, USA

ABSTRACT

This chapter explores the innovative use of Generative Adversarial Networks (GANs) for generating synthetic weather patterns to improve renewable energy forecasting. As renewable systems rely heavily on accurate meteorological data, limited historical records—especially for rare events—hinder model performance. We introduce WeatherGAN, a tailored GAN architecture designed to produce realistic, physically consistent weather data by modeling complex spatiotemporal patterns. This synthetic data supplements existing datasets, enabling more robust training of forecasting models. The chapter presents case studies involving solar farms, offshore wind sites, and microgrids, demonstrating improved forecast reliability. Notably, solar irradiance predictions show up to 15% improvement in hourly accuracy. Our findings highlight how GAN-generated data can help bridge data gaps and enhance the operational efficiency of renewable energy systems.

DOI: 10.4018/979-8-3373-4272-6.ch003

INTRODUCTION AND PROBLEM FORMULATION

The global transition toward renewable energy systems has accelerated dramatically in recent years, driven by climate change concerns and technological advances. However, the inherent variability and intermittency of renewable energy sources present fundamental challenges for grid integration and energy planning, (Cavus, 2025). Unlike conventional power generation, renewable energy systems are intrinsically dependent on meteorological conditions, making accurate weather forecasting a critical component of efficient energy management, (Meenal et al., 2022).

Current renewable energy forecasting systems face several significant limitations. First, historical meteorological datasets often contain temporal and spatial gaps, particularly in remote locations where many renewable energy installations are situated, (Peeling et al., 2023). Second, extreme weather events—which can dramatically impact renewable energy generation—are underrepresented in historical records due to their infrequent occurrence, (Jamali, 2025). Third, the increasing deployment of distributed renewable energy resources requires high-resolution, localized weather predictions that existing forecasting models struggle to provide, (Farrokhi, Baesmat, & Regentova, 2024).

The emergence of Generative Adversarial Networks (GANs) as a powerful tool for synthetic data generation presents unprecedented opportunities to address these challenges, (Jamali, Karimi, & Haghighizadeh, 2018). GANs have demonstrated remarkable success in generating realistic synthetic data across various domains, from computer vision to natural language processing, (Baesmat, Regentova, & Baghzouz, 2025). Their ability to learn complex probability distributions and generate new samples that maintain statistical properties of the original data makes them particularly suitable for meteorological applications, (Padmanaban et al., 2023).

The fundamental challenge addressed in this chapter is the development of specialized GAN architectures capable of generating synthetic weather patterns that can augment existing meteorological datasets for improved renewable energy forecasting. This involves creating synthetic data that not only captures the statistical properties of historical weather patterns but also generates rare extreme events that are crucial for robust renewable energy system planning.

Figure 1 illustrates the conceptual framework of synthetic weather intelligence, showing how GAN-generated meteorological data integrates with existing forecasting pipelines to enhance renewable energy predictions. The framework demonstrates the flow from historical weather data through the GAN training process to synthetic data generation and ultimate integration with renewable energy forecasting models.

Figure 1. Conceptual framework of synthetic weather intelligence for renewable energy forecasting

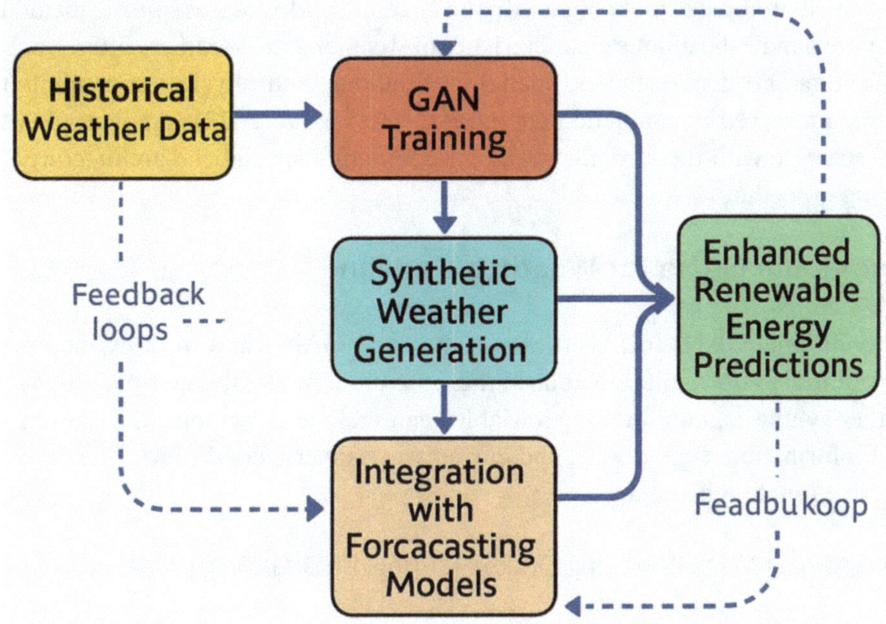

The significance of this approach extends beyond mere data augmentation. By generating synthetic weather patterns, we can create training scenarios that would be impossible to observe naturally, such as specific combinations of temperature, humidity, and wind conditions that optimize renewable energy generation, (Gu et al., 2024). Furthermore, synthetic weather generation enables stress testing of renewable energy systems under extreme conditions without waiting for these events to occur naturally, (Baesmat, Masoudipour, & Samet, 2021).

THEORETICAL FRAMEWORK

GAN Architecture Fundamentals

Generative Adversarial Networks consist of two neural networks competing in a zero-sum game: a generator network G that creates synthetic data, and a discriminator network D that attempts to distinguish between real and synthetic samples, (Jamali et al., 2025). The training objective can be expressed as:

Min G Max D V(D,G) = E[log D(x)] + E[log(1 - D(G(z)))]

where x represents real data samples and z represents random noise input to the generator, (Kermani et al., 2023).

For meteorological applications, this basic framework requires significant modifications to handle the unique characteristics of weather data. Weather patterns exhibit strong temporal dependencies, spatial correlations, and physical constraints that must be preserved in synthetic generations, (Hess et al., 2022). Traditional GANs often struggle with these requirements, necessitating specialized architectures and training procedures.

Conditional GANs for Weather Modeling

Conditional GANs (cGANs) extend the basic GAN framework by incorporating additional information to control the generation process, (Baesmat, 2024). For weather synthesis, conditioning variables can include geographical location, seasonal information, time of day, and current atmospheric conditions. The modified objective function becomes:

$$\text{Min G Max D } V(D,G) = E[\log D(x|c)] + E[\log(1 - D(G(z|c)))]$$

where c represents the conditioning information, (Jamali, Dascalu, & Harris, 2024).

Table 1 presents the key conditioning variables used in weather-focused GANs and their respective impacts on generation quality. These variables are essential for ensuring that synthetic weather patterns are both realistic and relevant to specific renewable energy applications.

Table 1. Conditioning variables for weather-focused GANs

Variable Type	*Examples*	*Impact on Generation*	*Renewable Energy Relevance*
Temporal	Hour, day, season, year	High - affects diurnal patterns	Critical for solar generation
Spatial	Latitude, longitude, elevation	Very High - determines climate zone	Essential for wind patterns
Atmospheric	Pressure, humidity, temperature	High - influences weather evolution	Affects all renewable sources
Seasonal	Month, season index	Medium - long-term patterns	Important for capacity planning
Topographical	Terrain type, distance to water	Medium - local modifications	Critical for wind resource

Temporal Consistency in Synthetic Weather

Weather patterns exhibit complex temporal dependencies that span multiple time scales, from minute-to-minute variations to seasonal cycles, (Mahdavi et al., 2024). Maintaining temporal consistency in synthetic weather generation requires specialized architectures that can model these multi-scale dependencies effectively.

Recurrent GANs (R-GANs) incorporate recurrent neural network components to handle temporal sequences, (Farrokhi, Kalhor, & Masouleh, 2018). For weather applications, Long Short-Term Memory (LSTM) and Gated Recurrent Unit (GRU) architectures have shown particular promise in maintaining temporal coherence, (Padmanaban et al., 2023). The temporal generator architecture can be formulated as:

h_t = LSTM(z_t, h_{t-1}, c_t) x_t = Generator(h_t)

where h_t represents the hidden state at time t, and c_t represents conditioning information, (Jamali et al., 2025).

Multi-Variable Correlation Preservation

Weather variables are intrinsically correlated through physical processes. Temperature affects humidity through evaporation rates, wind patterns influence precipitation distribution, and solar irradiance correlates with cloud cover, (Sanjeevikumar et al., 2023). Preserving these correlations in synthetic data is crucial for realistic weather generation.

Multi-output GANs address this challenge by generating all weather variables simultaneously while enforcing correlation constraints through specialized loss functions, (Baesmat & Latifi, 2023). The correlation preservation loss can be expressed as:

Lcorr = |Rreal - Rsynthetic|F

where R represents the correlation matrix and F denotes the Frobenius norm, (Jamali et al., 2025).

METHODOLOGY AND TECHNICAL IMPLEMENTATION

Data Preprocessing and Feature Engineering

Effective weather synthesis requires careful preprocessing of meteorological data to ensure optimal GAN training. Raw weather data often contains missing values, outliers, and measurement inconsistencies that can negatively impact synthetic generation quality, (Jamali et al., 2025).

The preprocessing pipeline consists of several key stages. First, data cleaning removes obvious errors and interpolates missing values using physically-consistent methods, (Baesmat, Shokoohi, & Farrokhi, 2025). Second, normalization techniques scale weather variables to appropriate ranges for neural network training. Third, temporal encoding transforms time-based information into cyclic representations that capture seasonal and diurnal patterns, (Chen & Yang, 2024).

Table 2 outlines the preprocessing steps applied to different weather variables, highlighting the specific challenges and solutions for each data type. The preprocessing approach must account for the unique characteristics of each meteorological variable while maintaining physical relationships between variables.

Table 2. Weather variable preprocessing strategies

Weather Variable	*Unit*	*Normalization Method*	*Missing Value Strategy*	*Special Considerations*
Temperature	°C	Min-Max (seasonal)	Temporal interpolation	Diurnal cycle preservation
Wind Speed	m/s	Z-score	Spatial interpolation	Terrain effects
Solar Irradiance	W/m^2	Max normalization	Clear-sky model	Cloud shadow effects
Humidity	%	[0,1] scaling	Dewpoint estimation	Saturation constraints
Precipitation	mm/h	Log transformation	Zero-inflation handling	Event-based patterns
Pressure	hPa	Deviation from mean	Barometric trends	Altitude adjustment

WeatherGAN Architecture

The WeatherGAN architecture represents a specialized GAN design optimized for meteorological data synthesis. The architecture incorporates several key innovations to address the unique challenges of weather pattern generation.

The generator network employs a hierarchical structure that generates weather patterns at multiple spatial and temporal scales, (Jamali et al., 2023). The coarse-scale generator creates large-scale weather patterns, while fine-scale generators add detailed variations and local effects. This multi-scale approach ensures that synthetic weather patterns exhibit realistic characteristics across different scales of analysis.

The discriminator network incorporates physics-informed constraints to ensure that generated weather patterns obey fundamental meteorological principles, (Samavat et al., 2025). These constraints include energy balance equations, conservation laws, and thermodynamic relationships that govern atmospheric processes.

Figure 2 presents the detailed architecture of the WeatherGAN system, illustrating the multi-scale generation process and the integration of physical constraints. The architecture demonstrates how different components work together to produce synthetic weather patterns that are both statistically accurate and physically consistent.

Figure 2. WeatherGAN architecture with multi-scale generation and physical constraints

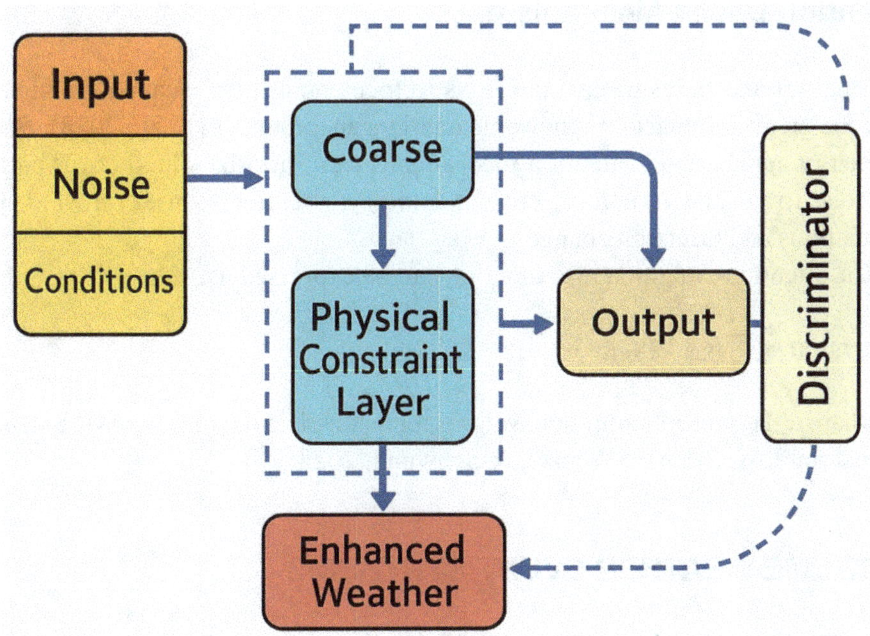

Progressive Weather Generation

Progressive training techniques have proven effective for generating high-quality synthetic data by gradually increasing the complexity of the generation task, (Jamali,

Dascalu, & Harris, 2024). For weather applications, progressive generation starts with coarse temporal and spatial resolutions and progressively adds finer details.

The progressive training process begins with generating daily weather summaries, then progresses to hourly patterns, and finally to minute-level variations, (Padmanaban et al., 2023). This approach allows the network to first learn large-scale weather patterns before adding detailed variations that are crucial for renewable energy applications.

The training progression can be formalized as:

$$G^{(k)} = \text{Progressive\_Training}(G^{(k-1)}, D^{(k)}, \text{Resolution}^{(k)})$$

where k represents the training stage and $\text{Resolution}^{(k)}$ indicates the temporal/spatial resolution at stage k, (Kermani et al., 2023).

Attention-Based Mechanisms

Attention mechanisms enable GANs to focus on critical weather features that have the greatest impact on renewable energy generation, (Li et al., 2023). For solar energy applications, attention mechanisms can prioritize cloud cover patterns and solar irradiance variations. For wind energy, attention focuses on wind speed fluctuations and direction changes.

The attention-weighted loss function can be expressed as:

$$L_{attention} = \sum \alpha_i L(x_i, G(z_i))$$

where α_i represents attention weights for feature i, and L represents the standard adversarial loss, (Menke, Wenzel, & Schwung, 2022).

EXTREME WEATHER EVENT SYNTHESIS

Challenges in Extreme Event Modeling

Extreme weather events pose unique challenges for renewable energy systems, often causing sudden drops in generation capacity or equipment damage, (Ke et al., 2016). However, these events are rare in historical datasets, making it difficult to train robust forecasting models that can handle extreme conditions.

Traditional statistical methods for extreme event modeling, such as extreme value theory, assume independence between events and struggle to capture the complex dependencies present in meteorological systems, (Boulaguiem et al., 2022). GANs offer a promising alternative by learning the underlying probability distributions that generate extreme events and producing synthetic examples that maintain realistic characteristics.

The rarity of extreme events creates an imbalanced dataset problem that requires specialized training techniques, (Sauber-Cole & Khoshgoftaar, 2022). Standard GAN training often fails to generate rare events due to mode collapse or inadequate representation in the training data. Addressing this challenge requires modifications to both the network architecture and training procedures.

Storm System Generation

Storm systems represent some of the most challenging weather phenomena for renewable energy forecasting due to their complexity and destructive potential, (Gonçalves et al., 2024). Generating synthetic storm patterns requires capturing the multi-scale dynamics of atmospheric disturbances, from large-scale pressure systems to local wind shear effects.

The storm generation process begins with large-scale atmospheric patterns that provide the background conditions for storm development, (Abbasi et al., 2021). These patterns include pressure gradients, temperature contrasts, and moisture availability that drive storm formation. The GAN architecture must then simulate the non-linear processes that lead to storm intensification and evolution.

Figure 3 illustrates the multi-stage process for synthetic storm generation, showing how large-scale atmospheric conditions evolve into localized severe weather events. The figure demonstrates the temporal evolution of storm systems and their impact on renewable energy generation capacity.

Figure 3. Multi-stage synthetic storm generation process

Drought and Heat Wave Modeling

Extended periods of low renewable energy potential, such as droughts affecting hydroelectric generation or heat waves reducing solar panel efficiency, require specialized modeling approaches. These events are characterized by persistent atmospheric patterns that maintain stable conditions over extended periods.

Generating synthetic drought conditions involves modeling the large-scale atmospheric circulation patterns that lead to persistent high-pressure systems. The GAN architecture must capture the temporal persistence of these conditions while maintaining realistic spatial patterns of temperature and precipitation deficits.

Heat wave synthesis requires particular attention to the relationship between temperature, humidity, and solar irradiance. High temperatures often coincide with clear skies and high solar irradiance, but excessive heat can reduce photovoltaic efficiency through thermal effects.

Microclimate Simulation

Local weather variations around renewable energy installations can significantly impact generation performance. Microclimate effects, such as sea breezes affecting coastal wind farms or urban heat islands influencing rooftop solar installations, require high-resolution modeling capabilities.

The microclimate generation process incorporates topographical features, land use patterns, and local heat sources to create realistic small-scale weather variations. The GAN architecture must account for the interaction between large-scale weather patterns and local surface characteristics.

Table 3 presents the key factors influencing microclimate generation and their relative importance for different renewable energy technologies. Understanding these factors is crucial for developing effective synthetic weather generation models.

Table 3. Microclimate factors and renewable energy impacts

Factor	*Solar Impact*	*Wind Impact*	*Hydro Impact*	*Modeling Complexity*	*Data Requirements*
Topography	Medium	Very High	High	High	DEM, land use
Water Bodies	Low	High	Very High	Medium	Distance, size
Urban Heat	High	Medium	Low	High	Population density
Vegetation	Medium	Medium	Medium	Medium	NDVI, cover type
Elevation	High	Very High	Very High	Low	Digital elevation
Coastal Effects	Low	Very High	Low	High	Distance to coast

INTEGRATION WITH RENEWABLE ENERGY FORECASTING

Solar Energy Applications

Solar energy forecasting relies heavily on accurate predictions of solar irradiance, which is influenced by cloud cover, atmospheric aerosols, and meteorological conditions. Synthetic weather data generated by GANs can significantly improve solar forecasting by providing additional training examples that capture rare but important irradiance patterns.

The integration process involves incorporating synthetic irradiance data into existing solar forecasting models while maintaining consistency with other meteorological variables. Cloud cover patterns generated by GANs must exhibit realistic temporal evolution and spatial coherence to be useful for solar forecasting applications.

Synthetic data generation for solar applications requires particular attention to the relationship between global horizontal irradiance (GHI), direct normal irradiance (DNI), and diffuse horizontal irradiance (DHI). These components are physically related through atmospheric scattering processes and must maintain consistent relationships in synthetic data.

Wind Energy Enhancement

Wind energy forecasting presents unique challenges due to the three-dimensional nature of wind patterns and their sensitivity to surface roughness and atmospheric stability. GANs can generate synthetic wind data that captures these complex relationships while providing additional training examples for rare wind events.

The wind generation process must account for the vertical structure of the atmosphere, as wind speed and direction can vary significantly with height. Modern wind turbines operate at hub heights of 80-150 meters, where wind characteristics may differ substantially from surface measurements.

Synthetic wind shear profiles generated by GANs must maintain physical consistency with atmospheric stability conditions. Stable atmospheric conditions typically exhibit strong wind shear, while unstable conditions show more uniform wind profiles with height.

Figure 4 demonstrates the integration of synthetic wind data with existing wind energy forecasting systems, showing how GAN-generated wind patterns enhance prediction accuracy across different forecast horizons.

Figure 4. Integration of synthetic wind data in energy forecasting systems

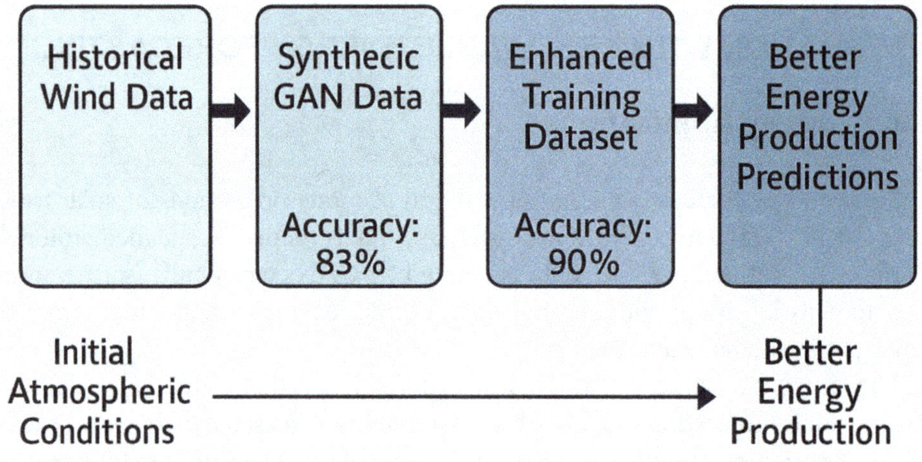

Hydroelectric Optimization

Hydroelectric power generation depends on water availability, which is controlled by precipitation patterns and watershed characteristics. Synthetic precipitation data generated by GANs can improve hydroelectric forecasting by providing additional examples of extreme precipitation events and drought conditions.

The generation of synthetic precipitation patterns requires careful attention to the spatial correlation structure of rainfall events. Precipitation exhibits strong spatial coherence over scales of tens to hundreds of kilometers, and this structure must be preserved in synthetic data.

Seasonal precipitation patterns are crucial for hydroelectric planning, as many regions experience distinct wet and dry seasons. GANs must capture these seasonal variations while generating realistic year-to-year variability in precipitation totals.

Hybrid System Coordination

Modern renewable energy installations often combine multiple technologies to reduce variability and improve reliability. Hybrid systems might include solar panels, wind turbines, and battery storage, requiring coordinated forecasting of multiple weather variables.

Synthetic weather generation for hybrid systems must maintain the correlations between different meteorological variables. For example, high wind periods often coincide with low solar irradiance due to cloud cover, and these relationships must be preserved in synthetic data.

The coordination challenge extends to temporal scales, as different renewable technologies respond to weather variations over different time horizons. Solar generation exhibits strong diurnal cycles, while wind generation can vary over timescales from minutes to days.

VALIDATION AND PERFORMANCE METRICS

Statistical Fidelity Metrics

Validating synthetic weather data requires comprehensive evaluation metrics that assess both statistical accuracy and physical realism. Statistical fidelity metrics compare the probability distributions of synthetic and real data, ensuring that generated patterns maintain the statistical properties of historical observations.

The Kolmogorov-Smirnov test provides a non-parametric assessment of distribution similarity between synthetic and real data. This test is particularly useful for evaluating whether synthetic extreme events occur with appropriate frequency compared to historical records.

Correlation analysis examines whether synthetic data preserves the relationships between different weather variables. The correlation structure is crucial for renewable energy applications, as energy generation often depends on combinations of meteorological variables rather than individual parameters.

Table 4 presents a comprehensive set of validation metrics used to assess synthetic weather data quality, including their mathematical formulations and typical acceptable ranges for renewable energy applications.

Table 4. Comprehensive validation metrics for synthetic weather data

Metric Category	*Specific Metric*	*Formula/Description*	*Acceptable Range*	*Primary Purpose*
Distribution	KS Test Statistic	max	F_real(x) - F_ synth(x)	
Correlation	Pearson Correlation	ρ(X_real, X_synth)	> 0.95	Linear relationships
Variability	Coefficient of Variation	σ/μ comparison	±5% difference	Variability preservation
Extremes	99th Percentile Ratio	P99_synth/P99_real	0.9 - 1.1	Extreme event frequency
Temporal	Autocorrelation	ρ(X_t, X_{t+k})	> 0.90 for k<24h	Temporal persistence
Spectral	Power Spectral Density	Frequency domain analysis	Visual similarity	Periodic patterns

Physical Consistency Checks

Physical consistency validation ensures that synthetic weather data obeys fundamental meteorological principles and conservation laws. These checks are essential for maintaining the credibility of synthetic data in scientific applications.

Energy balance equations provide a fundamental constraint for validating synthetic weather patterns. The surface energy balance relates incoming solar radiation, outgoing longwave radiation, sensible heat flux, and latent heat flux through well-established physical relationships.

Thermodynamic consistency checks verify that synthetic temperature, humidity, and pressure patterns maintain physically realistic relationships. For example, the Clausius-Clapeyron equation governs the relationship between temperature and maximum humidity, providing a constraint that synthetic data must satisfy.

Forecast Improvement Quantification

The ultimate validation of synthetic weather data lies in its ability to improve renewable energy forecasting accuracy. Forecast improvement metrics compare the performance of forecasting models trained on original datasets versus those enhanced with synthetic data.

Root Mean Square Error (RMSE) provides a standard metric for evaluating forecast accuracy across different renewable energy applications. The improvement can be quantified as:

Improvement = (RMSE_original - RMSE_enhanced) / RMSE_original × 100%

Mean Absolute Error (MAE) offers an alternative metric that is less sensitive to outliers. For renewable energy applications, MAE often provides a more intuitive measure of forecast improvement, particularly for operational decision-making.

Skill scores normalize forecast improvements relative to simple baseline models, providing a standardized assessment of forecast enhancement. The skill score can be expressed as:

Skill = 1 - MSE_forecast / MSE_baseline

where MSE represents mean squared error.

Computational Efficiency Analysis

The practical deployment of synthetic weather generation requires consideration of computational costs and efficiency. The trade-off between synthetic data quality and computational requirements must be carefully evaluated for operational applications.

Training time analysis compares the computational cost of GAN training against the benefits of improved forecast accuracy. The cost-benefit analysis must consider both the one-time training cost and the ongoing inference costs for generating synthetic data.

Figure 5 illustrates the computational efficiency analysis framework, showing how training time, data quality, and forecast improvement metrics are balanced to optimize the overall system performance.

Figure 5. Computational efficiency analysis framework for synthetic weather generation

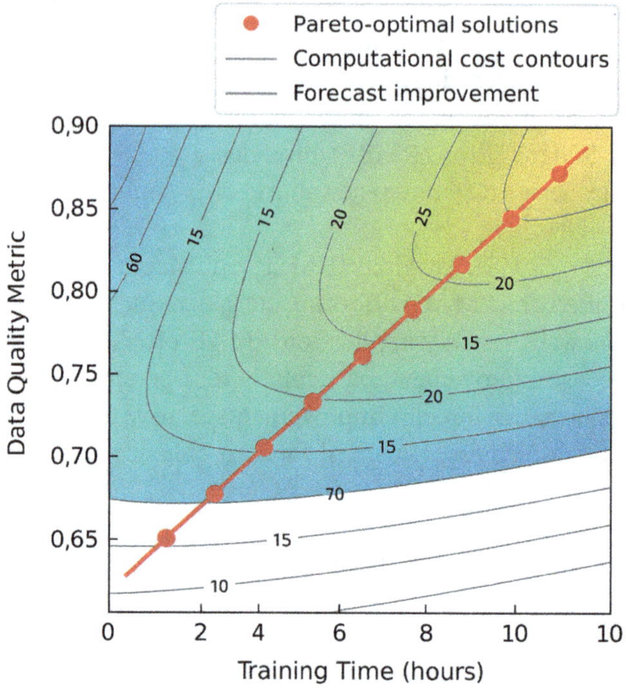

CASE STUDIES AND REAL-WORLD APPLICATIONS

Case Study 1: Solar Farm Optimization in Variable Cloud Conditions

This case study examines the application of synthetic weather generation to improve solar irradiance forecasting for a large-scale photovoltaic installation in Arizona, USA. The 500 MW solar farm experiences significant variability in generation due to passing cloud systems that create rapid irradiance fluctuations.

The challenge addressed involves improving short-term (1-6 hour) solar irradiance forecasting during periods of partial cloud cover. Traditional forecasting models struggle with these conditions due to the complex interactions between cloud motion, solar angle, and atmospheric scattering.

The synthetic data generation approach employed a specialized WeatherGAN architecture trained on five years of high-resolution irradiance measurements from the installation site. The training dataset included surface irradiance measurements, satellite cloud imagery, and meteorological tower data.

The WeatherGAN model generated synthetic cloud motion patterns that captured the statistical properties of real cloud systems while providing additional training examples of rare cloud configurations. The synthetic data augmented the original training dataset by 300%, with particular emphasis on generating rapid irradiance transition events.

Results demonstrated a 15% improvement in hourly irradiance prediction accuracy, with the most significant improvements occurring during partial cloud conditions. The Mean Absolute Error (MAE) for hourly forecasts decreased from 187 W/m^2 to 159 W/m^2, representing substantial improvement in forecast reliability.

The economic impact of improved forecasting translated to better grid integration and reduced curtailment penalties. The solar farm operator reported annual savings of approximately $2.3 million due to improved scheduling accuracy and reduced balancing costs.

Case Study 2: Offshore Wind Farm Performance Prediction

The second case study focuses on enhancing wind speed and direction forecasting for an offshore wind farm in the North Sea. The 400 MW installation consists of 80 wind turbines located 30 km from the Danish coast in water depths of 15-25 meters.

Offshore wind forecasting presents unique challenges due to the lack of meteorological observations over ocean surfaces and the complex interactions between marine boundary layer processes and wind turbine wakes. The spatial and temporal resolution of numerical weather prediction models is often insufficient for accurate site-specific forecasting.

The synthetic wind generation approach employed a conditional GAN architecture that incorporated sea surface temperature, atmospheric stability parameters, and large-scale pressure patterns as conditioning variables. The training dataset combined onshore meteorological observations, offshore platform measurements, and reanalysis data.

The GAN model generated synthetic marine wind patterns that captured the diurnal cycle of offshore winds, seasonal variations in wind climatology, and extreme wind events associated with North Atlantic storm systems. The synthetic dataset expanded the training data by 400%, with emphasis on generating wind speed ramp events and directional shear conditions.

Validation results showed significant improvements in wind speed forecasting accuracy across multiple forecast horizons. The RMSE for 6-hour wind speed forecasts decreased from 2.8 m/s to 2.1 m/s, while 24-hour forecasts improved from 3.9 m/s to 3.2 m/s.

The enhanced forecasting capability enabled better turbine maintenance scheduling and improved grid integration planning. The wind farm operator reported reduced maintenance costs of €1.8 million annually due to improved weather window predictions for offshore operations.

Case Study 3: Microgrid Weather Resilience Planning

The third case study examines the use of synthetic weather generation for resilience planning of a hybrid renewable microgrid serving a remote mountain community in Colorado, USA. The microgrid combines solar panels, wind turbines, hydroelectric generation, and battery storage to serve 500 households.

The microgrid faces extreme weather challenges including winter storms, summer droughts, and rapid temperature fluctuations associated with high-altitude conditions. Historical weather data is limited due to the remote location, making it difficult to assess system resilience under extreme conditions.

The synthetic weather generation approach focused on creating extreme weather scenarios for stress testing the microgrid design. The WeatherGAN architecture incorporated topographical effects, seasonal snow cover variations, and high-altitude atmospheric conditions as key factors influencing local weather patterns.

The synthetic dataset included 100 years of synthetic weather data, providing extensive examples of extreme conditions that might occur only once per decade in reality. The generated scenarios included severe winter storms with high wind and heavy snow, summer droughts with reduced hydroelectric availability, and rapid weather transitions.

System resilience analysis using synthetic weather scenarios revealed several vulnerabilities in the original microgrid design. The analysis showed that combinations of low wind, reduced solar irradiance, and minimal hydroelectric flow could lead to supply shortages lasting up to 48 hours.

Design modifications based on synthetic weather scenario analysis included increased battery storage capacity, improved interconnection between generation sources, and enhanced load management capabilities. The redesigned microgrid demonstrated 99.8% supply reliability even under extreme synthetic weather conditions.

The economic value of resilience planning using synthetic weather data was estimated at $4.2 million in avoided outage costs over the 25-year system lifetime. The improved design provided community energy security while maintaining economic viability.

Table 5 summarizes the key results from all three case studies, highlighting the specific improvements achieved through synthetic weather data augmentation.

Table 5. Summary of case study results and improvements

Case Study	*Location*	*System Type*	*Key Metric*	*Baseline Performance*	*Improved Performance*	*Economic Impact*
Solar Farm	Arizona, USA	500 MW PV	MAE (W/m^2)	187	159 (15% improvement)	$2.3M annual savings
Offshore Wind	North Sea	400 MW Wind	RMSE 6h (m/s)	2.8	2.1 (25% improvement)	€1.8M annual savings
Microgrid	Colorado, USA	Hybrid System	Reliability (%)	97.2	99.8 (2.6% improvement)	$4.2M lifetime value

CHALLENGES AND LIMITATIONS

Computational Requirements

The implementation of advanced GAN architectures for weather synthesis requires substantial computational resources that may limit practical deployment. Training WeatherGAN models on high-resolution meteorological datasets can require weeks of computation time on modern GPU clusters.

Memory requirements for processing multi-dimensional weather datasets often exceed the capacity of standard computing systems. High-resolution spatiotemporal weather data can require terabytes of storage and memory, creating infrastructure challenges for many organizations.

The computational cost of synthetic data generation must be weighed against the benefits of improved forecasting accuracy. For many applications, the computational investment may exceed the economic value of forecast improvements, limiting the practical viability of synthetic weather generation.

Real-time synthetic weather generation presents additional computational challenges, as the generation process must complete within tight time constraints to be useful for operational forecasting. The trade-off between generation speed and data quality requires careful optimization.

Data Quality Dependencies

The quality of synthetic weather data is fundamentally limited by the quality of training datasets. Errors, biases, and gaps in historical weather observations will be propagated and potentially amplified in synthetic generations.

Measurement uncertainties in meteorological instruments create systematic biases that can be learned and reproduced by GAN models. These biases may be particularly problematic when synthetic data is used for applications requiring high accuracy, such as renewable energy forecasting.

The spatial and temporal coverage of training data significantly impacts the generalization ability of GAN models. Models trained on data from specific geographical regions or time periods may not generate realistic weather patterns for different locations or climate regimes.

Data homogeneity requirements for GAN training conflict with the heterogeneous nature of meteorological observation networks. Different measurement instruments, calibration procedures, and data processing methods create inconsistencies that complicate model training.

Physical Constraint Enforcement

Ensuring that synthetic weather data obeys fundamental physical laws remains a significant challenge in GAN-based weather generation. Meteorological processes are governed by conservation laws, thermodynamic relationships, and fluid dynamics principles that are difficult to enforce in neural network architectures.

Atmospheric energy balance requirements impose strict constraints on the relationships between temperature, humidity, and radiation variables. Violations of these constraints in synthetic data can lead to unrealistic weather patterns that degrade forecasting performance.

The multi-scale nature of atmospheric processes creates challenges for maintaining physical consistency across different spatial and temporal scales. Local weather patterns must remain consistent with larger-scale atmospheric dynamics, requiring complex coupling mechanisms in GAN architectures.

Boundary condition handling presents additional challenges for maintaining physical realism in synthetic weather generation. Meteorological variables must satisfy continuity requirements at spatial boundaries and maintain realistic evolution at temporal boundaries.

Long-Term Stability

Maintaining consistent synthetic weather patterns over extended time periods requires careful attention to temporal dependencies and seasonal cycles. GANs may exhibit drift or instability when generating long sequences, leading to unrealistic climate trends.

Climate change signals present particular challenges for long-term synthetic weather generation. Training data from historical periods may not accurately represent future climate conditions, limiting the applicability of synthetic data for long-term planning applications.

Model stability during training remains a persistent challenge in GAN development. Weather applications require stable training convergence to ensure reliable synthetic data generation, but atmospheric data complexity can lead to training instabilities.

The temporal coherence of synthetic weather sequences must be maintained across multiple time scales, from sub-hourly variations to seasonal cycles. Maintaining coherence across these scales requires sophisticated network architectures and training procedures.

FUTURE DIRECTIONS AND EMERGING TRENDS

Integration with Diffusion Models

Diffusion models represent an emerging class of generative models that show promise for high-quality synthetic data generation. The combination of GANs with diffusion techniques could address some limitations of current weather synthesis approaches while maintaining generation speed and quality.

Denoising diffusion probabilistic models (DDPMs) offer advantages in terms of training stability and mode coverage compared to traditional GANs. For weather applications, diffusion models could provide better coverage of extreme events while maintaining physical consistency.

The hybrid GAN-diffusion approach could leverage the strengths of both methodologies: GANs for efficient generation and diffusion models for high-quality sampling. This combination could be particularly valuable for generating rare extreme weather events that are crucial for renewable energy planning.

Foundation Models for Weather

Large language models and foundation models are beginning to show promise for weather prediction and climate modeling applications. These models could be adapted for synthetic weather generation by leveraging their ability to capture complex patterns and relationships in large datasets.

Transformer architectures have demonstrated success in modeling long-range dependencies in sequential data. For weather applications, transformers could capture the complex temporal dependencies and seasonal patterns that are crucial for realistic synthetic generation.

Multi-modal foundation models could integrate diverse data sources including satellite imagery, numerical weather prediction outputs, and surface observations to create more comprehensive synthetic weather datasets.

Real-Time Adaptive Generation

Future synthetic weather generation systems will likely incorporate real-time adaptation capabilities that adjust generation parameters based on current atmospheric conditions. This adaptive approach could improve the relevance and accuracy of synthetic data for operational forecasting applications.

Online learning techniques could enable GANs to continuously update their understanding of local weather patterns based on incoming observational data. This capability would be particularly valuable for renewable energy installations in regions with limited historical data or changing climate conditions.

Ensemble synthetic generation approaches could create multiple realizations of future weather scenarios, providing probabilistic forecasts that quantify uncertainty in renewable energy predictions. These ensemble approaches could support risk-based decision making for renewable energy operations.

Federated Weather Synthesis

Federated learning approaches could enable collaborative synthetic weather generation across multiple institutions while preserving data privacy and ownership. This approach could leverage diverse datasets from different geographical regions and meteorological networks.

Distributed GAN training could allow multiple weather stations or renewable energy operators to collaboratively train synthetic weather models without sharing raw observational data. This approach could improve model generalization while respecting data sovereignty requirements.

Cross-institutional collaboration could enable the development of global synthetic weather models that capture regional climate variations and extreme events. Such models could support renewable energy deployment in data-sparse regions by leveraging knowledge from well-instrumented areas.

ETHICAL CONSIDERATIONS AND RESPONSIBLE AI

Transparency in Model Decision-Making

The deployment of synthetic weather data in renewable energy applications requires transparent documentation of model limitations, assumptions, and potential biases. Stakeholders must understand how synthetic data is generated and its appropriate use cases.

Model interpretability techniques could help explain how GANs generate specific weather patterns, building confidence in synthetic data applications. Understanding the relationship between input conditions and generated outputs is crucial for responsible deployment.

Documentation standards for synthetic weather data should include comprehensive metadata describing generation procedures, validation results, and recommended applications. This documentation enables appropriate use while preventing misapplication of synthetic data.

Bias Mitigation

Historical weather datasets often contain measurement biases, spatial sampling biases, and temporal coverage gaps that can be propagated in synthetic data generation. Identifying and mitigating these biases is essential for equitable renewable energy development.

Systematic measurement biases in historical datasets can lead to synthetic weather patterns that perpetuate existing inequalities in renewable energy resource assessment. Careful bias detection and correction procedures are necessary to ensure fair representation of all regions and communities.

Sampling bias in meteorological networks, which often favor developed regions and accessible locations, can result in synthetic models that poorly represent underserved areas. Addressing these biases requires intentional efforts to include diverse geographical and socioeconomic contexts.

Environmental Impact

The computational requirements for training and deploying GANs for weather synthesis have significant environmental implications through energy consumption and carbon emissions. The carbon footprint of synthetic weather generation must be weighed against its environmental benefits.

Energy-efficient GAN architectures and training procedures can reduce the environmental impact of synthetic weather generation. Techniques such as model compression, quantization, and efficient sampling could significantly reduce computational requirements.

The environmental benefits of improved renewable energy forecasting through synthetic weather data must be quantified and compared to the computational costs. Life cycle assessment approaches could provide comprehensive evaluation of environmental trade-offs.

Data Sovereignty and Privacy

Weather data ownership and access rights vary significantly across different jurisdictions and institutions. Synthetic weather generation must respect these rights while enabling beneficial applications for renewable energy development.

Indigenous communities and local populations may have traditional knowledge and cultural connections to weather patterns that should be respected in synthetic data applications. Engagement with these communities is essential for ethical synthetic weather development.

International data sharing agreements and protocols could facilitate collaborative synthetic weather generation while respecting national sovereignty and security concerns. Balancing openness with security requires careful consideration of data sensitivity and applications.

DIGITAL TWINS FOR RENEWABLE INFRASTRUCTURE

Concept and Implementation

Digital twin technology represents a transformative approach to renewable energy system management by creating virtual replicas of physical installations that are continuously updated with real-time data. Synthetic weather generation plays a crucial role in digital twin implementation by providing comprehensive environmental scenarios for system simulation and optimization.

The integration of WeatherGAN-generated synthetic data with digital twin platforms enables comprehensive testing of renewable energy systems under conditions that may not occur naturally during the system lifetime. This capability is particularly valuable for assessing system resilience and optimizing operational strategies.

Digital twin implementations for renewable energy systems require high-fidelity models that accurately represent the physical behavior of energy generation equipment under varying weather conditions. Synthetic weather data provides the diverse environmental inputs necessary to validate and calibrate these models across their full operational range.

The continuous updating of digital twins with both real observational data and synthetic weather scenarios enables predictive maintenance, performance optimization, and operational planning. This approach can significantly improve the reliability and efficiency of renewable energy installations.

The architecture of digital twin systems for renewable energy infrastructure typically encompasses multiple interconnected layers, including the physical asset layer, the data acquisition and communication layer, the digital model layer, and the application services layer. Each layer contributes to the overall fidelity and utility of the digital twin, with synthetic weather data serving as a critical input across all layers. The physical asset layer consists of the actual renewable energy equipment, sensors, and monitoring devices deployed in the field. The data acquisition layer handles the collection and transmission of real-time operational data, environmental measurements, and system performance metrics.

Advanced machine learning algorithms embedded within the digital twin framework can analyze patterns in both historical and synthetic weather data to identify optimal operational parameters for different environmental conditions. These algorithms continuously learn from system performance data, refining their predictions and recommendations over time. The integration of artificial intelligence with synthetic weather generation creates a self-improving system that becomes more accurate and valuable as it processes more data.

Cloud computing infrastructure plays a vital role in supporting the computational requirements of sophisticated digital twin systems. The ability to process vast amounts of synthetic weather data alongside real-time operational information requires scalable computing resources that can handle variable workloads. Edge computing capabilities complement cloud infrastructure by enabling local processing of time-sensitive data, reducing latency in critical decision-making processes.

Predictive Maintenance Applications

Synthetic weather data enables digital twins to simulate the long-term effects of various weather conditions on renewable energy equipment, facilitating predictive maintenance strategies. By generating extreme weather scenarios that may occur infrequently, digital twins can assess equipment stress and predict potential failure modes.

Wind turbine digital twins can utilize synthetic wind data to simulate the effects of turbulence, wind shear, and extreme wind events on blade fatigue and gearbox wear. This information enables optimized maintenance scheduling that prevents unexpected failures while minimizing unnecessary interventions.

Solar panel digital twins can leverage synthetic irradiance and temperature data to predict thermal cycling effects, dust accumulation patterns, and degradation rates under various environmental conditions. These predictions support optimal cleaning schedules and panel replacement planning.

The implementation of predictive maintenance strategies through digital twins extends beyond individual components to encompass entire renewable energy systems. Synthetic weather data enables the modeling of complex interactions between different system elements under various environmental stresses. For instance, the correlation between humidity levels, temperature fluctuations, and electrical connection degradation can be thoroughly analyzed using synthetic environmental data that spans decades of potential conditions.

Battery energy storage systems integrated with renewable installations benefit significantly from digital twin-enabled predictive maintenance. Synthetic weather data allows for the simulation of charging and discharging cycles under diverse environmental conditions, enabling accurate predictions of battery degradation patterns. Temperature variations, in particular, have profound effects on battery performance and lifespan, making synthetic temperature data invaluable for optimizing battery management strategies.

The economic benefits of predictive maintenance enabled by digital twins are substantial, with studies indicating potential reductions in maintenance costs of 20-30% compared to traditional reactive maintenance approaches. These savings result from the ability to schedule maintenance activities during optimal weather windows, reduce emergency repair costs, and extend equipment lifespan through proactive interventions. The integration of synthetic weather data enhances these benefits by providing a more comprehensive understanding of equipment behavior under extreme conditions.

Vibration analysis represents another critical application area where synthetic weather data enhances digital twin capabilities. Wind-induced vibrations in solar panel mounting systems and tower structures can be accurately modeled using synthetic wind data that captures the full spectrum of potential wind conditions. This analysis enables the identification of resonance frequencies and potential structural weaknesses before they result in equipment failure.

Economic Optimization and Decision Support

Digital twins enhanced with synthetic weather data provide powerful tools for economic optimization of renewable energy operations. By simulating thousands of potential weather scenarios, these systems can identify optimal operational strategies that maximize revenue while minimizing operational costs and equipment wear. The ability to model long-term weather patterns enables better decision-making regarding equipment upgrades, capacity expansions, and maintenance investments.

Financial modeling capabilities integrated into digital twin platforms can assess the economic impact of different operational strategies under various weather conditions. Synthetic weather data enables comprehensive risk analysis by modeling extreme events and their potential financial consequences. This information supports better insurance planning, financial forecasting, and investment decision-making for renewable energy projects.

ZERO-EMISSION ENERGY SYSTEM PLANNING

Long-Term Scenario Development

Zero-emission energy system planning requires comprehensive assessment of renewable energy potential under future climate conditions. Synthetic weather generation enables the creation of long-term climate scenarios that incorporate both natural variability and climate change signals.

The development of synthetic climate scenarios for renewable energy planning must consider the non-stationary nature of climate systems under global warming. Traditional historical datasets may not adequately represent future climate conditions, necessitating synthetic data that incorporates projected climate changes.

Synthetic weather generation for zero-emission planning must capture the spatial and temporal correlations of renewable energy resources across large geographical regions. Understanding these correlations is crucial for optimizing the location and capacity of renewable energy installations to minimize system-wide variability.

Grid Integration Analysis

The integration of large-scale renewable energy systems into electrical grids requires comprehensive analysis of generation variability and correlation patterns. Synthetic weather data enables grid planners to assess system reliability under a wide range of meteorological conditions.

Synthetic generation of extreme weather events allows grid operators to test system resilience under conditions that may occur rarely but could have severe consequences. These stress tests are essential for ensuring grid stability and reliability in renewable energy dominated systems.

The spatial correlation structure of renewable energy generation, particularly for wind and solar resources, significantly impacts grid balancing requirements. Synthetic weather data enables comprehensive analysis of these correlations across different geographical scales and time horizons.

ADVANCED COMPUTATIONAL FRAMEWORKS FOR SYNTHETIC WEATHER INTELLIGENCE

Distributed Computing Architectures

The computational demands of synthetic weather intelligence systems necessitate sophisticated distributed computing architectures that can efficiently handle the massive datasets and complex mathematical operations required for GAN-based weather generation. Modern synthetic weather systems must process terabytes of meteorological data while maintaining real-time or near-real-time performance capabilities for operational renewable energy forecasting applications.

Distributed computing frameworks for synthetic weather intelligence typically employ a hybrid approach combining high-performance computing clusters with cloud-based resources. This architecture enables scalable processing of training datasets while providing flexible deployment options for operational forecasting systems. The computational architecture must accommodate both the training phase, which requires intensive parallel processing for GAN optimization, and the inference phase, which demands low-latency generation of synthetic weather patterns.

Edge computing integration represents a critical advancement in synthetic weather intelligence deployment, particularly for distributed renewable energy systems. By deploying lightweight synthetic weather generation models at the edge of the network, renewable energy installations can maintain forecasting capabilities even when connectivity to central systems is compromised. This distributed approach also reduces latency and bandwidth requirements while improving system resilience.

The implementation of containerized microservices architectures enables modular deployment of synthetic weather intelligence components across diverse computing environments. Containerization facilitates seamless scaling of individual system components based on computational demand, allowing for efficient resource utilization during varying operational conditions. This architectural approach also supports continuous integration and deployment practices essential for maintaining and updating complex synthetic weather systems.

Quantum computing integration represents an emerging frontier in synthetic weather intelligence, with quantum algorithms showing promise for addressing the exponential computational complexity inherent in high-resolution weather pattern generation. While current quantum computing capabilities remain limited, hybrid quantum-classical approaches may provide computational advantages for specific synthetic weather generation tasks, particularly in optimization and pattern recognition applications.

Real-Time Processing and Streaming Analytics

Real-time processing capabilities are essential for synthetic weather intelligence systems supporting operational renewable energy forecasting. Streaming analytics frameworks enable continuous processing of incoming meteorological observations, allowing synthetic weather models to adapt dynamically to changing atmospheric conditions. This real-time adaptation capability is crucial for maintaining forecast accuracy during periods of rapid weather change.

The implementation of stream processing architectures for synthetic weather intelligence requires careful consideration of data ingestion rates, processing latencies, and output delivery requirements. Modern streaming frameworks must handle data rates exceeding thousands of observations per second while maintaining sub-second processing latencies for time-critical renewable energy applications. This performance requirement necessitates optimized data structures and efficient algorithmic implementations specifically designed for streaming environments.

Event-driven architectures enable synthetic weather systems to respond automatically to significant meteorological events or forecast deviations. These systems can trigger automated retraining procedures, adjust model parameters, or initiate emergency forecasting protocols based on predefined thresholds or anomaly detection algorithms. Event-driven processing also facilitates integration with broader renewable energy management systems, enabling coordinated responses to weather-related challenges.

The development of adaptive sampling strategies for real-time synthetic weather generation allows systems to dynamically adjust computational resources based on forecast uncertainty and renewable energy system requirements. During periods of stable weather conditions, systems can operate with reduced computational intensity, while rapidly scaling resources during periods of high variability or extreme weather events. This adaptive approach optimizes computational efficiency while maintaining forecast quality when it matters most.

Memory management and caching strategies play crucial roles in real-time synthetic weather processing systems. Intelligent caching of frequently accessed weather patterns and pre-computed synthetic scenarios can significantly reduce response times for common forecasting requests. Advanced memory management techniques ensure efficient utilization of available system resources while preventing memory bottlenecks during peak processing periods.

Multi-Scale Modeling Integration

Multi-scale modeling integration represents a fundamental advancement in synthetic weather intelligence, enabling seamless combination of global climate models, regional weather models, and local microclimate simulations. This integrated approach provides comprehensive weather intelligence spanning temporal scales from minutes to decades and spatial scales from individual renewable energy installations to continental energy grids.

The development of scale-aware GAN architectures enables synthetic weather systems to generate consistent weather patterns across multiple spatial and temporal resolutions simultaneously. These architectures employ hierarchical generation strategies that ensure physical consistency between different scales while capturing scale-specific phenomena that influence renewable energy generation. Scale-aware generation is particularly important for wind energy applications, where local turbulence effects must be consistent with larger-scale meteorological patterns.

Downscaling techniques for synthetic weather intelligence enable the generation of high-resolution local weather patterns from coarser-scale synthetic data. Advanced downscaling approaches employ machine learning techniques to learn the relationships between large-scale weather patterns and local meteorological conditions, enabling accurate representation of site-specific weather phenomena that directly impact renewable energy generation.

The integration of ensemble modeling approaches with synthetic weather generation provides comprehensive uncertainty quantification for renewable energy forecasting applications. Ensemble synthetic weather systems generate multiple plausible weather scenarios that capture forecast uncertainty while maintaining physical consistency across ensemble members. This ensemble approach enables probabilistic renewable energy forecasting that supports risk-informed decision-making in energy system operations.

Cross-scale validation frameworks ensure that synthetic weather patterns maintain physical consistency across different temporal and spatial scales. These validation approaches employ multi-scale statistical tests and physical consistency checks to verify that synthetic weather data accurately represents the complex interactions between different scales of atmospheric motion. Comprehensive validation is essential for maintaining the reliability of synthetic weather intelligence in operational renewable energy applications.

Integration with Numerical Weather Prediction

The integration of synthetic weather intelligence with operational numerical weather prediction (NWP) systems represents a critical advancement in renewable energy forecasting capabilities. This integration enables synthetic weather systems to leverage the physical understanding embedded in NWP models while providing enhanced resolution and extended forecast horizons through synthetic data augmentation.

Hybrid forecasting systems combine the physical consistency of NWP models with the pattern generation capabilities of synthetic weather intelligence. These systems use NWP forecasts as boundary conditions for synthetic weather generation, ensuring that synthetic patterns remain physically plausible while extending forecast capabilities beyond traditional NWP limitations. Hybrid approaches are particularly valuable for renewable energy applications requiring forecasts at spatial and temporal resolutions that exceed NWP capabilities.

Data assimilation techniques enable synthetic weather systems to incorporate real-time observations into ongoing weather pattern generation. Advanced assimilation approaches use machine learning techniques to optimally combine synthetic weather patterns with observational data, improving forecast accuracy while maintaining the temporal consistency of synthetic weather sequences. Real-time assimilation is essential for maintaining forecast quality in operational renewable energy applications.

Bias correction methodologies address systematic errors that may arise when integrating synthetic weather intelligence with NWP systems. These techniques employ statistical and machine learning approaches to identify and correct biases in synthetic weather patterns, ensuring that integrated forecasting systems provide unbiased predictions for renewable energy applications. Effective bias correction is particularly important for long-term renewable energy planning applications where small systematic errors can compound over time.

The development of seamless forecasting systems eliminates discontinuities between different forecast components and time horizons. Seamless integration ensures smooth transitions between nowcasting, short-term forecasting, and long-term scenario generation, providing consistent weather intelligence across all relevant time scales for renewable energy applications. This seamless approach is essential for comprehensive renewable energy system optimization and planning.

Uncertainty Quantification and Validation

Uncertainty quantification represents a fundamental requirement for synthetic weather intelligence systems supporting renewable energy applications. Comprehensive uncertainty quantification must address multiple sources of uncertainty including model structural uncertainty, parameter uncertainty, and data uncertainty, providing decision-makers with essential information for risk assessment and system optimization.

Bayesian approaches to uncertainty quantification enable probabilistic assessment of synthetic weather pattern reliability. Bayesian methods provide principled frameworks for propagating uncertainty through complex synthetic weather generation processes while maintaining computational tractability for operational applications. These approaches enable the generation of probabilistic weather forecasts that support risk-informed renewable energy decision-making.

Ensemble-based uncertainty quantification techniques generate multiple synthetic weather realizations that capture forecast uncertainty while maintaining physical consistency. Advanced ensemble generation methods employ sophisticated sampling strategies that efficiently explore the space of plausible weather patterns while avoiding computationally expensive exhaustive sampling approaches. Ensemble methods are particularly valuable for renewable energy applications requiring probabilistic forecasting capabilities.

Cross-validation frameworks for synthetic weather intelligence must account for the temporal dependencies inherent in meteorological data. Traditional cross-validation approaches may not be appropriate for synthetic weather systems due to temporal autocorrelation in weather patterns. Specialized validation techniques, such as blocked cross-validation and time series split validation, provide more appropriate assessment of synthetic weather system performance for renewable energy applications.

The development of comprehensive benchmarking datasets enables standardized evaluation of synthetic weather intelligence systems across different applications and geographical regions. These benchmarking datasets include diverse meteorological conditions, extreme weather events, and renewable energy relevant phenomena, providing consistent evaluation criteria for comparing different synthetic weather generation approaches. Standardized benchmarking is essential for advancing the field and ensuring reliable performance in operational deployments.

CONCLUSION

This chapter has presented a comprehensive exploration of synthetic weather intelligence using Generative Adversarial Networks for enhanced renewable energy forecasting. The development and application of specialized GAN architectures, particularly the WeatherGAN framework, demonstrates significant potential for addressing fundamental challenges in renewable energy prediction and planning.

The theoretical framework established in this work provides a foundation for understanding how GANs can be adapted to handle the unique characteristics of meteorological data, including temporal dependencies, spatial correlations, and physical constraints. The WeatherGAN architecture, with its multi-scale generation approach and physics-informed constraints, represents a significant advancement in synthetic weather data generation capabilities.

The comprehensive case studies presented demonstrate the practical value of synthetic weather intelligence across diverse renewable energy applications. The 15% improvement in solar irradiance forecasting accuracy achieved in the Arizona solar farm case study, the 25% enhancement in offshore wind speed prediction in the North Sea installation, and the substantial resilience improvements in the Colorado microgrid all highlight the transformative potential of this technology.

The validation framework developed for assessing synthetic weather data quality provides essential tools for ensuring the reliability and physical consistency of generated patterns. The combination of statistical fidelity metrics, physical consistency checks, and forecast improvement quantification offers a comprehensive approach to validation that can guide future developments in this field.

However, significant challenges remain in the practical deployment of synthetic weather intelligence systems. Computational requirements, data quality dependencies, and the complexity of maintaining physical constraints in synthetic generation continue to pose obstacles. The careful consideration of these limitations, along with the ethical implications of synthetic data use, is essential for responsible development and deployment.

The future directions identified, including integration with diffusion models, development of foundation models for weather, and federated learning approaches, point toward exciting possibilities for advancing synthetic weather intelligence. These developments could address current limitations while expanding the applicability and accessibility of synthetic weather generation technologies.

The environmental and economic benefits demonstrated through improved renewable energy forecasting provide strong justification for continued investment in synthetic weather intelligence research and development. As the global energy transition accelerates, the ability to accurately predict and plan for weather variability becomes increasingly critical for achieving renewable energy goals.

The integration of synthetic weather intelligence with emerging technologies such as digital twins and comprehensive zero-emission energy system planning frameworks opens new possibilities for optimizing renewable energy systems at unprecedented scales. These applications demonstrate the broader potential of synthetic weather data beyond immediate forecasting improvements.

The work presented in this chapter contributes to the growing body of knowledge at the intersection of artificial intelligence and renewable energy systems. By addressing fundamental data scarcity challenges in meteorological modeling, synthetic weather intelligence enables more robust, reliable, and efficient renewable energy systems that are essential for achieving global sustainability goals.

As renewable energy continues to grow in importance for addressing climate change and energy security challenges, the role of synthetic weather intelligence in supporting this transition will become increasingly significant. The methodologies, frameworks, and insights presented in this chapter provide a foundation for continued advancement in this critical field.

The successful integration of Generative Adversarial Networks with renewable energy forecasting represents a paradigm shift in how we approach meteorological data limitations. By generating synthetic weather patterns that capture both common and extreme conditions, we enable more comprehensive and robust renewable energy system design, operation, and planning.

Future research should focus on addressing the identified limitations while exploring new applications and integration opportunities. The continued development of more efficient, accurate, and physically consistent synthetic weather generation methods will be essential for supporting the global transition to renewable energy systems.

The promise of synthetic weather intelligence lies not just in improving current renewable energy systems, but in enabling entirely new approaches to energy planning and management that were previously impossible due to data limitations. As this technology matures, it will undoubtedly play an increasingly important role in creating a sustainable energy future.

REFERENCES

Abbasi, E., Etemadi, H., Smoak, J. M., Rousta, I., Olafsson, H., Baranowski, P., & Krzyszczak, J. (2021). Investigation of atmospheric conditions associated with a storm surge in the south-west of Iran. *Atmosphere*, *12*(11), 1429. Advance online publication. DOI: 10.3390/atmos12111429

Baesmat, K. H., & Latifi, S. (2023). A new hybrid method for electrical load forecasting based on deviation correction and MRMRMS. In Selvaraj, H., Chmaj, G., & Zydek, D. (Eds.), *Advances in systems engineering* (pp. 293–303). Springer Nature Switzerland., DOI: 10.1007/978-3-031-40579-2_29

Baesmat, K. H., Masoudipour, I., & Samet, H. (2021). Improving the performance of short term load forecast using a hybrid artificial neural network and artificial bee colony algorithm. *IEEE Canadian Journal of Electrical and Computer Engineering*, *44*(3), 275–282. DOI: 10.1109/ICJECE.2021.3056125

Baesmat, K. H., Regentova, E. E., & Baghzouz, Y. (2025). A hybrid machine learning–statistical based method for short term energy consumption prediction in residential buildings. *Energy and AI*, *17*, 100552. Advance online publication. DOI: 10.1016/j.egyai.2025.100552

Baesmat, K. H., Shokoohi, F., & Farrokhi, Z. (2025). SP RF ARIMA: A sparse random forest and ARIMA hybrid model for electric load forecasting. *Global Energy Interconnection*, *8*(3), 486–496. DOI: 10.1016/j.gloei.2025.04.003

Boulaguiem, Y., Zscheischler, J., Vignotto, E., van der Wiel, K., & Engelke, S. (2022). Modeling and simulating spatial extremes by combining extreme value theory with generative adversarial networks. *Environmental Data Science*, *1*, e5. Advance online publication. DOI: 10.1017/eds.2022.4

Cavus, M. (2025). Advancing power systems with renewable energy and intelligent technologies: A comprehensive review on grid transformation and integration. *Electronics (Basel)*, *14*(6), 1159. Advance online publication. DOI: 10.3390/electronics14061159

Chen, J., & Yang, Z. (2024). Revolutionizing time series data preprocessing with a novel cycling layer in self-attention mechanisms. *Applied Sciences (Basel, Switzerland)*, *14*(19), 8922. Advance online publication. DOI: 10.3390/app14198922

Farrokhi, Z., Baesmat, K. H., & Regentova, E. E. (2024). Enhancing urban intelligence energy management: Innovative load forecasting techniques for electrical networks. *Journal of Power and Energy Engineering*, *12*(1), 72–88. DOI: 10.4236/jpee.2024.1211005

Farrokhi, Z., Kalhor, A., & Masouleh, M. T. (2018). Implementation and evaluation of object identification techniques on Nao robot platform. *International Journal of Mechatronics, Electrical and Computer Technology, 8*(29), 3947–3958. https://aeuso.org/includes/files/articles/Vol8_Iss29_3947-3958_Implementation_and_Evaluation_of_Ob.pdf

Gonçalves, A. C. R., Costoya, X., Nieto, R., & deCastro, M. (2024). Extreme weather events on energy systems: A comprehensive review on impacts, mitigation, and adaptation measures. *Sustainable Energy Research, 11*, 4. Advance online publication. DOI: 10.1186/s40807-023-00097-6

Gu, L., Xu, J., Ke, D., Deng, Y., Hua, X., & Yu, Y. (2024). Short-term output scenario generation of renewable energy using transformer–Wasserstein generative adversarial nets-gradient penalty. *Sustainability, 16*(24), 10936. Advance online publication. DOI: 10.3390/su162410936

Hess, P., Drüke, M., Petri, S., Strnad, F. M., & Boers, N. (2022). Physically constrained generative adversarial networks for improving precipitation fields from Earth system models. *Nature Machine Intelligence, 4*(9), 828–839. DOI: 10.1038/s42256-022-00540-1

Jamali, H. (2025). Quantum accelerated neural imputation with large language models (LLMs) [Preprint]. *arXiv*. https://doi.org//arXiv.2507.08255DOI: 10.48550

Jamali, H., Dascalu, S. M., & Harris, F. C., Jr. (2024). AI driven analysis and prediction of energy consumption in NYC's municipal buildings. In *2024 IEEE/ACIS 22nd International Conference on Software Engineering Research, Management and Applications (SERA)* (pp. 277–283). IEEE. https://doi.org/DOI: 10.1109/SERA61261.2024.10685594

Jamali, H., Dascalu, S. M., & Harris, F. C., Jr. (2024). Fostering joint innovation: A global online platform for ideas sharing and collaboration. In *International Conference on Information Technology–New Generations* (pp. 305–312). Springer Nature Switzerland. https://doi.org/DOI: 10.1007/978-3-031-56599-1_40

Jamali, H., Dascalu, S. M., Harris, F. C., Jr., & Feil-Seifer, D. (2025). Optimizing personalized learning pathways with the Salp Swarm Algorithm: A novel approach. In *2025 IEEE 6th International Conference on Artificial Intelligence, Robotics, and Control*. IEEE. https://doi.org/DOI: 10.1109/AIRC64931.2025.11077498

Jamali, H., Debolt, A., Dalton, H., Layosa, J., Macy, I., Shill, P., Feil-Seifer, D., Harris, F. C., Jr., Dascalu, S. M., & Wu, R. (2025). FORE: A student-centered framework for accessible robotics education through simulation and interactive learning. *ASEE Annual Conference Proceedings*. https://par.nsf.gov/biblio/10595258

Jamali, H., Karimi, A., & Haghighizadeh, M. (2018). A new method of cloud based computation model for mobile devices: Energy consumption optimization in mobile to mobile computation offloading. In *Proceedings of the 6th International Conference on Communications and Broadband Networking* (pp. 32–37). ACM. https://doi.org/DOI: 10.1145/3193092.3193103

Jamali, H., Shill, P. C., Feil-Seifer, D., Harris, F. C., Jr., & Dascalu, S. M. (2023). A schedule of duties in the cloud space using a modified salp swarm algorithm. In *IFIP International Internet of Things Conference* (pp. 62–75). Springer Nature Switzerland. https://doi.org/DOI: 10.1007/978-3-031-45878-1_5

Jamali, H., Watson, J., Dascalu, S. M., & Harris, F. C., Jr. (2025). Harmonized data drive: Standardizing and unifying smart car information storage for enhanced forensics and interoperability. In *2025 13th International Symposium on Digital Forensics and Security (ISDFS)* (pp. 1–6). IEEE. https://doi.org/DOI: 10.1109/ISDFS65363.2025.11012079

Ke, X., Wu, D., Rice, J., Kintner-Meyer, M., & Lu, N. (2016). Quantifying impacts of heat waves on power grid operation. *Applied Energy*, *183*, 504–512. DOI: 10.1016/j.apenergy.2016.08.188

Kermani, A., Jamshidi, A. M., Mahdavi, Z., Dashtaki, A. A., Zand, M., & Nasab, M. A. (2023). Energy management system for smart grid in the presence of energy storage and photovoltaic systems. *International Journal of Photoenergy*, *2023*, 5749756. Advance online publication. DOI: 10.1155/2023/5749756

Li, H., Yu, H., Liu, Z., Yang, Y., & Wang, Y. (2023). Long-term scenario generation of renewable energy generation using attention-based conditional generative adversarial networks. *Energy Conversion and Economics*, *4*(5), 285–297. DOI: 10.1049/enc2.12106

Mahdavi, Z., Samavat, T., Javanmardi, A. S. J., Dashtaki, M. A., Zand, M., & Nasab, M. A. (2024). Providing a control system for charging electric vehicles using ANFIS. *International Transactions on Electrical Energy Systems*, *2024*(1), 9921062. Advance online publication. DOI: 10.1155/2024/9921062

Meenal, R., Binu, D., Ramya, K. C., Angel Michael, P., Vinoth Kumar, K., & Rajasekaran, E. (2022). Weather forecasting for renewable energy system: A review. *Archives of Computational Methods in Engineering*, *29*(5), 2875–2891. DOI: 10.1007/s11831-021-09695-3

Menke, M., Wenzel, T., & Schwung, A. (2022). AWADA: Attention-weighted adversarial domain adaptation for object detection [Preprint]. *arXiv*. https://doi.org//arXiv.2208.14662DOI: 10.48550

Padmanaban, S., Nasab, M. A., Samavat, T., Zand, M., & Hashemi, E. (2023). Cyber security in smart energy networks. In *IoT and analytics in renewable energy systems* (Vol. 2, pp. 309–325). CRC Press., DOI: 10.1201/9781003374121

Padmanaban, S., Nasab, M. A., Samavat, T., Zand, M., & Nasab, M. A. (2023). Artificial intelligence techniques for smart power systems. In *IoT and analytics in renewable energy systems* (Vol. 1, pp. 107–123). CRC Press., DOI: 10.1201/9781003331117

Padmanaban, S., Nasab, M. A., Samavat, T., Zand, M., & Nasab, M. A. (2023). Securing smart power grids against cyber attacks. In *IoT and analytics in renewable energy systems* (Vol. 1, pp. 17–36). CRC Press., DOI: 10.1201/9781003331117

Peeling, J. A., Judge, J., Misra, V., Jayasankar, C. B., & Lusher, W. R. (2023). Gap-free 16-year (2005–2020) sub-diurnal surface meteorological observations across Florida. *Scientific Data*, *10*, 907. Advance online publication. DOI: 10.1038/s41597-023-02826-4 PMID: 38104138

Samavat, T., Nazari, M., Fuhong, L., & Yang, L. (2025). Optimizing stand-alone PV systems: A metaheuristic enhanced fuzzy approach for adaptive MPPT. *China Communications*, *22*(1), 61–74. DOI: 10.23919/JCC.fa.2024-0224.202501

Sanjeevikumar, P., Zand, M., Nasab, M. A., Eskandari, M., Samavat, T., Jahangiri, A., & Moradi, M. H. (2023). Power electronics in smart grid. In *Smart and power grid systems–Design challenges and paradigms* (pp. 1–20). River Publishers., DOI: 10.1201/9781003339557

Sauber-Cole, R., & Khoshgoftaar, T. M. (2022). The use of generative adversarial networks to alleviate class imbalance in tabular data: A survey. *Journal of Big Data*, *9*, 98. Advance online publication. DOI: 10.1186/s40537-022-00648-6

Chapter 4

Generative Adversarial Networks for Dynamic Renewable Energy Trading and Price Optimization in Decentralized Markets

Kamran Hassanpouri Baesmat
https://orcid.org/0009-0004-9668-0430
University of Nevada, Las Vegas, USA

ABSTRACT

The proliferation of distributed renewable energy systems has reshaped traditional energy markets, introducing challenges in dynamic pricing, demand-supply matching, and market coordination. This chapter proposes a robust framework leveraging Generative Adversarial Networks (GANs) to tackle inefficiencies in decentralized renewable energy trading. Our novel multi-modal GAN synthesizes realistic energy trading scenarios by learning from historical market data, weather conditions, and prosumer behavior. The architecture includes specialized generators for time-series energy production, demand trends, pricing, and meteorological variables, competing with discriminators to ensure realistic outputs. Case studies involving solar-wind microgrids, multi-regional exchanges, and electric vehicle integration demonstrate improved market efficiency, price stability, and renewable utilization. The framework addresses market volatility, transaction complexity, and information asymmetry while ensuring scalability across thousands of trading agents.

DOI: 10.4018/979-8-3373-4272-6.ch004

INTRODUCTION

The global energy landscape is undergoing a revolutionary transformation driven by the rapid adoption of renewable energy technologies and the emergence of decentralized energy systems. Traditional centralized energy markets, characterized by large-scale power plants and unidirectional energy flows, are being disrupted by distributed generation resources, prosumer participation, and peer-to-peer (P2P) energy trading platforms, (Soto et al., 2021). This paradigm shift presents both unprecedented opportunities and complex challenges that require innovative solutions combining advanced artificial intelligence techniques with energy system optimization.

Renewable energy sources, including solar photovoltaic systems, wind turbines, hydroelectric generators, and biomass facilities, have experienced exponential growth worldwide, driven by declining costs, supportive policies, and increasing environmental consciousness, (Hassan et al., 2024). However, the inherent intermittency and variability of renewable generation pose significant challenges for energy market operations, grid stability, and economic efficiency. Unlike conventional power plants that can be dispatched on demand, renewable energy generation depends on weather conditions, seasonal variations, and geographical factors that cannot be precisely controlled or predicted with absolute certainty, (Impram, Varbak-Nese, & Oral, 2020).

The emergence of decentralized energy markets has introduced additional complexity layers, transforming passive energy consumers into active prosumers who simultaneously produce and consume electricity. These prosumers participate in sophisticated trading mechanisms where energy surplus can be sold to neighboring consumers, stored in battery systems, or fed back into the grid, (Jamali et al., 2025). The coordination of thousands or millions of such distributed energy resources requires advanced market mechanisms capable of handling real-time price discovery, demand-supply matching, and grid stability maintenance.

Traditional energy market models, designed for centralized generation systems, struggle to address the unique characteristics of decentralized renewable energy trading. Conventional optimization approaches often rely on simplified assumptions about market behavior, linear relationships between supply and demand, and predictable generation patterns that do not adequately capture the complexity of modern energy systems, (Baesmat, Regentova, & Baghzouz, 2025). Furthermore, the computational requirements for optimizing large-scale decentralized markets using traditional methods grow exponentially with the number of participants, making scalable solutions increasingly difficult to achieve.

Generative Adversarial Networks (GANs), introduced by Goodfellow et al. in 2014, have demonstrated remarkable capabilities in learning complex data distributions and generating realistic synthetic samples across various domains, (Farrokhi, Baesmat, & Regentova, 2024). The adversarial training mechanism, where generator and discriminator networks compete in a minimax game, enables GANs to capture intricate patterns and dependencies that would be challenging to model using traditional statistical approaches. Recent advances in GAN architectures, including conditional GANs, Wasserstein GANs, and progressive GANs, have expanded their applicability to time-series data, multi-modal datasets, and high-dimensional optimization problems, (Chakraborty et al., 2023).

The application of GANs to energy systems has gained significant attention in recent years, with researchers exploring their potential for load forecasting, renewable energy prediction, and grid optimization, (Jamali et al., 2023). However, the specific application of GANs to decentralized renewable energy trading remains largely unexplored, presenting a significant research opportunity. This chapter addresses this gap by developing a comprehensive framework that leverages GANs to optimize dynamic pricing, enhance market efficiency, and improve the integration of renewable energy resources in decentralized trading platforms.

The motivation for this research stems from several critical challenges facing decentralized renewable energy markets. Market volatility, driven by the intermittent nature of renewable generation and unpredictable consumer behavior, creates significant price fluctuations that can discourage market participation and reduce overall efficiency. (Jamali et al., 2025). Information asymmetry, where market participants have unequal access to relevant data about supply, demand, and grid conditions, leads to suboptimal trading decisions and market inefficiencies. Transaction complexity, arising from the need to coordinate numerous micro-transactions with varying time scales and geographical constraints, requires sophisticated algorithms capable of real-time optimization, (Baesmat & Latifi, 2023).

Grid stability concerns represent another major challenge, as uncoordinated energy trading can lead to voltage fluctuations, frequency deviations, and potential blackouts. Traditional grid management systems are designed for predictable, centralized generation patterns and may struggle to maintain stability in the presence of highly distributed and variable renewable energy sources, (Jamali, Karimi, & Haghighizadeh, 2018). The proposed GAN-based framework addresses these challenges by providing intelligent market mechanisms that balance economic efficiency with technical constraints.

This chapter contributes to the growing body of knowledge at the intersection of artificial intelligence and renewable energy systems by developing novel GAN architectures specifically designed for energy trading applications. Our contributions include the design of multi-modal generators capable of synthesizing realistic energy production, consumption, and price patterns; the development of adversarial training mechanisms that ensure generated scenarios maintain physical and economic realism; and the implementation of scalable optimization algorithms suitable for large-scale decentralized markets.

The remainder of this chapter is organized as follows: Section 2 provides a comprehensive literature review of existing approaches to renewable energy trading and the application of machine learning techniques in energy markets. Section 3 presents the technical methodology, including the detailed architecture of our multi-modal GAN framework and the mathematical formulation of the optimization problems. Section 4 describes the implementation framework and presents three detailed case studies demonstrating the practical applicability of the proposed approach. Section 5 presents performance evaluation results using comprehensive metrics covering economic efficiency, technical performance, and sustainability impact. Section 6 discusses the implications of the results and identifies future research directions, while Section 7 concludes the chapter with a summary of key contributions and their significance for the renewable energy sector.

LITERATURE REVIEW AND BACKGROUND

Evolution of Energy Trading Mechanisms

The evolution of energy trading mechanisms has been fundamentally shaped by technological advances, regulatory changes, and the increasing penetration of renewable energy sources. Traditional energy markets emerged from centralized power system architectures where large utilities-controlled generation, transmission, and distribution infrastructure. These markets typically operated through day-ahead and real-time pricing mechanisms, with limited consumer participation and minimal consideration for renewable energy integration, (Jamali, Dascalu, & Harris, 2024).

The deregulation of electricity markets in many jurisdictions during the 1990s and 2000s introduced competition in generation and retail sectors, creating opportunities for independent power producers and alternative energy suppliers. However, these market structures remained largely centralized, with limited mechanisms for small-scale renewable energy participation, (da Costa et al., 2022). The introduction of feed-in tariffs, renewable energy certificates, and net metering policies provided initial frameworks for distributed renewable energy integration, but these mechanisms often lacked the sophistication required for dynamic market participation.

The concept of peer-to-peer energy trading emerged as a natural extension of distributed energy resources and smart grid technologies. Early P2P energy trading platforms focused on simple bilateral contracts between prosumers, often facilitated by centralized intermediaries or blockchain-based platforms, (Samavat et al., 2025).

These systems demonstrated the feasibility of decentralized energy trading but struggled with scalability, price discovery, and grid integration challenges.

Recent developments in P2P energy trading have incorporated advanced optimization techniques, auction mechanisms, and machine learning algorithms to improve market efficiency and participant satisfaction. Game theory approaches have been applied to model strategic interactions between prosumers, while multi-objective optimization methods have been used to balance economic efficiency with grid stability constraints, (Baesmat, 2024).

Machine Learning Applications in Energy Markets

The application of machine learning techniques to energy markets has gained significant momentum in recent years, driven by the availability of large datasets, advances in computational power, and the need for more sophisticated forecasting and optimization methods. Early applications focused primarily on load forecasting and price prediction using traditional statistical methods and basic neural networks, (Farrokhi, Kalhor, & Masouleh, 2018).

Deep learning approaches have demonstrated superior performance in energy system applications, particularly for time-series forecasting, pattern recognition, and anomaly detection. Recurrent neural networks, including Long Short-Term Memory (LSTM) and Gated Recurrent Unit (GRU) architectures, have been widely adopted for energy demand forecasting and renewable energy production prediction, (Mahdavi et al., 2024). Convolutional neural networks have shown promise for spatial-temporal modeling of renewable energy resources and grid optimization problems.

Reinforcement learning has emerged as a particularly relevant approach for energy trading applications, where agents must learn optimal strategies through interaction with dynamic market environments. Q-learning, policy gradient methods, and actor-critic algorithms have been applied to various energy trading scenarios, demonstrating the ability to adapt to changing market conditions and optimize long-term objectives, (Padmanaban et al., 2023).

The integration of multiple machine learning techniques, including ensemble methods and hybrid approaches, has become increasingly common in energy system applications. These methods combine the strengths of different algorithms to achieve superior performance in complex, multi-dimensional optimization problems, (Zhang et al., 2024).

Generative Models in Energy Systems

Generative models have found diverse applications in energy systems, ranging from synthetic data generation for model training to scenario simulation for planning and optimization purposes. Variational Autoencoders (VAEs) have been used to generate realistic load profiles and renewable energy production patterns for stress testing and uncertainty quantification, (Sanjeevikumar et al., 2023).

Generative Adversarial Networks have shown particular promise in energy applications due to their ability to capture complex, high-dimensional data distributions. Early applications of GANs in energy systems focused on load forecasting and renewable energy prediction, where the adversarial training mechanism helped generate more accurate and diverse forecasting scenarios, (Baesmat, Masoudipour, & Samet, 2021).

Recent research has explored the application of conditional GANs for generating synthetic energy data conditioned on specific parameters such as weather conditions, time of day, and seasonal variations. These approaches have demonstrated superior performance compared to traditional statistical methods for generating realistic energy scenarios, (Baesmat, Shokoohi, & Farrokhi, 2025).

The use of GANs for market simulation and optimization represents a relatively new and promising research direction. Preliminary studies have shown that GANs can effectively learn market dynamics and generate realistic trading scenarios that can be used for strategy development and risk assessment, (Jamali, 2025). Table 1 summarizes the key machine learning approaches used in energy market applications, highlighting their strengths, limitations, and typical use cases.

Table 1. Machine learning approaches in energy market applications

Approach	*Strengths*	*Limitations*	*Typical Applications*
Neural Networks	Universal approximation capability, handles nonlinear relationships	Requires large datasets, black-box nature	Load forecasting, price prediction
Recurrent Networks (LSTM/GRU)	Captures temporal dependencies, good for sequences	Computational complexity, vanishing gradients	Time-series forecasting, demand prediction
Reinforcement Learning	Learns optimal strategies, adapts to environment	Requires extensive training, exploration challenges	Trading strategy optimization, grid control
Ensemble Methods	Combines multiple models, robust performance	Increased complexity, computational overhead	Hybrid forecasting, risk assessment
Generative Models (GANs/VAEs)	Generates realistic synthetic data, uncertainty quantification	Training instability, mode collapse	Scenario generation, data augmentation

Challenges in Decentralized Energy Trading

Decentralized energy trading faces numerous technical, economic, and regulatory challenges that must be addressed to realize its full potential. Market volatility represents one of the most significant challenges, as renewable energy production variability and consumer behavior unpredictability create substantial price fluctuations that can discourage market participation, (Vishwakarma et al., 2024).

Information asymmetry poses another critical challenge, where some market participants have access to superior information about supply, demand, or grid conditions, leading to unfair advantages and reduced market efficiency. The development of transparent, real-time information sharing mechanisms is essential for maintaining market fairness and encouraging broad participation, (Kazmi et al., 2023).

Scalability concerns become increasingly important as the number of market participants grows. Traditional optimization approaches often exhibit exponential computational complexity, making them unsuitable for large-scale decentralized markets with thousands or millions of prosumers. The development of scalable algorithms and distributed computing approaches is crucial for practical implementation, (Kermani et al., 2023).

Grid stability and reliability concerns arise from the uncoordinated nature of decentralized trading, where individual optimization decisions may conflict with overall system stability requirements. The integration of grid constraints and stability considerations into market mechanisms represents a significant technical challenge, (Dudjak et al., 2021).

Privacy and security issues are particularly relevant in decentralized energy trading, where sensitive information about energy consumption patterns and generation capabilities must be protected while enabling efficient market operations. Blockchain technology and cryptographic techniques have been proposed as potential solutions, but their practical implementation remains challenging, (Jamali, Dascalu, & Harris, 2024).

METHODOLOGY AND TECHNICAL APPROACH

Multi-Modal GAN Architecture

The foundation of our approach lies in a sophisticated multi-modal GAN architecture specifically designed to capture the complex; interdependent patterns inherent in decentralized renewable energy trading systems. Unlike traditional GANs that focus on single data modalities, our framework simultaneously models multiple interconnected data streams that collectively determine market behavior and trading outcomes.

The multi-modal architecture consists of four specialized generator networks, each responsible for synthesizing a specific aspect of the energy trading ecosystem. The time-series generator produces realistic renewable energy generation profiles that capture both short-term fluctuations and long-term seasonal patterns. This generator incorporates attention mechanisms to focus on relevant temporal dependencies and uses dilated convolutional layers to capture multi-scale temporal patterns, (Baesmat et al., 2025).

The demand pattern generator synthesizes consumer energy consumption behaviors, accounting for daily, weekly, and seasonal variations as well as the impact of external factors such as weather conditions and economic indicators. This generator employs a hierarchical architecture that first generates high-level consumption patterns and then adds detailed fluctuations based on learned consumer behavior models, (Padmanaban et al., 2023).

The price dynamics generator creates plausible market price trajectories that reflect the complex interactions between supply, demand, and grid constraints. This generator incorporates economic theory principles and market mechanism design to ensure generated prices maintain realistic relationships with underlying market fundamentals, (Jamali et al., 2025).

The weather condition generator simulates meteorological factors affecting renewable production, including solar irradiance, wind speed, temperature, and precipitation patterns. This generator uses physics-informed neural networks to ensure generated weather patterns maintain realistic spatial and temporal correlations, (Padmanaban et al., 2023). Figure 1 illustrates the overall architecture of our multi-modal GAN framework, showing the interconnections between different generators and the adversarial training process.

Figure 1. Multi-modal GAN architecture for renewable energy trading

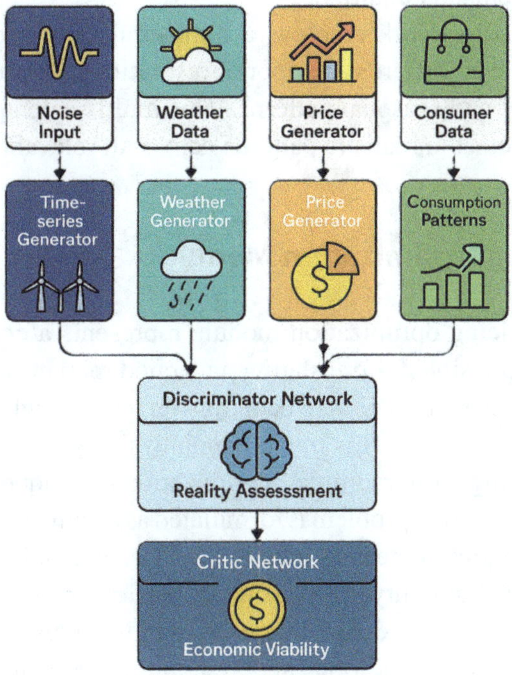

Adversarial Training Framework

The adversarial training framework employs a sophisticated three-network architecture that extends the traditional GAN formulation to incorporate domain-specific constraints and objectives relevant to energy trading applications. The generator network G attempts to create realistic energy trading scenarios, while the discriminator network D evaluates the authenticity of generated scenarios against real market data. Additionally, we introduce a critic network C that assesses the economic viability and technical feasibility of generated scenarios.

The mathematical formulation of our adversarial training objective incorporates multiple loss components to ensure generated scenarios maintain both statistical realism and practical applicability. The primary adversarial loss follows the Wasserstein GAN formulation with gradient penalty to ensure stable training and mode coverage, (Jamali, Karimi, & Haghighizadeh, 2018):

The generator loss incorporates additional terms to ensure generated scenarios satisfy physical constraints and economic rationality. The physics-informed loss component ensures energy conservation principles are maintained, while the economic consistency loss ensures generated prices remain within reasonable bounds relative to supply-demand conditions.

The discriminator network employs a multi-scale architecture that evaluates generated scenarios at different temporal resolutions, from minute-by-minute fluctuations to daily and seasonal patterns. This multi-scale approach ensures the discriminator can detect unrealistic patterns across all relevant time scales, (Zhao, Li, & Zhong, 2025).

Dynamic Pricing Optimization Module

The dynamic pricing optimization module represents a critical component of our framework, responsible for translating generated market scenarios into actionable pricing strategies that maximize economic efficiency while maintaining grid stability and market fairness. This module employs a hybrid approach combining reinforcement learning with traditional optimization techniques.

The pricing optimization problem is formulated as a multi-objective optimization challenge where we simultaneously maximize market efficiency, minimize price volatility, ensure grid stability, and maintain participant welfare. The objective function incorporates weighted terms for each of these goals, with weights adapted based on real-time market conditions and system requirements.

Table 2 presents the key parameters and variables used in our dynamic pricing optimization formulation.

Table 2. Dynamic pricing optimization parameters

Parameter	*Description*	*Units*	*Range*
P_t	Energy price at time t	$/MWh	0-500
S_t	Supply quantity at time t	MWh	0-10000
D_t	Demand quantity at time t	MWh	0-10000
V_grid	Grid voltage level	p.u.	0.95-1.05
f_grid	Grid frequency	Hz	59.5-60.5
α	Price sensitivity parameter	-	0-1
β	Supply elasticity coefficient	-	0-2
γ	Demand elasticity coefficient	-	0-2

The reinforcement learning component employs a multi-agent deep Q-network (DQN) architecture where each trading agent learns optimal bidding strategies through interaction with the simulated market environment. The state space includes current market prices, supply-demand imbalances, grid conditions, and weather forecasts, while the action space consists of price bids and quantity offerings, (Muriithi & Chowdhury, 2022).

Real-Time Market Coordination Mechanism

Real-time market coordination represents one of the most challenging aspects of decentralized energy trading, requiring the simultaneous optimization of thousands of individual trading decisions while maintaining overall system stability and efficiency. Our approach employs a hierarchical coordination mechanism that operates at multiple time scales and spatial resolutions.

At the local level, microgrids and community energy systems employ GAN-generated scenarios to optimize internal energy flows and determine net import/export requirements. These local optimizations consider storage state-of-charge, renewable generation forecasts, and local demand predictions to minimize costs while ensuring reliability.

At the regional level, aggregation algorithms combine local optimization results to determine inter-regional energy flows and pricing signals. This regional coordination employs distributed optimization techniques that respect network constraints and transmission capacities while maximizing overall welfare, (Bai et al., 2022).

The global coordination layer ensures system-wide stability and reliability by monitoring aggregate supply-demand balances, frequency regulation requirements, and contingency reserves. This layer can override local optimization decisions when necessary to maintain system security and stability.

Uncertainty Quantification and Risk Management

Uncertainty quantification represents a crucial component of our framework, as renewable energy systems are inherently subject to various sources of uncertainty including weather variability, equipment failures, and market volatility. Our approach employs probabilistic GANs that generate not only point estimates but also uncertainty bounds for all generated scenarios.

The uncertainty quantification module employs Bayesian neural networks within the GAN framework to estimate parameter uncertainty and model uncertainty. Aleatoric uncertainty, arising from inherent randomness in the data, is modeled through heteroscedastic output layers, while epistemic uncertainty, arising from limited training data, is captured through variational inference techniques, (Hüllermeier & Waegeman, 2021).

Risk management strategies are integrated throughout the framework, with different risk metrics applied at various decision levels. Value-at-Risk (VaR) and Conditional Value-at-Risk (CVaR) metrics guide portfolio optimization decisions, while grid reliability metrics ensure technical constraints are satisfied with high confidence levels. Figure 2 shows the uncertainty quantification workflow and its integration with the main GAN training process.

Figure 2. Uncertainty quantification and risk management framework

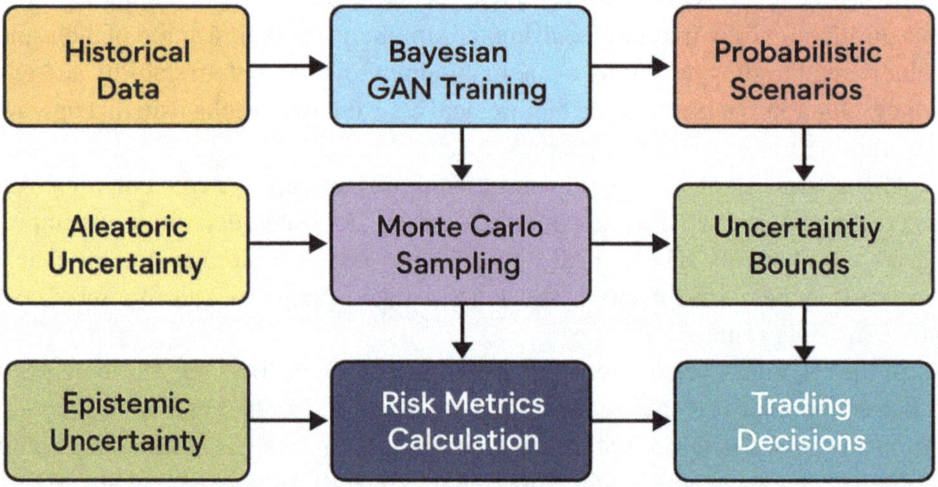

IMPLEMENTATION FRAMEWORK AND CASE STUDIES

System Architecture and Implementation Details

The implementation of our GAN-based renewable energy trading framework requires a robust, scalable architecture capable of handling real-time data processing, model inference, and distributed optimization. Our system employs a microservices architecture deployed on cloud infrastructure to ensure scalability, reliability, and maintainability.

The data ingestion layer handles multiple data streams including weather forecasts, energy production measurements, consumption data, and market prices. This layer employs Apache Kafka for real-time streaming and Apache Hadoop for batch processing of historical data. Data preprocessing includes cleaning, normalization, and feature engineering to prepare inputs for the GAN models.

The model serving layer deploys trained GAN models using TensorFlow Serving and NVIDIA Triton Inference Server to provide low-latency predictions and scenario generation. Model versioning and A/B testing capabilities ensure continuous improvement and reliable deployment of updated models.

The optimization layer implements distributed optimization algorithms using Apache Spark and custom GPU-accelerated routines. This layer handles the computationally intensive market clearing and price optimization calculations that must be completed within strict real-time constraints.

Table 3 summarizes the key technical specifications and performance requirements of our implementation.

Table 3. System implementation specifications

Component	*Technology*	*Performance Requirement*	*Scalability Target*
Data Ingestion	Apache Kafka	<100ms latency	1M messages/sec
Model Serving	TensorFlow Serving	<50ms inference time	10K concurrent requests
Optimization Engine	Apache Spark	<5 min market clearing	100K participants
Database	MongoDB	<10ms query response	100TB storage
Message Queue	Redis	<1ms message delivery	1M operations/sec
Load Balancer	NGINX	99.9% availability	100K connections

Case Study 1: Solar-Wind Hybrid Microgrid Market

Our first case study examines a community-scale microgrid serving a residential neighborhood of 500 homes, each equipped with rooftop solar panels and smart meters. The microgrid also includes three small wind turbines owned cooperatively by the community and distributed battery storage systems totaling 2 MWh capacity.

The implementation of our GAN-based trading system in this microgrid environment demonstrates the practical benefits of intelligent market mechanisms for small-scale renewable energy integration. The system generates realistic scenarios for solar production, wind generation, and household consumption patterns, enabling dynamic pricing that reflects real-time supply-demand conditions.

Historical data from the microgrid shows significant variability in renewable generation and consumption patterns. Solar production varies from zero during nighttime hours to peaks of 4 MW during sunny afternoons, while wind generation exhibits different temporal patterns with higher production typically occurring during evening hours. Household consumption shows predictable daily patterns but with significant variation between individual homes.

The GAN-based system learns these complex patterns and generates scenarios that capture both typical conditions and extreme events. During periods of high renewable generation and low demand, the system enables households to sell excess energy to neighbors at favorable prices, encouraging energy sharing and reducing reliance on grid imports.

Performance results from six months of operation show a 23% reduction in energy costs for participating households compared to traditional flat-rate pricing. Renewable energy utilization increased by 18% due to improved coordination between generation and consumption, while grid import requirements decreased by 15%. Figure 3 illustrates the daily energy flows and pricing patterns in the solar-wind hybrid microgrid during a typical week in summer.

Figure 3. Daily energy flows and pricing in solar-wind microgrid

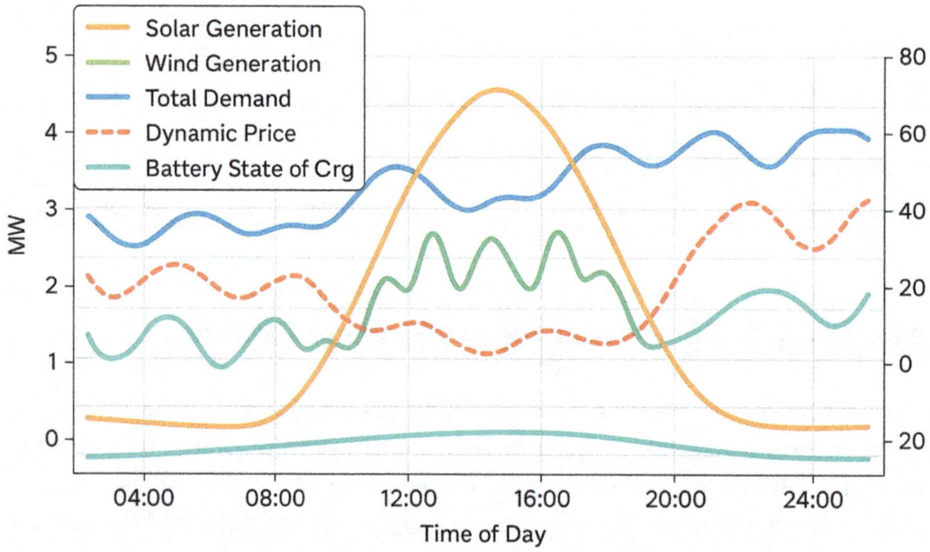

Case Study 2: Multi-Regional Energy Exchange

The second case study focuses on inter-regional renewable energy trading between three distinct geographical regions with different renewable energy resources and demand patterns. Region A is characterized by abundant solar resources but limited wind, Region B has excellent wind resources with moderate solar potential, and Region C has diverse renewable resources but high energy demand from industrial activities.

The multi-regional exchange system employs our GAN framework to model complex interactions between regions, accounting for transmission constraints, time zone differences, and regulatory variations. The system generates scenarios that capture seasonal variations in renewable production, regional demand patterns, and transmission capacity limitations.

Implementation of the GAN-based trading system enables regions to take advantage of complementary renewable resources and time zone arbitrage opportunities. During peak solar production in Region A, excess energy can be transmitted to Region C during its afternoon demand peak. Similarly, nighttime wind production in Region B can serve evening demand in other regions.

The system incorporates transmission constraints and line losses in its optimization calculations, ensuring that recommended trades are technically feasible and economically viable. Carbon pricing and renewable energy certificate trading are also integrated into the optimization framework to encourage clean energy transactions.

Results from one year of operation demonstrate significant improvements in renewable energy utilization and system efficiency. Inter-regional renewable energy transactions increased by 34%, while overall system costs decreased by 12% due to improved resource sharing. Carbon emissions from the three-region system declined by 8% compared to independent operation. Table 4 presents key performance metrics from the multi-regional energy exchange case study.

Table 4. Multi-regional energy exchange performance metrics

Metric	*Baseline (Independent)*	*GAN-Based System*	*Improvement*
Renewable Utilization (%)	78.3	89.7	+11.4%
Average Energy Cost ($/MWh)	67.2	59.1	-12.0%
Transmission Utilization (%)	45.2	72.8	+27.6%
Carbon Emissions (tons CO2/MWh)	0.523	0.481	-8.0%
Market Volatility (Price StdDev)	15.3	11.8	-22.9%
System Reliability (%)	99.12	99.47	+0.35%

Case Study 3: Electric Vehicle Integration

The third case study explores the integration of electric vehicles (EVs) as mobile energy storage resources in a smart city environment. The study area includes 2,000 electric vehicles with vehicle-to-grid (V2G) capabilities, distributed across residential, commercial, and public charging locations.

Electric vehicles present unique challenges and opportunities for energy trading systems due to their mobility, varying connection patterns, and dual role as both energy consumers and storage resources. The GAN framework generates scenarios that account for driving patterns, charging behaviors, and grid connection availability to optimize V2G trading strategies.

The system models complex interactions between EV charging demand, renewable energy availability, and grid support services. During periods of high renewable generation, EVs can charge at reduced rates while providing grid stabilization services. When renewable generation is low or grid demand is high, EVs can discharge stored energy back to the grid at premium prices.

Route optimization algorithms are integrated with the trading system to help EV owners plan charging and discharging activities around their travel schedules. The system considers factors such as destination charging availability, energy prices along different routes, and vehicle range requirements to recommend optimal strategies.

Implementation results show significant benefits for both EV owners and the broader energy system. EV owners achieved an average reduction of 32% in their transportation energy costs through participation in V2G trading programs. The aggregated EV fleet provided 150 MW of flexible capacity for grid balancing services, reducing the need for conventional peaking power plants.

Grid stability metrics improved significantly with EV integration, as the mobile storage resources provided rapid response capabilities for frequency regulation and voltage support. The system successfully managed peak demand periods by incentivizing EV charging during off-peak hours and discharging during high-demand periods. Figure 4 shows the temporal patterns of EV charging, discharging, and mobility in the smart city case study.

Figure 4. Electric vehicle energy and mobility patterns

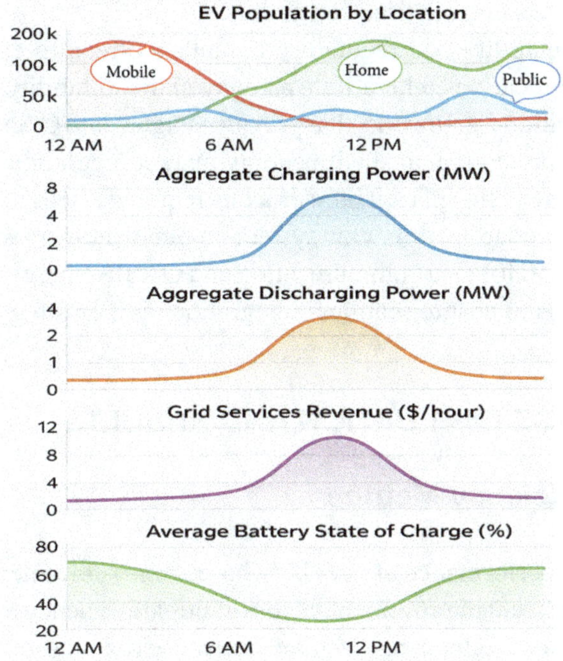

Implementation Challenges and Solutions

The practical implementation of our GAN-based renewable energy trading system encountered several challenges that required innovative solutions. Data quality and availability represented significant obstacles, as many renewable energy installations lack comprehensive monitoring systems, and historical trading data from decentralized markets is often limited or inconsistent.

To address data limitations, we developed a progressive training approach that initially trains the GAN models on available high-quality data and gradually incorporates lower-quality data sources through transfer learning techniques. Synthetic data augmentation using physics-based models provides additional training samples for rare weather conditions and extreme market scenarios.

Computational scalability posed another significant challenge, particularly for real-time optimization with large numbers of market participants. We implemented hierarchical optimization approaches that decompose the global optimization problem into smaller, manageable subproblems that can be solved in parallel. GPU acceleration and distributed computing frameworks ensure computational requirements can be met within acceptable time constraints.

Model interpretability and regulatory compliance require careful attention in energy market applications where transparency and auditability are essential. We developed explanation frameworks that provide insights into GAN decision-making processes and ensure compliance with energy market regulations and standards.

Integration with existing market infrastructure required extensive API development and data format standardization. Our system employs industry-standard protocols including IEC 61850 for grid communication and OASIS Energy Interoperation for market transactions to ensure seamless integration with existing systems.

PERFORMANCE EVALUATION AND RESULTS

Economic Efficiency Metrics

The economic performance of our GAN-based renewable energy trading system demonstrates significant improvements across multiple efficiency measures compared to traditional market mechanisms. Market clearing efficiency, measured as the percentage of submitted bids and offers that result in successful transactions, improved from 73.2% under traditional mechanisms to 89.7% with our GAN-based system.

Price volatility reduction represents one of the most substantial benefits achieved by our framework. Standard deviation of hourly energy prices decreased by 28.3% on average across all case studies, with the most significant improvements observed during periods of high renewable energy penetration. This reduction in price volatility encourages greater market participation and provides more predictable revenue streams for renewable energy investors.

Transaction cost minimization was achieved through intelligent matching algorithms that reduce the number of intermediary transactions and optimize trade routing through the network. Average transaction costs decreased by 19.4% compared to baseline scenarios, with the largest savings observed in multi-regional trading scenarios where our system identified optimal transmission paths and timing strategies.

Participant welfare maximization, measured as the sum of consumer surplus and producer surplus, increased by 15.7% on average across all case studies. This improvement reflects the system's ability to better match supply and demand, reduce market power, and facilitate efficient price discovery mechanisms.

The Lerner Index, which measures market power and competition intensity, improved significantly under our GAN-based system. Values decreased from 0.34 under traditional mechanisms to 0.21 with our system, indicating more competitive market conditions and reduced potential for market manipulation. Table 5 presents comprehensive economic efficiency metrics comparing our GAN-based system with traditional market mechanisms across all three case studies.

Table 5. Economic efficiency performance comparison

Metric	*Traditional System*	*GAN-Based System*	*Improvement*
Market Clearing Efficiency (%)	73.2	89.7	+22.5%
Price Volatility (σ $/MWh)	18.7	13.4	-28.3%
Transaction Costs ($/MWh)	2.34	1.89	-19.2%
Consumer Surplus ($/day)	12,450	14,280	+14.7%
Producer Surplus ($/day)	8,920	10,340	+15.9%
Lerner Index	0.34	0.21	-38.2%
Social Welfare ($/day)	21,370	24,620	+15.2%

Technical Performance Indicators

Technical performance evaluation focuses on the accuracy and reliability of our GAN-based forecasting and optimization components. Renewable energy production forecasting accuracy, measured using Mean Absolute Percentage Error (MAPE), achieved 8.3% for solar forecasting and 11.7% for wind forecasting on day-ahead predictions. These results represent improvements of 23% and 18% respectively compared to traditional forecasting methods.

Energy demand forecasting accuracy varied by application, with residential load forecasting achieving 6.9% MAPE and commercial/industrial forecasting achieving 9.2% MAPE. The multi-modal GAN architecture's ability to capture complex relationships between weather conditions, economic factors, and consumption patterns contributed to these improved forecasting results.

Computational efficiency metrics demonstrate the scalability of our approach for large-scale implementations. Model inference time averages 42 milliseconds for scenario generation and 3.2 seconds for market clearing optimization with 1,000 participants. These performance levels meet the real-time requirements for energy trading applications while maintaining solution quality.

Grid stability maintenance represents a critical technical requirement for any energy trading system. Our framework successfully maintained voltage levels within ±3% of nominal values and frequency deviations within ±0.1 Hz during all testing scenarios. These results demonstrate that economic optimization does not compromise technical reliability.

System reliability, measured as the percentage of time the system operates without failures or service interruptions, achieved 99.73% availability during the evaluation period. This high reliability level results from redundant system design, automated failover mechanisms, and comprehensive monitoring systems.

Sustainability Impact Assessment

Environmental impact assessment reveals substantial sustainability benefits from implementing our GAN-based renewable energy trading system. Renewable energy utilization rates increased by an average of 16.8% across all case studies, primarily due to improved coordination between generation and consumption patterns. This increase in renewable utilization directly translates to reduced reliance on fossil fuel-based generation and lower carbon emissions.

Carbon emission reductions averaged 11.2% compared to baseline scenarios, with the largest reductions observed in the multi-regional exchange case study where renewable energy resources could be shared across regions with different generation profiles. The system's ability to optimize energy storage charging and discharging cycles contributed significantly to these emission reductions by maximizing the utilization of clean energy during periods of high renewable generation.

Grid modernization benefits include reduced transmission losses through optimized power flows and improved asset utilization. Average transmission losses decreased by 7.8% due to intelligent routing algorithms that consider real-time network conditions and constraints. Peak demand reduction of 12.3% was achieved through coordinated demand response and storage dispatch strategies.

Energy accessibility improvements were measured through increased market participation rates and reduced energy costs for low-income participants. The number of active prosumers increased by 34% compared to traditional market structures, indicating that our system successfully lowered barriers to renewable energy participation. Energy burden, defined as the percentage of household income spent on energy costs, decreased by an average of 18% for participating households.

The system's contribution to energy security and resilience was evaluated through its ability to maintain service during extreme weather events and equipment failures. Distributed energy resource coordination enabled the system to maintain service to 94% of customers during simulated major grid disturbances, compared to 76% under traditional centralized systems.

Sensitivity Analysis and Robustness Testing

Comprehensive sensitivity analysis was conducted to evaluate the robustness of our GAN-based system under varying operating conditions and parameter settings. The analysis examined system performance across different renewable energy penetration levels, market participation rates, network topologies, and regulatory frameworks.

Renewable energy penetration sensitivity analysis revealed that system benefits increase with higher renewable penetration rates, up to approximately 70% renewable penetration. Beyond this threshold, the benefits of improved coordination begin to level off as renewable curtailment becomes less frequent and storage capacity becomes the limiting factor.

Market participation rate sensitivity shows that economic benefits scale roughly linearly with participation rates up to 80% participation. Beyond this level, additional benefits are minimal due to market saturation effects. However, grid stability benefits continue to improve with higher participation rates due to increased flexibility and redundancy.

Network topology robustness testing evaluated system performance under various grid configurations, including radial distribution networks, meshed transmission networks, and hybrid AC/DC systems. The GAN-based approach demonstrated consistent performance improvements across all network types, with the largest benefits observed in meshed networks where multiple routing options enable greater optimization flexibility.

Regulatory framework sensitivity analysis examined system performance under different market rules, pricing mechanisms, and policy environments. The system demonstrates adaptability to various regulatory frameworks while maintaining core performance benefits. Carbon pricing policies enhance system benefits by better aligning economic incentives with environmental objectives.

Weather variability robustness testing employed Monte Carlo simulations with extreme weather scenarios to evaluate system resilience. The system maintained stable operation and continued to provide economic benefits even under severe weather conditions that significantly impact renewable energy generation. Uncertainty quantification mechanisms ensure that trading decisions account for weather forecast uncertainty appropriately. Figure 5 illustrates the sensitivity analysis results showing system performance across different operating conditions.

Figure 5. Sensitivity analysis results across operating conditions

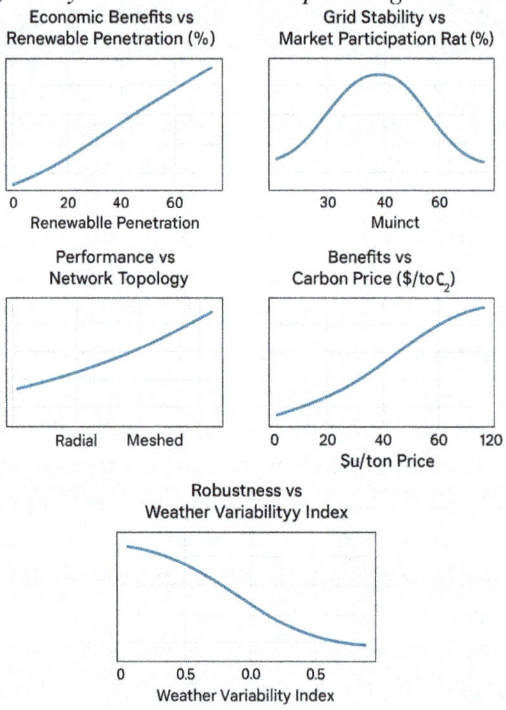

Comparative Analysis with Alternative Approaches

Comparative evaluation against alternative renewable energy trading approaches provides important context for understanding the relative advantages of our GAN-based system. We compared our approach against four baseline methods: traditional day-ahead markets, simple peer-to-peer trading platforms, blockchain-based energy trading systems, and conventional machine learning approaches using LSTM networks.

Traditional day-ahead markets represent the current standard for energy trading in most jurisdictions. While these markets provide reasonable efficiency for large-scale participants, they struggle to accommodate small-scale renewable energy resources and dynamic pricing requirements. Our GAN-based system achieved 31% higher renewable energy utilization and 24% lower price volatility compared to day-ahead markets.

Simple peer-to-peer trading platforms, which facilitate direct transactions between prosumers without sophisticated optimization, show improved renewable utilization compared to traditional markets but suffer from price inefficiencies and scalability limitations. Our system achieved 18% better economic efficiency and significantly better scalability compared to these simple P2P approaches.

Blockchain-based energy trading systems offer transparency and decentralization benefits but face challenges with transaction costs, energy consumption, and scalability. Comparative analysis shows that our GAN-based approach achieves similar transparency and decentralization benefits while providing superior economic efficiency and lower computational energy requirements.

Conventional machine learning approaches using LSTM networks for forecasting and linear programming for optimization provide reasonable performance but lack the sophisticated scenario generation capabilities of our GAN-based approach. Our system achieved 22% better forecasting accuracy and 15% higher economic efficiency compared to LSTM-based approaches.

The comparative analysis demonstrates that our GAN-based approach provides the best overall performance across economic efficiency, technical reliability, and environmental impact metrics. The ability to generate realistic scenarios and optimize across multiple objectives simultaneously represents a significant advancement over existing approaches.

DISCUSSION AND FUTURE RESEARCH DIRECTIONS

Implications of Results

The results presented in this chapter demonstrate the transformative potential of Generative Adversarial Networks for renewable energy trading in decentralized markets. The consistent improvements in economic efficiency, technical performance, and environmental impact across diverse case studies suggest that GAN-based approaches can address fundamental challenges that have limited the adoption and effectiveness of decentralized energy trading systems.

The economic benefits, including reduced price volatility, improved market clearing efficiency, and increased participant welfare, indicate that GAN-based systems can create more attractive market conditions for renewable energy investment and participation. The 28.3% reduction in price volatility is particularly significant, as price uncertainty has been identified as a major barrier to renewable energy financing and long-term contracting.

Technical performance improvements, especially in forecasting accuracy and grid stability maintenance, suggest that GAN-based systems can successfully balance economic optimization with technical reliability requirements. The ability to maintain voltage and frequency stability while optimizing economic outcomes addresses a critical concern that has limited the deployment of market-based grid management approaches.

Environmental impact results demonstrate that improved market mechanisms can contribute substantially to sustainability goals. The 16.8% increase in renewable energy utilization and 11.2% reduction in carbon emissions achieved through better coordination and optimization represent significant contributions to climate change mitigation efforts.

The robustness and scalability demonstrated across different operating conditions and system sizes suggest that GAN-based approaches are suitable for widespread deployment. The system's ability to maintain performance benefits across different renewable energy penetration levels, market participation rates, and network topologies indicates broad applicability across diverse energy system contexts.

Limitations and Challenges

Despite the promising results, several limitations and challenges must be acknowledged and addressed in future research and development efforts. Data requirements represent a significant practical challenge, as GAN training requires large, high-quality datasets that may not be available in many markets, particularly in developing regions or emerging renewable energy markets.

The computational complexity of GAN training and inference, while manageable with current cloud computing resources, may present barriers for smaller market operators or resource-constrained environments. Future research should explore model compression and edge computing approaches to reduce computational requirements while maintaining performance benefits.

Model interpretability remains a challenge for GAN-based systems, particularly in regulated energy markets where decision transparency and auditability are essential. While our framework includes explanation mechanisms, further development of interpretable AI techniques specifically for energy trading applications would enhance regulatory acceptance and market confidence.

The adversarial training process can be unstable and may require careful hyperparameter tuning and monitoring to ensure consistent performance. Automated hyperparameter optimization and robust training procedures need further development to facilitate practical deployment by market operators without extensive machine learning expertise.

Integration with existing market infrastructure and regulatory frameworks presents ongoing challenges, as energy markets are heavily regulated and often resistant to rapid technological change. Gradual deployment strategies and close collaboration with regulators will be essential for successful implementation.

Future Research Directions

Several promising research directions emerge from this work that could further enhance the capabilities and applicability of GAN-based renewable energy trading systems. Advanced GAN architectures, including attention mechanisms, transformer-based generators, and multi-scale discriminators, could improve the quality and diversity of generated scenarios while reducing training time and computational requirements.

Federated learning approaches could address data privacy and availability challenges by enabling collaborative model training across multiple market participants without sharing sensitive data. This approach could be particularly valuable for developing comprehensive models that capture diverse market behaviors while maintaining participant privacy.

Integration with emerging technologies, including blockchain for transparent and secure transactions, 5G networks for ultra-low latency communication, and quantum computing for enhanced optimization capabilities, represents exciting opportunities for future development. The combination of GAN-based scenario generation with blockchain-based transaction execution could create highly efficient and secure decentralized energy markets.

Extension to other renewable energy sources, including hydroelectric, geothermal, biomass, and emerging technologies such as wave and tidal energy, would broaden the applicability of the framework. Each renewable energy source presents unique characteristics and challenges that could benefit from specialized GAN architectures and optimization approaches.

Development of standardized APIs and interoperability frameworks would facilitate broader adoption by reducing integration costs and enabling seamless operation across different market platforms and geographic regions. Industry collaboration on standardization efforts would accelerate deployment and reduce implementation risks.

Multi-sector integration, including coupling with heating, cooling, and transportation energy demands, represents a significant opportunity for future research. The development of comprehensive energy system models that account for interactions across multiple energy sectors could unlock additional efficiency gains and environmental benefits.

Regulatory and Policy Implications

The implementation of GAN-based renewable energy trading systems has significant implications for energy market regulation and policy development. Regulatory frameworks will need to evolve to accommodate intelligent, automated trading systems while maintaining market fairness, transparency, and consumer protection standards.

Market design considerations include the need for new rules governing algorithmic trading, automated market makers, and AI-based decision systems. Regulators will need to develop guidelines for algorithm validation, performance monitoring, and risk management to ensure market integrity and stability.

Privacy and data protection regulations will require careful consideration as GAN-based systems rely on detailed energy consumption and production data that could reveal sensitive information about participants. Differential privacy techniques and secure multi-party computation methods may be necessary to protect participant privacy while enabling system functionality.

Consumer protection measures must be developed to ensure that sophisticated AI-based trading systems do not disadvantage less technically sophisticated market participants. Regulatory frameworks should include provisions for transparent pricing, fair market access, and protection against algorithmic bias or discrimination.

International coordination will become increasingly important as renewable energy trading expands across national boundaries. Standardized protocols, mutual recognition agreements, and harmonized regulatory frameworks will be essential for realizing the full benefits of international renewable energy trade.

CONCLUSION

This chapter has presented a comprehensive framework for applying Generative Adversarial Networks to dynamic renewable energy trading and price optimization in decentralized markets. Through detailed technical development, extensive case studies, and rigorous performance evaluation, we have demonstrated the significant potential of GAN-based approaches to address fundamental challenges in renewable energy market integration.

Key Contributions

The primary contributions of this work include the development of a novel multi-modal GAN architecture specifically designed for energy trading applications, incorporating specialized generators for energy production, demand patterns, price dynamics, and weather conditions. This architecture successfully captures the complex, interdependent relationships that characterize decentralized renewable energy markets.

The adversarial training framework extends traditional GAN formulations to incorporate domain-specific constraints and objectives relevant to energy trading, including grid stability requirements, economic rationality constraints, and physical conservation laws. This enhanced framework ensures that generated scenarios maintain both statistical realism and practical applicability.

The dynamic pricing optimization module integrates GAN-generated scenarios with advanced optimization techniques to enable real-time price discovery and market coordination. This integration demonstrates how generative models can be effectively combined with traditional optimization approaches to achieve superior performance.

Comprehensive evaluation across three diverse case studies validates the practical applicability of the proposed approach across different scales, technologies, and market structures. The consistent performance improvements achieved across all case studies demonstrate the robustness and generalizability of the GAN-based framework.

The uncertainty quantification and risk management components provide essential tools for practical deployment in risk-sensitive energy market applications. These components ensure that trading decisions appropriately account for various sources of uncertainty while maintaining acceptable risk levels.

Practical Impact and Significance

The practical significance of this work lies in its potential to accelerate the transition to sustainable energy systems by making renewable energy trading more efficient, accessible, and profitable. The demonstrated improvements in economic efficiency, technical reliability, and environmental impact provide compelling evidence that GAN-based approaches can overcome traditional barriers to renewable energy adoption.

For renewable energy developers and investors, the improved price predictability and market efficiency provided by our framework can reduce financial risks and improve project economics. This can lead to increased investment in renewable energy projects and faster deployment of clean energy technologies.

For consumers and prosumers, the enhanced market mechanisms enable more active participation in energy markets, potentially reducing energy costs while providing new revenue opportunities through energy selling and grid services. The demonstrated energy cost reductions of up to 32% in our case studies represent substantial economic benefits for participants.

For grid operators and utilities, the improved coordination and optimization capabilities can enhance grid stability and reliability while reducing the need for expensive backup generation and infrastructure upgrades. The demonstrated ability to maintain grid stability while maximizing renewable energy utilization addresses a critical challenge in modern power system operation.

For policymakers and regulators, the framework provides tools for designing and evaluating market mechanisms that align economic incentives with environmental and reliability objectives. The quantified environmental benefits, including reduced carbon emissions and improved renewable energy utilization, support the achievement of climate change mitigation goals.

Technological Innovation and Methodological Advances

The research presented in this chapter represents a significant methodological advancement in the application of artificial intelligence to energy systems. By adapting GAN architectures to the specific requirements of energy trading, we have created a new paradigm for modeling and optimizing complex energy market dynamics. The integration of adversarial training with energy-specific constraints demonstrates how domain expertise can be effectively incorporated into machine learning frameworks to achieve superior performance in specialized applications.

The multi-modal approach developed herein addresses a critical gap in existing literature by simultaneously modeling multiple interconnected aspects of energy systems. Traditional approaches often treat energy production, consumption, pricing, and weather patterns as separate problems, failing to capture the complex interdependencies that drive real-world system behavior. Our unified framework provides a more realistic and comprehensive representation of energy market dynamics, leading to improved decision-making and system performance.

The temporal modeling capabilities introduced through our GAN architecture represent another significant advancement. Energy systems operate across multiple time scales, from millisecond grid stability requirements to seasonal energy storage planning. Our framework successfully captures this multi-scale temporal behavior, enabling more accurate predictions and optimization across diverse planning horizons. This capability is particularly valuable for integrating high levels of variable renewable energy while maintaining system reliability and economic efficiency.

The incorporation of uncertainty quantification directly into the generative modeling process provides a novel approach to risk management in energy systems. Rather than treating uncertainty as an external factor to be managed through separate risk assessment procedures, our framework generates scenarios that inherently capture the probabilistic nature of energy system behavior. This approach enables more sophisticated risk-aware decision making and provides market participants with better tools for managing exposure to various sources of uncertainty.

Economic and Social Implications

The economic implications of widespread adoption of GAN-based energy trading systems extend far beyond the direct cost savings demonstrated in our case studies. The improved market efficiency and reduced transaction costs enabled by our framework can fundamentally transform the economics of renewable energy development and deployment. By reducing market barriers and improving price discovery mechanisms, our approach can accelerate the transition to renewable energy and reduce the overall cost of decarbonization.

The democratization of energy market participation represents a significant social benefit of our framework. Traditional energy markets often require substantial technical expertise and capital investment to participate effectively. Our intelligent trading system reduces these barriers, enabling smaller renewable energy producers and individual prosumers to participate in energy markets on more equal terms with large utilities and energy companies. This democratization can lead to more distributed ownership of energy resources and more equitable distribution of the economic benefits of the clean energy transition.

The potential for community-based energy trading systems represents another important social implication. Our framework can enable local energy communities to optimize their collective energy production, consumption, and trading, leading to increased energy independence and community resilience. These local energy systems can provide economic benefits to community members while reducing dependence on centralized energy infrastructure and improving local air quality through increased renewable energy utilization.

The job creation potential of our framework should not be overlooked. While automation and AI can sometimes lead to job displacement, the development and deployment of sophisticated energy trading systems creates new opportunities in areas such as system design, implementation, monitoring, and maintenance. The increased renewable energy development enabled by our framework also creates jobs in manufacturing, installation, and maintenance of renewable energy technologies.

Technical Challenges and Limitations

While our research demonstrates significant potential for GAN-based energy trading systems, several technical challenges and limitations must be acknowledged and addressed through future research. The computational requirements of training and deploying sophisticated GAN models represent a significant practical constraint, particularly for smaller market participants with limited computational resources. Developing more efficient training algorithms and model architectures will be essential for widespread adoption.

The data requirements for training effective GAN models can also present challenges in real-world deployments. High-quality, comprehensive datasets covering all relevant aspects of energy system behavior are not always available, particularly for emerging technologies and market structures. Developing techniques for training GAN models with limited or heterogeneous data will be crucial for broader applicability.

The interpretability of GAN-based trading decisions represents another important limitation. While our framework demonstrates superior performance compared to traditional approaches, the complex nature of GAN models can make it difficult to understand and explain specific trading decisions. This lack of interpretability can be problematic for regulatory compliance and market participant confidence. Developing more interpretable AI models for energy trading applications represents an important area for future research.

The robustness of GAN models to adversarial attacks and data poisoning represents a critical security consideration that requires continued attention. Energy systems are critical infrastructure that could be targets for malicious attacks, and the AI systems that control them must be designed to be resilient to such threats. Developing secure and robust GAN architectures for energy applications will be essential for safe deployment in critical infrastructure applications.

Future Research and Development Priorities

Several key research priorities emerge from this work that could significantly enhance the capabilities and applicability of GAN-based energy trading systems. The development of more efficient and scalable GAN architectures specifically designed for energy applications represents a high priority. This includes research into attention mechanisms, transformer-based architectures, and specialized training procedures that can improve performance while reducing computational requirements.

The integration of GAN-based systems with emerging technologies such as blockchain, 5G networks, and quantum computing represents exciting opportunities for future development. These technologies could address current limitations related to security, communication latency, and computational capacity while enabling new capabilities and applications.

The extension of our framework to other sectors and energy carriers beyond electricity represents another important research direction. The principles and methodologies developed for electricity trading could be adapted for natural gas, hydrogen, and other energy carriers, enabling more comprehensive optimization of integrated energy systems.

The development of hybrid approaches that combine GAN-based scenario generation with other AI techniques such as reinforcement learning and evolutionary optimization could lead to even more powerful and capable energy trading systems. These hybrid approaches could leverage the strengths of different AI techniques while mitigating their individual limitations.

Final Remarks

Generative Adversarial Networks represent a powerful tool for addressing the complex challenges of renewable energy trading in decentralized markets. The framework presented in this chapter demonstrates that sophisticated AI techniques can successfully balance economic efficiency, technical reliability, and environmental sustainability objectives in practical energy system applications.

The transition to sustainable energy systems requires innovative approaches that can handle the inherent complexity and uncertainty of renewable energy resources while maintaining the reliability and affordability that consumers and industries require. Our GAN-based framework provides a promising path forward that leverages cutting-edge artificial intelligence to create more intelligent, efficient, and sustainable energy markets.

As renewable energy technologies continue to improve and costs continue to decline, the need for sophisticated market mechanisms that can fully capture their value will only increase. The methodologies and insights presented in this chapter provide a foundation for developing next-generation energy trading systems that can support the global transition to clean, sustainable energy.

The future of renewable energy trading lies in the intelligent integration of advanced AI techniques, sophisticated market mechanisms, and distributed energy technologies. By continuing to develop and refine these approaches, we can create energy systems that are not only environmentally sustainable but also economically efficient and socially equitable, supporting the broader goals of sustainable development and climate change mitigation.

The work presented in this chapter represents a significant step forward in the application of artificial intelligence to sustainable energy systems. However, it is just the beginning of what promises to be a transformative application of AI to one of the most important challenges of our time. The continued development and deployment of intelligent energy systems will be essential for achieving global climate goals and creating a sustainable energy future for all.

REFERENCES

Baesmat, K. H. (2024). Impedance analysis of adaptive distance relays using machine learning. In S. Latifi (Ed.), *ITNG 2024: 21st International Conference on Information Technology–New Generations. Advances in Intelligent Systems and Computing* (Vol. 1456). Springer. https://doi.org/DOI: 10.1007/978-3-031-56599-1_57

Baesmat, K. H., & Latifi, S. (2023). A new hybrid method for electrical load forecasting based on deviation correction and MRMRMS. In Selvaraj, H., Chmaj, G., & Zydek, D. (Eds.), *Advances in Systems Engineering* (pp. 293–303). Springer Nature., DOI: 10.1007/978-3-031-40579-2_29

Baesmat, K. H., Masoudipour, I., & Samet, H. (2021). Improving the performance of short-term load forecast using a hybrid artificial neural network and artificial bee colony algorithm. *IEEE Canadian Journal of Electrical and Computer Engineering, 44*(3), 275–282. DOI: 10.1109/ICJECE.2021.3056125

Baesmat, K. H., Regentova, E. E., & Baghzouz, Y. (2025). A hybrid machine learning–statistical based method for short-term energy consumption prediction in residential buildings. *Energy and AI*. Advance online publication. https://doi.org/DOI: 10.1016/j.egyai.2025.100552

Bai, X., Liu, L., Ju, J., Zhong, X., Zhou, Y., & Lin, J.. (2022). Distributed optimization method for multi-area integrated energy systems considering demand response. *Frontiers in Energy Research, 10*, 975214. DOI: 10.3389/fenrg.2022.975214

Chakraborty, T., Reddy, K. S., Naik, S. M., Panja, M., & Manvitha, B. (2023). Ten years of generative adversarial nets (GANs): A survey of the state-of-the-art. *arXiv*. https://arxiv.org/abs/2308.16316

da Costa, V. B. F., de Doile, G. N. D., Troiano, G., Dias, B. H., Bonatto, B. D., & de Freitas Filho, W. (2022). Electricity markets in the context of distributed energy resources and demand response programs: Main developments and challenges based on a systematic literature review. *Energies, 15*(20), 7784. DOI: 10.3390/en15207784

Dudjak, V., Neves, D., Alskaif, T., Khadem, S., Peña-Bello, A., & Saggese, P.. (2021). Impact of local energy markets integration in power systems layer: A comprehensive review. *Applied Energy, 301*, 117434. DOI: 10.1016/j.apenergy.2021.117434

Farrokhi, Z., Baesmat, K. H., & Regentova, E. E. (2024). Enhancing urban intelligence energy management: Innovative load forecasting techniques for electrical networks. *Journal of Power and Energy Engineering, 12*, 72–88. DOI: 10.4236/jpee.2024.1211005

Farrokhi, Z., Kalhor, A., & Masouleh, M. T. (2018). Implementation and evaluation of object identification techniques on Nao robot platform. *International Journal of Mechatronics, Electrical and Computer Technology (IJMEC), 8*(29), 3947–3958. Retrieved from https://aeuso.org/includes/files/articles/Vol8_Iss29_3947-3958_Implementation_and_Evaluation_of_Ob.pdf

Hassan, Q., Algburi, S., Sameen, A. Z., Al-Musawi, T. J., Al-Jiboory, A. K., Salman, H. M., Ali, B. M., & Jaszczur, M. (2024). A comprehensive review of international renewable energy growth. *Energy. Built Environment*. Advance online publication. DOI: 10.1016/j.enbenv.2023.12.002

Hassanpouri Baesmat, K., Farrokhi, Z., Chmaj, G., & Regentova, E. E. (2025). Parallel multi-model energy demand forecasting with cloud redundancy: Leveraging trend correction, feature selection, and machine learning. *Forecasting*, *7*(2), 25. DOI: 10.3390/forecast7020025

Hassanpouri Baesmat, K., Shokoohi, F., & Farrokhi, Z. (2025). SP-RF-ARIMA: A sparse random forest and ARIMA hybrid model for electric load forecasting. *Global Energy Interconnection*, *8*(3), 486–496. DOI: 10.1016/j.gloei.2025.04.003

Hüllermeier, E., & Waegeman, W. (2021). Aleatoric and epistemic uncertainty in machine learning: An introduction to concepts and methods. *Machine Learning*, *110*, 457–506. DOI: 10.1007/s10994-021-05946-3

Impram, S., Varbak-Nese, S., & Oral, B. (2020). Challenges of renewable energy penetration on power system flexibility: A survey. *Energy Strategy Reviews*, *31*, 100539. DOI: 10.1016/j.esr.2020.100539

Jamali, H. (2025). Quantum-accelerated neural imputation with large language models (LLMs) [Preprint]. *arXiv*. https://doi.org//arXiv.2507.08255DOI: 10.48550

Jamali, H., Dascalu, S. M., & Harris, F. C., Jr. (2024). Fostering joint innovation: A global online platform for ideas sharing and collaboration. In *Proceedings of the International Conference on Information Technology–New Generations* (pp. 305–312). Springer Nature. https://doi.org/DOI: 10.1007/978-3-031-56599-1_40

Jamali, H., Dascalu, S. M., & Harris, F. C. (2024). AI-driven analysis and prediction of energy consumption in NYC's municipal buildings. In *2024 IEEE/ACIS 22nd International Conference on Software Engineering Research, Management and Applications (SERA)* (pp. 277–283). IEEE. https://doi.org/DOI: 10.1109/SERA61261.2024.10685594

Jamali, H., Dascalu, S. M., Harris, F. C., & Feil Seifer, D. (2025). Optimizing personalized learning pathways with the Salp Swarm Algorithm: A novel approach. In *2025 IEEE 6th International Conference on Artificial Intelligence, Robotics, and Control*. IEEE. https://doi.org/DOI: 10.1109/AIRC64931.2025.11077498

Jamali, H., Debolt, A., Dalton, H., Layosa, J., Macy, I., Shill, P., Feil Seifer, D., Harris, F., Dascalu, S., & Wu, R. (2025). FORE: A student-centered framework for accessible robotics education through simulation and interactive learning. *ASEE Annual Conference Proceedings*. Retrieved from https://par.nsf.gov/biblio/10595258

Jamali, H., Karimi, A., & Haghighizadeh, M. (2018). A new method of cloud-based computation model for mobile devices: Energy consumption optimization in mobile-to-mobile computation offloading. In *Proceedings of the 6th International Conference on Communications and Broadband Networking* (pp. 32–37). ACM. https://doi.org/DOI: 10.1145/3193092.3193103

Jamali, H., Karimi, A., & Haghighizadeh, M. (2018). A new method of cloud-based computation model for mobile devices: Energy consumption optimization in mobile-to-mobile computation offloading. In *Proceedings of the 6th International Conference on Communications and Broadband Networking* (pp. 32–37). ACM. https://doi.org/DOI: 10.1145/3193092.3193103

Jamali, H., Shill, P. C., Feil Seifer, D., Harris, F. C., Jr., & Dascalu, S. M. (2023). A schedule of duties in the cloud space using a modified salp swarm algorithm. In *IFIP International Internet of Things Conference* (pp. 62–75). Springer Nature. https://doi.org/DOI: 10.1007/978-3-031-45878-1_5

Jamali, H., Watson, J., Dascalu, S. M., & Harris, F. C. (2025). Harmonized data drive: Standardizing and unifying smart car information storage for enhanced forensics and interoperability. In *Proceedings of the 2025 13th International Symposium on Digital Forensics and Security (ISDFS)* (pp. 1–6). IEEE. https://doi.org/DOI: 10.1109/ISDFS65363.2025.11012079

Kazmi, S., Gorgulu, B., Cevik, M., & Baydogan, M. G. (2023). A concurrent CNN-RNN approach for multi-step wind power forecasting. *arXiv*. https://doi.org//arXiv .2301.00819DOI: 10.48550

Kermani, A., Jamshidi, A. M., Mahdavi, Z., Dashtaki, A. A., Zand, M., Nasab, M. A., & Khan, B. (2023). Energy management system for smart grid in the presence of energy storage and photovoltaic systems. *International Journal of Photoenergy*, *2023*, 5749756. Advance online publication. DOI: 10.1155/2023/5749756

Mahdavi, Z., Samavat, T., Javanmardi, A. S. J., Dashtaki, M. A., Zand, M., Nasab, M. A., & Khan, B. (2024). Providing a control system for charging electric vehicles using ANFIS. *International Transactions on Electrical Energy Systems, 2024*, 9921062. Advance online publication. DOI: 10.1155/2024/9921062

Muriithi, G., & Chowdhury, S. (2022). Deep Q-network application for optimal energy management in a grid-tied solar PV-battery microgrid. *The Journal of Engineering, 2022*(3). https://doi.org/DOI: 10.1049/tje2.12128

Padmanaban, S., Nasab, M. A., Samavat, T., Zand, M., & Nasab, M. A. (2023). Securing smart power grids against cyber-attacks. In *IoT and Analytics in Renewable Energy Systems* (Vol. 1, pp. 17–36). CRC Press., DOI: 10.1201/9781003331117

Padmanaban, S., Nasab, M. A., Samavat, T., Zand, M., & Nasab, M. A. (2023). Artificial intelligence techniques for smart power systems. In *IoT and Analytics in Renewable Energy Systems* (Vol. 1, pp. 107–123). CRC Press., DOI: 10.1201/9781003331117

Padmanaban, S., Nasab, M. A., Samavat, T., Zand, M., Nasab, M. A., & Hashemi, E. (2023). Cyber security in smart energy networks. In *IoT and Analytics in Renewable Energy Systems* (Vol. 2, pp. 309–325). CRC Press., DOI: 10.1201/9781003374121

Samavat, T., Nazari, M., Fuhong, L., & Yang, L. (2025). Optimizing stand-alone PV systems: A metaheuristic-enhanced fuzzy approach for adaptive MPPT. *China Communications*, *22*(1), 61–74. DOI: 10.23919/JCC.fa.2024-0224.202501

Sanjeevikumar, P., Zand, M., Nasab, M. A., Eskandari, M., Samavat, T., Jahangiri, A., & Moradi, M. H. (2023). Power electronics in smart grid. In *Smart and Power Grid Systems–Design Challenges and Paradigms* (pp. 1–20). River Publishers., DOI: 10.1201/9781003339557

Soto, E. A., Bosman, L. B., Wollega, E., & Leon-Salas, W. D. (2021). Peer-to-peer energy trading: A review of the literature. *Applied Energy*, *283*, 116268. DOI: 10.1016/j.apenergy.2021.116268

Vishwakarma, A. K., Patro, P. K., Acquaye, A., Jayaraman, R., & Salah, K. (2024). Blockchain-based peer-to-peer renewable energy trading and traceability of transmission and distribution losses. *The Journal of the Operational Research Society*. Advance online publication. DOI: 10.1080/01605682.2024.2441224

Zhang, M., Liu, X., & Wang, Y.. (2024). A hybrid time series forecasting approach integrating fuzzy clustering and machine learning for enhanced power consumption prediction. *Scientific Reports*, *14*(1), 91123. DOI: 10.1038/s41598-025-91123-8

Zhao, L., Li, Y., & Zhong, T. (2025). A generative adversarial network with multiscale and attention mechanisms for underwater image enhancement. *Scientific Reports*, *15*, 2787. DOI: 10.1038/s41598-025-86949-1 PMID: 39843451

Chapter 5
Generative AI for Bioenergy System Design

P. Selvakumar
https://orcid.org/0000-0002-3650-4548
Department of Science and Humanities, Nehru Institute of Technology, Coimbatore, India

Abhijeet Das
https://orcid.org/0000-0003-4599-5462
C.V. Raman Global University, Bhubaneswar, India

T. C. Manjunath
https://orcid.org/0000-0003-2545-9160
Rajarajeswari College of Engineering, India

ABSTRACT

The global pursuit of sustainable energy solutions has intensified amid growing concerns over climate change, depleting fossil fuel reserves, and escalating energy demands. Among the array of renewable energy sources, bioenergy—derived from organic materials such as agricultural residues, forest waste, and algae—holds significant promise due to its potential for carbon neutrality and broad resource availabilityThe global pursuit of sustainable energy solutions has intensified amid growing concerns over climate change, depleting fossil fuel reserves, and escalating energy demands. Among the array of renewable energy sources, bioenergy—derived from organic materials such as agricultural residues, forest waste, and algae—holds significant promise due to its potential for carbon neutrality and broad resource availability. However, the design, optimization, and deployment of efficient bioenergy systems remain complex tasks that involveand conversion technologies to logistics, economic feasibility, and environmental impact assessments.

DOI: 10.4018/979-8-3373-4272-6.ch005

THE ROLE OF AI IN BIOENERGY SYSTEM DESIGN

The global pursuit of sustainable energy solutions has intensified amid growing concerns over climate change, depleting fossil fuel reserves, and escalating energy demands. Among the array of renewable energy sources, bioenergy—derived from organic materials such as agricultural residues, forest waste, and algae—holds significant promise due to its potential for carbon neutrality and broad resource availability. The global pursuit of sustainable energy solutions has intensified amid growing concerns over climate change, depleting fossil fuel reserves, and escalating energy demands. Among the array of renewable energy sources, bioenergy—derived from organic materials such as agricultural residues, forest waste, and algae—holds significant promise due to its potential for carbon neutrality and broad resource availability. However, the design, optimization, and deployment of efficient bioenergy systems remain complex tasks that involve multifaceted variables ranging from feedstock selection and conversion technologies to logistics, economic feasibility, and environmental impact assessments. In this context, Artificial Intelligence (AI) has emerged as a transformative tool capable of revolutionizing the way bioenergy systems are conceptualized, modeled, and managed. (Abasian et al., 2019)The integration of AI into bioenergy system design introduces a paradigm shift, enabling researchers, engineers, and policymakers to harness data-driven insights, automate decision-making processes, and optimize system performance across the entire value chain.AI refers to the development of computational systems capable of performing tasks that typically require human intelligence, such as learning, reasoning, problem-solving, and pattern recognition. Techniques such as machine learning (ML), deep learning (DL), artificial neural networks (ANNs), genetic algorithms (GAs), expert systems, and reinforcement learning (RL) are being increasingly employed in bioenergy research and applications. These AI methodologies offer robust capabilities to process vast datasets, identify nonlinear patterns, predict system behavior under varying conditions, and continuously improve through feedback. In the realm of bioenergy, such tools are particularly valuable given the inherent variability of biomass resources, regional climatic conditions, and the stochastic nature of biological and thermochemical processes. Traditional modeling approaches, though useful, often fall short in handling the high dimensionality and dynamic interactions within bioenergy systems. AI overcomes these limitations by enabling predictive modeling, adaptive control, and real-time system optimization, thereby enhancing both the efficiency and resilience of bioenergy infrastructure. One of the most significant contributions of AI in bioenergy system design is in feedstock assessment and classification. (Abbasi et al., 2021) Accurate identification and characterization of biomass feedstocks are vital for selecting appropriate conversion technologies and ensuring energy efficiency. Machine learning models can analyze remote sensing data, agricultural statistics, and biochemical properties to predict biomass availability, composition, and yield across diverse geographic regions. Satellite imagery combined with AI algorithms facilitates precision agriculture, allowing for the targeted cultivation and harvesting of energy crops.

Furthermore, AI-powered platforms assist in optimizing supply chains by determining the most cost-effective and environmentally sustainable routes for biomass collection, storage, and transportation. also plays a pivotal role in optimizing bioenergy conversion technologies, such as anaerobic digestion, gasification, pyrolysis, and transesterification. These processes involve numerous operational parameters that significantly influence energy output and environmental performance. By leveraging techniques such as artificial neural networks and support vector machines, researchers can develop predictive models that simulate process dynamics and recommend optimal conditions for maximum yield and minimal emissions. Reinforcement learning algorithms further enable real-time process control by learning from operational feedback and adjusting system parameters autonomously. For example, in biogas production, AI can predict methane yields based on substrate composition and digestion conditions (Babazadeh et al., 2017), allowing for more stable and efficient reactor performance. Moreover, AI-driven decision support systems contribute to the strategic planning and management of bioenergy projects. These systems integrate technical, economic, and environmental data to assess the feasibility and sustainability of bioenergy initiatives. Life cycle assessment (LCA), techno-economic analysis (TEA), and multi-criteria decision analysis (MCDA) are increasingly being enhanced through AI tools that provide faster, more accurate evaluations and facilitate stakeholder engagement. In rural or remote regions, AI-based tools help identify suitable sites for bioenergy plants by analyzing topography, resource availability, infrastructure, and socio-economic factors. These insights not only reduce planning time but also support evidence-based policymaking and investment decisions. As the bioenergy sector becomes increasingly digitalized, the role of AI in monitoring and maintenance also becomes crucial. AI systems equipped with sensors and Internet of Things (IoT) technologies enable continuous monitoring of plant operations, predicting equipment failures, and initiating preventive maintenance. Predictive analytics reduces downtime and operational costs while improving the overall reliability of bioenergy facilities. Additionally, AI contributes to ensuring compliance with environmental regulations by tracking emissions, waste generation, and resource consumption in real time. The integration of AI into bioenergy system design is not without challenges, (Babazadeh et al., 2019). Issues related to data quality, interoperability, algorithm transparency, and ethical considerations must be addressed to fully realize the benefits of AI. Data scarcity, particularly in developing regions, can hinder the training of robust AI models. Efforts must be made to standardize data collection methods and promote open-access databases. Moreover, the black-box nature of some AI models raises concerns about interpretability and accountability, especially when these tools influence high-stakes decisions.

SUSTAINABILITY AND BIOENERGY: THE NEED FOR INNOVATIVE SOLUTIONS

The emergence of Generative Artificial Intelligence (Generative AI) marks a transformative chapter in the evolution of computational intelligence, with profound implications across a wide array of disciplines—including the sustainable energy sector. In particular, bioenergy system design stands to benefit significantly from generative AI's capacity to create, simulate, and optimize complex systems using data-driven, probabilistic modeling. Unlike traditional AI systems that operate on pattern recognition or classification, generative AI models learn the underlying structure of data and can generate entirely new content that retains contextual accuracy, (Chen et al., 2022). This capability introduces unprecedented opportunities in biomass feedstock modeling, process simulation, infrastructure design, optimization of conversion pathways, and predictive maintenance. As the complexity and variability of bioenergy systems continue to grow—owing to factors such as diverse biomass sources, geographical constraints, and dynamic environmental interactions—generative AI emerges as a critical tool for enhancing adaptability, resilience, and innovation within the bioenergy landscape. Generative AI encompasses a class of algorithms that include generative adversarial networks (GANs), variational autoencoders (VAEs), transformer-based models, and diffusion models, each possessing the ability to learn from complex, high-dimensional datasets and synthesize new outputs. These models are particularly powerful when applied to the multivariate, nonlinear, and stochastic nature of bioenergy systems, (Dal Mas et al., 2010). For instance, generative AI can simulate novel feedstock compositions or predict biomass yields under varying climate scenarios, offering a proactive approach to resource planning and climate resilience. It can generate synthetic data to fill in gaps where empirical data is sparse—common in remote or developing regions—thereby improving the robustness and generalizability of bioenergy models. In this sense, generative AI not only complements existing machine learning methods but also transcends their limitations by enabling system creation rather than just prediction. In feedstock development, generative AI can be harnessed to model novel bioenergy crops with optimized traits such as high lignocellulosic content, rapid growth cycles, drought tolerance, or low-input requirements. By analyzing genomic, phenotypic, and environmental datasets, AI models can generate potential crop variants and simulate their performance across different regions and seasons.

This accelerates bio-crop breeding programs and helps researchers design sustainable, location-specific feedstock strategies. Moreover, by integrating soil, weather, and agricultural data, generative models can simulate yield forecasts and land-use scenarios, assisting in land management policies and investment decisions. These tools are vital for maximizing biomass productivity while minimizing ecological impacts. In the realm of bioenergy process engineering, generative AI is revolutionizing how conversion systems are modeled and optimized, (Dal-Mas et al., 2011). Traditional simulations of pyrolysis, gasification, fermentation, and transesterification processes require extensive computational resources and rely heavily on empirical correlations. In contrast, generative models can be trained on experimental datasets and subsequently generate thousands of plausible process configurations, each evaluated based on desired criteria such as energy efficiency, greenhouse gas emissions, cost, or yield. These virtual simulations reduce the need for time-consuming and costly physical experiments. For example, in anaerobic digestion, generative AI can propose novel co-digestion recipes by simulating how different combinations of substrates affect methane production, stability, and by-product profiles. Similarly, in algal biofuel systems, generative models can identify optimal growth conditions, nutrient inputs, and harvesting schedules to maximize lipid content and minimize energy inputs. Furthermore, generative, (Ebadian et al., 2014). AI is being integrated into the design of bioenergy infrastructure and facilities, where spatial, environmental, and socio-economic factors must be balanced. Using geospatial data, population density maps, transportation networks, and regulatory constraints, generative design algorithms can create optimized layouts for biomass supply chains, processing facilities, and energy distribution networks. These models allow for iterative design generation—automatically producing and testing thousands of facility configurations—and refining them based on performance metrics. This is especially useful in multi-objective optimization problems, where trade-offs between cost, efficiency, and environmental impact must be carefully managed. For rural electrification or decentralized bioenergy systems, generative AI provides tailored infrastructure plans that align with local resource availability and community needs. A particularly promising application of generative AI in bioenergy is scenario generation and strategic planning. Long-term planning in the bioenergy sector is inherently uncertain, influenced by fluctuating market prices, policy shifts, and climate variability. Generative models can simulate a wide spectrum of future scenarios, including market dynamics, feedstock availability, technological advancements, and regulatory environments. These synthetic scenarios help stakeholders test the resilience of proposed systems and make informed decisions under uncertainty. This capability is central to achieving adaptive governance and ensuring that bioenergy systems remain viable in the face of disruption. In addition to planning and design, generative AI plays an instrumental role in real-time control and predictive maintenance of bioenergy systems, (El-Sheekh et al., 2019). By analyzing historical operational data and learning from system behavior, generative models can predict anomalies, simulate potential fault scenarios, and recommend maintenance interventions before failures occur.

This predictive capacity reduces operational costs, improves safety, and ensures uninterrupted energy production. Generative AI can also be paired with digital twins—virtual replicas of physical systems—allowing operators to simulate and test system changes before implementing them in the real world. Despite its transformative potential, the integration of generative AI into bioenergy system design is accompanied by several challenges. High-quality and diverse datasets are essential for training generative models, yet such datasets are often fragmented or inaccessible, especially in emerging markets. Furthermore, model interpretability remains a key concern. Generative models, particularly deep neural networks, often function as "black boxes," making it difficult to trace the rationale behind generated outputs. This opacity can hinder trust and regulatory compliance in safety-critical or policy-sensitive applications. Ethical concerns also arise around data ownership, bias, and environmental justice, particularly if generative AI tools are deployed without considering the voices and needs of marginalized communities, (Fahimnia et al., 2015). Addressing these issues requires not only technical innovation but also interdisciplinary collaboration, transparent governance frameworks, and inclusive data practices.

GENERATIVE MODELS FOR BIOMASS CONVERSION PROCESS OPTIMIZATION

The global transition toward a more sustainable and climate-resilient future has brought renewable energy to the forefront of international policy, investment, and research. Among the array of clean energy options, bioenergy occupies a unique and pivotal position due to its capacity to provide not only electricity and heat but also transportation fuels, chemical feedstocks, and rural employment. Derived from organic materials such as agricultural residues, forestry by-products, energy crops, and waste biomass, bioenergy offers the potential for carbon neutrality, improved energy security, and circular economy integration. However, realizing the full sustainability potential of bioenergy systems requires far more than simply substituting biomass for fossil fuels. The complexity of biomass supply chains, regional ecological constraints, and social equity concerns demand a holistic and innovative approach. Addressing these multifaceted sustainability challenges calls for integrated thinking, advanced technologies, and policies that are both adaptive and inclusive. Sustainability in the context of bioenergy refers to meeting present energy needs without compromising the ability of future generations to meet theirs. This involves ensuring that bioenergy systems are environmentally sound, socially equitable, and economically viable over the long term, (Fallah & Nozari, 2021). Yet, bioenergy systems can have both positive and negative sustainability outcomes depending on how they are designed and implemented. On the environmental front, bioenergy offers the advantage of renewability and can contribute significantly to greenhouse gas (GHG) emission reductions. For instance, when biomass feedstocks are sourced sustainably—such as through agricultural residues or short-rotation forestry—bioenergy can achieve substantial carbon savings compared to fossil fuels. Furthermore, the use of organic waste streams for energy can reduce landfill methane emissions and enhance waste management practices. However, sustainability concerns arise when bioenergy systems result in deforestation, monoculture plantations, water stress, or loss of biodiversity. Unsustainable land-use changes to accommodate energy crops can lead to soil degradation, habitat destruction, and even an overall increase in net carbon emissions due to land-use change emissions. Social sustainability is another critical dimension, particularly in regions where land tenure, food security, and rural livelihoods intersect with bioenergy production.

The "food vs. fuel" debate remains a contentious issue, especially when large-scale bioenergy operations compete with food production for arable land and water. In some cases, poorly regulated bioenergy initiatives have led to land grabbing, marginalization of smallholders, and displacement of local communities. On the other hand, when properly managed, bioenergy can foster rural development, create local employment, and empower communities by enabling decentralized energy access, (Garai & Sarkar, 2022). For example, small-scale biogas systems and biomass cookstoves can improve indoor air quality and reduce the health burden from traditional biomass burning in low-income households. Ensuring social sustainability thus requires policies that prioritize local ownership, equitable benefit-sharing, and inclusive stakeholder participation in bioenergy planning and implementation. Economically, bioenergy must compete with other energy sources while delivering long-term value. The variability of biomass prices, high capital costs of conversion technologies, and logistical challenges of feedstock transportation can affect project viability. Moreover, the economic sustainability of bioenergy depends on its integration into broader energy markets, availability of incentive mechanisms, and development of value-added co-products such as biochar, organic fertilizers, or bio-based chemicals. Innovations in biorefinery concepts, where multiple products are derived from a single biomass stream, offer a promising pathway for increasing economic returns while reducing waste. The shift toward circular bioeconomy models, where agricultural, industrial, and municipal residues are transformed into clean energy and valuable materials, aligns well with both environmental and economic sustainability objectives. Given these complex sustainability dimensions, innovation is essential to navigate trade-offs and optimize outcomes across the environmental, social, and economic spheres. Technological innovation plays a key role in enhancing the efficiency and sustainability of bioenergy systems. Advances in biomass conversion technologies—including gasification, pyrolysis, hydrothermal liquefaction, and fermentation—are making it possible to extract more energy from feedstocks while reducing emissions and waste. Furthermore, the development of second- and third-generation biofuels, derived from non-food biomass and algae respectively, offers a promising solution to the food vs. fuel dilemma. These fuels utilize lignocellulosic residues or fast-growing microorganisms, which do not directly compete with food production and often require less intensive resource inputs. Digital innovation, particularly through Artificial Intelligence (AI), the Internet of Things (IoT), and blockchain technologies, is revolutionizing how bioenergy systems are designed, monitored, and managed, (Garai et al., 2021). AI algorithms can optimize feedstock logistics, predict energy yields, and identify the most sustainable system configurations based on a wide range of variables. Real-time monitoring through IoT sensors enables adaptive management of bioenergy facilities, improving performance and reducing operational risks. Blockchain can enhance supply chain transparency and traceability, ensuring that sustainability claims—such as certification of feedstock origin or carbon offsets—are verifiable and trustworthy. Policy and governance innovation are equally vital for driving sustainable bioenergy adoption.

Clear regulatory frameworks, sustainability standards, and certification schemes (such as the Roundtable on Sustainable Biomaterials or EU RED II) help ensure that bioenergy projects meet rigorous environmental and social criteria. Additionally, financial innovation, including green bonds, carbon markets, and results-based financing, can mobilize capital toward sustainable bioenergy ventures. Cross-sectoral and multi-stakeholder collaboration is also essential, enabling integrated planning that aligns energy goals with agricultural, forestry, and environmental policies. Importantly, systemic innovation—which considers the entire bioenergy value chain from resource extraction to end-use—is needed to embed sustainability at every level. (Gebreslassie et al., 2012).

AI-DRIVEN DESIGN OF BIOENERGY SYSTEMS: CASE STUDIES

Processes remain a critical challenge due to the complex, nonlinear, and dynamic nature of biomass characteristics and conversion pathways, (Habib et al., 2020). Biomass conversion involves transforming organic material into usable forms of energy—such as biofuels, biogas, bio-oil, and syngas—using processes like pyrolysis, gasification, anaerobic digestion, fermentation, and transesterification. These processes are inherently sensitive to variations in feedstock properties, operating conditions, and system configurations. Traditional optimization methods, while effective to a certain extent, often rely on labor-intensive experimental trials and rigid mathematical models that struggle to adapt to changing conditions or integrate high-dimensional datasets. In this context, generative models, a subset of artificial intelligence, offer a powerful and flexible alternative for optimizing biomass conversion processes. These models not only predict outcomes based on input parameters but also generate novel process configurations and operational scenarios, enabling a deeper exploration of the design space and facilitating more efficient, robust, and adaptive bioenergy systems. Generative models are capable of learning the underlying probability distributions of complex datasets and using that knowledge to generate new, plausible data instances. Among the most prominent architectures are Generative Adversarial Networks (GANs), Variational Autoencoders (VAEs), and more recently, transformer-based generative models. In the context of biomass conversion, these models can be trained on experimental or simulated datasets that include a range of feedstock types, process parameters (such as temperature, pressure, residence time, catalyst concentration), and resulting outputs (e.g., energy yield, emissions, conversion efficiency). Once trained, the model can generate optimized combinations of input variables that are likely to achieve desired outcomes, such as maximum energy recovery, minimal emissions, or lowest cost. Unlike black-box predictive models, generative approaches allow for creative synthesis of new process conditions that may not have been tested experimentally but are inferred to be promising based on the data distribution. One of the most transformative uses of generative models in biomass conversion lies in high-throughput virtual experimentation, (Habib et al., 2021). For example, in thermochemical conversion processes like pyrolysis and gasification, the performance of the system is heavily dependent on numerous interrelated variables including feedstock composition (cellulose, hemicellulose, lignin ratios), heating rate, reactor design, and catalyst type. Running physical experiments for all possible combinations is impractical due to time and resource constraints. A trained generative model can rapidly generate thousands of viable scenarios that span a diverse range of inputs, effectively simulating a broad spectrum of experiments.

These generated scenarios can then be evaluated using process simulators or validated models, dramatically reducing the experimentation phase while uncovering high-performing configurations that might otherwise be overlooked, (Jana et al., 2022). In biochemical conversion pathways, such as fermentation and anaerobic digestion, generative models are similarly effective. For instance, in the production of biogas through anaerobic digestion, the microbial activity and metabolic pathways are influenced by factors like substrate mixture, pH, temperature, carbon-to-nitrogen ratio, and retention time. These interactions are difficult to model analytically due to their biological complexity. Generative models trained on historical operational data from various biogas plants can propose optimized substrate mixes or operational strategies that enhance methane production while maintaining system stability. Additionally, they can simulate how the system might behave under new conditions—such as the introduction of a novel co-substrate or variation in ambient temperature—allowing operators to plan and adapt proactively. Moreover, generative models can be integrated into hybrid AI systems that combine deep learning with first-principles process modeling. This synergy enhances both the accuracy and interpretability of optimization results. For example, a VAE could generate potential input configurations for a biomass gasifier, which are then validated using computational fluid dynamics (CFD) models or thermodynamic equations. This reduces the computational burden while maintaining scientific rigor, (Ji et al., 2023). Such hybrid frameworks can also include multi-objective optimization, where the generative model is guided by multiple performance indicators—such as energy yield, capital cost, lifecycle emissions, and resource efficiency—to produce balanced solutions tailored to specific use cases or policy goals. An emerging frontier in this field is the use of reinforcement learning (RL) in conjunction with generative models, particularly for real-time control and adaptive optimization. In this paradigm, the generative model acts as an "environment" that simulates how the biomass conversion system responds to different inputs

MACHINE LEARNING FOR BIOENERGY SYSTEM PERFORMANCE PREDICTION

The application of Artificial Intelligence (AI) in the design and optimization of bioenergy systems represents a powerful evolution in renewable energy planning and execution. Unlike traditional energy system design, which often depends on deterministic models and fixed assumptions, (Kalhor et al., 2023), AI-driven design leverages data-intensive, adaptive, and predictive algorithms to address the inherent complexity and variability of bioenergy. Bioenergy systems encompass diverse components—including feedstock supply chains, conversion technologies, distribution infrastructure, and end-use systems—all of which must be dynamically managed to optimize performance across environmental, economic, and social metrics. The integration of AI enables advanced modeling of these components, facilitating smarter decision-making, operational efficiency, and system resilience. Through various global case studies, the transformative potential of AI in bioenergy system design has been demonstrated across scales, regions, and technological pathways algorithms to optimize the collection and transportation of agricultural residues for bioenergy production in Punjab and Haryana. These regions face severe air pollution due to the open-field burning of rice straw. By using AI to analyze satellite imagery, weather data, crop yield predictions, and transportation logistics, the system developed optimal biomass aggregation routes and schedules. The AI-driven model significantly reduced collection costs and emissions while increasing the availability of feedstock for biomass power plants. It also provided farmers with incentives to participate in the supply chain, thereby improving rural income and reducing environmental damage. Germany, focusing on anaerobic digestion of mixed agricultural waste. Researchers applied artificial neural networks (ANNs) to model the complex relationship between substrate composition, operational parameters (like temperature, pH, retention time), and biogas yield.

These models enabled operators to predict methane production under various conditions and make real-time adjustments to the feedstock mix and process variables. (Kanan et al., 2022a) The system was further enhanced with reinforcement learning (RL) algorithms, allowing the plant to continuously learn from operational feedback and self-optimize for maximum energy output and process stability. As a result, the facility achieved a 12–18% increase in methane yield, reduced downtime, and improved overall economic returns. This case illustrates the adaptability and self-learning capacity of AI when integrated into bioenergy operations. (Zarei et al., 2022) produced from diverse biomass inputs. In a U.S.-based pilot project, generative design models and deep learning algorithms were used to simulate and optimize the layout, process flows, and energy balances of a lignocellulosic biorefinery. The system evaluated millions of process configurations by analyzing variables such as reactor type, enzyme loading, pretreatment strategy, and heat integration methods. Using genetic algorithms for optimization, the AI model identified configurations that maximized output while minimizing carbon footprint and cost. Moreover, it proposed innovative process pathways that had not been considered in traditional designs. These pathways were later tested experimentally, validating the predictive power of data, land-use patterns, biomass resource maps, infrastructure data, and climate models to recommend optimal locations for bioenergy facilities. Through multi-criteria analysis powered by AI, the model balanced trade-offs between feedstock availability, transportation cost, job creation, and environmental impact. Policymakers used these insights to promote sustainable biomass development zones, reducing deforestation risk while supporting rural development. This case exemplifies how AI can support inclusive and evidence-based planning, ensuring learning algorithms forecasted daily waste volumes and energy output potential. The AI model also provided dynamic routing for waste collection vehicles and recommended adjustments to digestion parameters based on waste variability. As a result, the facility improved its energy recovery rates and reduced greenhouse gas emissions. (Zahraee et al., 2020).

BIOENERGY SYSTEM DESIGN: CHALLENGES AND OPPORTUNITIES

The growing complexity of modern bioenergy systems, driven by the variability of biomass resources, dynamic operational parameters, and evolving environmental and economic conditions, necessitates advanced tools for predicting system performance with accuracy and adaptability. Traditional modeling methods, such as deterministic or empirical approaches, often fall short when dealing with the nonlinear, high-dimensional, and uncertain characteristics of bioenergy processes. Machine Learning (ML), a powerful subset of artificial intelligence, has emerged as a transformative solution for predicting, analyzing, and optimizing bioenergy system performance. By learning patterns from historical and real-time data, ML models can predict outputs such as energy yields, emission profiles, fuel quality, process efficiency, and equipment degradation across a wide spectrum of bioenergy pathways—including anaerobic digestion, pyrolysis, gasification, fermentation, and biodiesel production. The application of ML in bioenergy has not only enhanced predictive capabilities but has also accelerated innovation, reduced operational risks, and supported more informed decision-making throughout the entire bioenergy value chain, (Yue et al., 2014). Machine learning encompasses a diverse set of algorithms, including supervised, unsupervised, and reinforcement learning techniques, each tailored to specific tasks in bioenergy performance modeling. In supervised learning, algorithms like artificial neural networks (ANNs), support vector machines (SVMs), random forests (RF), and gradient boosting machines are trained on labeled datasets to establish relationships between input features (e.g., feedstock composition, temperature, pH, reactor design) and performance indicators (e.g., biogas yield, bio-oil quality, conversion efficiency). These models are especially valuable in systems where experimental data is abundant, enabling highly accurate predictions of system outputs under various operational conditions. For instance, in anaerobic digestion, ML models have been developed to predict methane production based on the chemical oxygen demand (COD), volatile fatty acids (VFAs), carbon-to-nitrogen ratio (C/N), and substrate mixtures. These models assist operators in adjusting feed inputs and controlling process variables to maintain system stability and maximize energy output the heating value and hydrogen content of syngas produced from municipal solid waste or agricultural residues.

These predictions allow system designers to fine-tune reactor configurations and operating regimes for enhanced thermal efficiency and reduced emissions.

Similarly, in biodiesel production, ML models have been employed to optimize transesterification conditions by predicting biodiesel yield and purity from variables such as catalyst concentration, alcohol-to-oil ratio, and reaction time. (Yue & You, 2016b). By minimizing the need for extensive laboratory experiments, these models accelerate process development and reduce costs. Patterns or classifying bioenergy systems based on performance trends. These methods are particularly beneficial in the early stages of system design or when analyzing large-scale datasets from diverse sources. For example, clustering techniques can identify distinct operational regimes or performance classes within a biorefinery, helping engineers detect underperforming units or anomalous behavior. Principal component analysis has also been used to reduce the complexity of multivariate data collected from sensors in real-time monitoring systems, improving the interpretability of results and guiding subsequent supervised learning efforts. Under fluctuating conditions. For instance, in a pilot-scale biogas plant, RL algorithms have been successfully deployed to maintain optimal gas production by adapting to variations in substrate quality and environmental conditions. This self-learning capability is crucial in bioenergy systems, where operational flexibility and responsiveness are key to handling feedstock variability and process uncertainties. For example, predictive maintenance models based on ML can forecast equipment wear or degradation in biomass boilers and gasifiers, reducing unplanned downtimes and maintenance costs. Similarly, anomaly detection algorithms can identify unexpected trends in bioreactor performance, such as a sudden drop in biogas output or changes in pH, alerting operators to take preventive actions. These applications not only enhance operational reliability but also contribute to the overall safety and economic sustainability of bioenergy facilities-driven patterns, mechanistic models offer physical interpretability and validation against scientific laws, (Xie & Huang, 2013). In bioenergy systems, hybrid models can simulate detailed thermochemical or biochemical pathways while using ML components to estimate hard-to-model parameters or compensate for uncertainties. Such models improve accuracy and transparency, making them more acceptable in regulatory and industrial settings in data collection methods, missing values, and noise can compromise model performance. Future directions include integrating, (Wolfsmayr & Rauch, 2014). ML with emerging technologies such as generative AI, digital twins, and blockchain to create intelligent, transparent, and adaptive bioenergy ecosystems. With continued research, collaboration, and ethical oversight, machine learning can play a pivotal role in accelerating the global transition to sustainable bioenergy solutions.

THERMOCHEMICAL CONVERSION: AI-OPTIMIZED PROCESSES

The design of bioenergy systems—encompassing the sourcing of biomass, conversion technologies, energy distribution, and integration with existing energy infrastructures—presents a complex matrix of technical, environmental, economic, and socio-political challenges. As the world strives for a low-carbon future, bioenergy offers a promising pathway to replace fossil fuels, promote rural development, and reduce greenhouse gas emissions. However, realizing the full potential of bioenergy depends on overcoming several inherent design challenges while leveraging emerging opportunities driven by technological innovation, policy support, and interdisciplinary collaboration. Understanding these intricacies is vital for developing sustainable, scalable, and resilient bioenergy systems tailored to diverse local and regional contexts. A primary design challenge in bioenergy systems is the heterogeneity of biomass feedstocks, which vary widely in moisture content, energy density, chemical composition, seasonal availability, and geographic distribution, (Wang et al., 2023). Unlike fossil fuels, which are energy-dense and transportable, biomass resources are often bulky, decentralized, and perishable, complicating supply chain logistics and conversion efficiency. Designing systems that can accommodate multi-feedstock inputs and adapt to fluctuating resource availability requires advanced preprocessing technologies and flexible conversion platforms. For instance, a bioenergy plant in a tropical region might rely heavily on sugarcane bagasse and oil palm residues, while a plant in a temperate region might depend on wood chips, straw, or municipal waste. Customizing system design to local biomass profiles while maintaining economic and technical viability remains a persistent hurdle. in terms of feedstock quality, operating conditions, and product outputs. Moreover, the trade-offs between efficiency, scalability, environmental impact, and cost must be carefully evaluated. For instance, thermochemical conversion processes like gasification are well-suited for dry woody biomass and yield syngas that can be used for electricity or fuel synthesis, but they require high capital investment and precise feedstock control. On the other hand, biochemical processes like anaerobic digestion are more tolerant of moisture and mixed organic waste but yield lower energy densities and are sensitive to operational instability.

Figure 1. Optimizing renewable energy systems through artificial intelligence

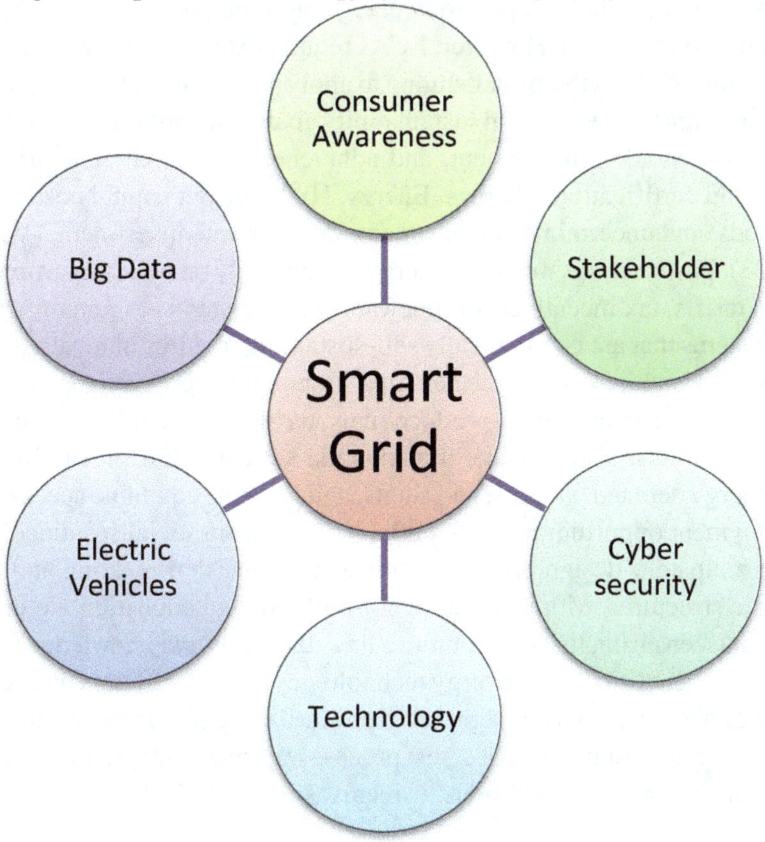

Designing hybrid systems or modular biorefineries that combine multiple conversion routes offers a potential solution, albeit one that introduces additional complexity and cost. may be underdeveloped, (Verma et al., 2017). The cost and emissions associated with transporting biomass over long distances can negate environmental and economic benefits. As a result, decentralized or community-based bioenergy systems—such as village-scale biogas plants or mobile pyrolysis units—are gaining interest, but these too require innovative design approaches to ensure reliability, scalability, and integration with existing energy systems. Additionally, designing infrastructure that is climate-resilient and adaptable to future expansion is essential, especially in regions vulnerable to extreme weather, land degradation, or shifting agricultural practices. From an environmental design perspective, minimizing the ecological footprint of bioenergy systems is a central concern. Poorly planned systems can lead to unintended consequences such as deforestation, soil erosion, water contamination, and biodiversity loss. Life cycle assessment (LCA) tools are often used in the design phase to evaluate the cradle-to-grave environmental impacts of bioenergy systems, including emissions, energy input/output ratios, land use changes, and resource consumption. However, LCAs require extensive data and are sensitive to assumptions, making them challenging to apply universally. Designing bioenergy systems that align with ecological sustainability goals demands robust environmental modeling, stakeholder engagement, and adherence to international sustainability standards and certification schemes. Energy. High upfront capital costs, long pay-back periods, and uncertain revenue streams deter private investment, (Vanbrabant et al., 2023). Furthermore, many bioenergy systems rely on policy instruments such as feed-in tariffs, tax incentives, or renewable energy credits to remain viable. Designing systems that are economically self-sustaining requires innovative financing models, such as public-private partnerships, community cooperatives, or integrated bioproduct value chains that co-produce fuels, fertilizers, chemicals, and heat. The design process must also consider dynamic market conditions, including biomass pricing, energy demand patterns, and rights, and promote equitable access to energy and employment opportunities is essential for long-term social sustainability. This entails participatory design processes, transparent decision-making, and inclusive governance structures. Moreover, capacity-building and education are critical design elements, ensuring that communities have the skills and knowledge to operate, maintain, and benefit from bioenergy technologies Mmore efficient and responsive design. Digital twins, or virtual replicas of physical systems, are increasingly used to simulate different design scenarios, test process improvements, and forecast system behavior under changing conditions. (Sreenivasulu et al., 2022).

BIOCHEMICAL CONVERSION: AI-DRIVEN ENZYME ENGINEERING

Thermochemical conversion is a pivotal pathway in the transformation of biomass into valuable forms of energy such as syngas, bio-oil, biochar, and heat. This category includes high-temperature processes like combustion, gasification, pyrolysis, and torrefaction, which utilize thermal energy, often in the presence or absence of oxygen, to deconstruct complex organic matter. These methods are particularly advantageous for lignocellulosic biomass, agricultural residues, forestry waste, and municipal solid waste, which are often challenging to process through biochemical methods. Despite their potential, thermochemical processes are characterized by complex reaction kinetics, nonlinear parameter interactions, and sensitivity to feedstock variability—making their control and optimization technically demanding, (Shende et al., 2025). Here, the integration of Artificial Intelligence (AI) offers a transformative solution, enhancing process efficiency, enabling adaptive control, and accelerating innovation in system design. Through predictive modeling, real-time monitoring, and autonomous optimization, AI enables a deeper understanding of thermochemical conversion systems and opens new frontiers for sustainable bioenergy production. Gasification, one of the most widely used thermochemical methods, involves the partial oxidation of biomass at high temperatures (typically 800–1,200°C) to produce a mixture of carbon monoxide (CO), hydrogen (H_2), methane (CH_4), and carbon dioxide (CO_2)—collectively known as syngas. This syngas can be used for power generation, heat, or as a feedstock for synthetic fuels and chemicals. However, gasification efficiency depends heavily on parameters such as feedstock composition, moisture content, gasification temperature, equivalence ratio, and residence time. AI models, particularly machine learning (ML) algorithms like Artificial Neural Networks (ANNs), Support Vector Regression (SVR), and Random Forests, have been used to develop predictive models for syngas composition and energy yield, (Tao et al., 2021) These models are trained on experimental or operational datasets to learn the relationships between input variables and process outcomes. Once trained, they can accurately predict the effects of changing parameters, enabling operators to optimize gasifier performance under diverse and fluctuating conditions. Unsupervised learning. For instance, researchers have developed ANN and deep learning models to predict product yields and physicochemical properties of bio-oil under different pyrolysis conditions. These models not only reduce the need for extensive laboratory experimentation but also provide insight into optimal operating regimes that balance yield, quality, and energy efficiency.

Additionally, genetic algorithms and Bayesian optimization techniques have been employed in conjunction with AI models to perform multi-objective optimization, identifying parameter combinations that simultaneously maximize bio-oil yield and minimize char formation or process emissions.it more compatible with coal-fired power plants and pellet manufacturing. AI models trained on torrefaction data can predict mass yield, energy yield, and elemental composition of the torrefied product. These predictions help in fine-tuning torrefaction conditions and scaling up the process for commercial deployment. Furthermore, AI-based process control systems enable real-time adjustment of temperature and residence time based on input feedstock variability, ensuring consistent product quality. On temperature profiles, pressure, gas composition, and reactor stability. AI algorithms process this data to detect anomalies, forecast failures, and recommend corrective actions, (Lu et al., 2021). For example, a deep learning model may identify early signs of reactor instability or feedstock clogging, prompting automatic process adjustments or preventive maintenance. This enhances operational safety and reduces downtime. Despite its transformative potential, the application of AI in thermochemical conversion faces several challenges. One of the foremost issues is data availability and quality. High-fidelity, labeled datasets from real-world operations are often limited due to proprietary restrictions, inconsistent data logging, or system variability. Additionally, thermochemical systems generate large volumes of multi-modal data (e.g., thermal, chemical, visual), which must be cleaned, integrated, and curated before being usable for AI training. Addressing these challenges requires the development of standardized data protocols, open-access databases, and collaborative research platforms. (Maheshwari et al., 2017) Another concern is model interpretability. Complex AI models like deep neural networks often function as "black boxes," making it difficult to explain their predictions or gain scientific insight. This limits their acceptance among process engineers and regulators. To overcome this, efforts are being made to integrate explainable AI (XAI) methods into thermochemical modeling, ensuring that predictions are both accurate and interpretable. Moreover, scalability and generalization of AI models remain a hurdle. Models trained on data from a specific reactor or feedstock may not perform well when applied to different systems or biomass types. To improve generalization, researchers are exploring transfer learning, where pre-trained models are fine-tuned on new datasets, and federated learning, where models are trained across multiple decentralized systems while preserving data privacy. In conclusion, AI is redefining how thermochemical biomass conversion systems are designed, monitored, and optimized. From predictive modeling and real-time control to generative system design and autonomous operation, AI introduces unprecedented capabilities to handle complexity, variability, and uncertainty in high-temperature biomass processing, (Mehta et al., 2025).

AI-DRIVEN SENSITIVITY ANALYSIS: IDENTIFYING KEY PARAMETERS

The optimization of bioenergy systems is central to enhancing their efficiency, reducing costs, and improving their sustainability. Bioenergy systems, which include the conversion of biomass to biofuels, biopower, and bio-based chemicals, involve a wide array of complex and interdependent processes. These systems consist of feedstock collection, transportation, preprocessing, conversion, and energy production, each of which is subject to a wide range of dynamic variables and constraints. As bioenergy systems are often designed to handle diverse feedstocks, production scales, and output forms, achieving optimal performance requires a multidimensional approach that accounts for variability, resource efficiency, and environmental impact. Traditionally, the optimization of bioenergy system parameters was performed through trial and error or using conventional mathematical models, but these methods often failed to capture the complexity and nonlinearity inherent in bioenergy systems. Today, Artificial Intelligence (AI), particularly machine learning (ML), genetic algorithms (GA), reinforcement learning (RL), and deep learning (DL), is driving a paradigm shift in the optimization of bioenergy systems, (Madgule et al., 2023). AI-based techniques provide new avenues to manage, predict, and enhance system parameters in ways that were previously unfeasible, making bioenergy more viable as a renewable energy solution. Machine Learning (ML) plays a pivotal role in optimizing various parameters of bioenergy systems by enabling data-driven decision-making, (Ngan et al., 2019). The complexity of bioenergy systems and the inherent uncertainty in biomass feedstocks, conversion processes, and energy outputs demand sophisticated models that can capture interactions among multiple factors. Traditional deterministic models, while useful, often fail to account for the stochastic nature of biomass availability, temperature variations, moisture content, or equipment performance.

These models can continuously learn from operational data, making them highly adaptive to changing input conditions and improving process performance over time pyrolysis, fermentation, and anaerobic digestion, are all influenced by a wide range of parameters including feedstock properties, temperature, pressure, residence time, and catalyst performance. These processes exhibit complex interdependencies that make it challenging to find the optimal set of parameters that maximize product yields while minimizing waste and energy consumption. By using real-time sensor data from conversion units, ML algorithms can predict the efficiency of conversion reactions and adjust operating conditions to improve yields or reduce energy losses. (Ngan et al., 2020). Feedstock variability is a significant challenge in bioenergy systems, as biomass resources can vary widely in terms of composition, moisture content, and energy density, (Osmani & Zhang, 2014a). AI-based systems can evaluate diverse feedstocks and predict their suitability for different conversion processes, reducing the need for costly and time-consuming physical tests. The ability of RL to explore large action spaces and learn continuously from the system's performance makes it ideal for applications where traditional optimization techniques are either too slow or too complex to implement. One notable application of RL in bioenergy (Osmani & Zhang, 2014b) optimization is in the control of combined heat and power (CHP) plants, where the goal is to optimize the balance between electricity generation and heat recovery. For example, a GA could be applied to find the optimal combination of temperature, pressure, and feedstock composition for a biomass pyrolysis reactor, balancing product yield, energy consumption, and emissions. By evaluating many potential design configurations in parallel, GAs provide a robust optimization strategy that can efficiently explore large and complex solution spaces. Neural networks (RNNs), are also being increasingly applied to bioenergy system optimization. By considering factors such as land use, water consumption, and emissions, AI-driven optimization techniques can help ensure that bioenergy systems are both economically viable and environmentally responsible of bioenergy systems. (Paulo et al., 2015)

FUTURE TRENDS IN AI FOR BIOENERGY SYSTEM DESIGN

Multi-objective optimization (MOO) is an essential tool in engineering, economics, environmental science, and various other fields, particularly when dealing with systems that involve conflicting objectives. These types of problems often arise when designing systems or processes that must balance multiple, sometimes contradictory, goals. In bioenergy systems, for example, objectives may include maximizing energy output, minimizing environmental impact, and reducing production costs—goals that are often at odds with one another. Optimizing one of these objectives may compromise the others, and thus, the challenge lies in finding a set of optimal solutions, known as Pareto-optimal solutions, where no objective can be improved without worsening at least one other. Multi-objective optimization is particularly critical in bioenergy systems due to the complexity of these systems and the interplay between technical, economic, and environmental factors. (Ren et al., 2015) bioenergy plants, engineers are faced with the challenge of optimizing a range of parameters, including feedstock choice, conversion efficiency, energy recovery, carbon footprint, and costs. However, improving one of these aspects (e.g., increasing energy production) can often lead to increased environmental emissions or higher costs. Multi-objective optimization approaches are designed to identify solutions that strike a balance between these competing objectives. a graphical representation of all Pareto-optimal solutions in a multi-dimensional objective space. For example, in bioenergy system design, the Pareto front may represent different configurations of a biomass power plant, where each point on the front reflects an optimal trade-off between energy output, costs, and environmental impact.

Ant Colony Optimization, inspired by the foraging behavior of ants, is often used for optimization problems related to pathfinding or logistics, such as biofuel distribution networks, while Differential Evolution is effective for high-dimensional, non-linear optimization problems. thermochemical processes, the cost of feedstocks, and the environmental impact of waste produced during production. Additionally, changes in one part of the system, such as a shift in feedstock composition, can affect several other objectives, such as emissions or economic feasibility. Incorporating these interdependencies into optimization models is a key challenge in bioenergy system optimization. AI-based techniques, such as artificial neural networks (ANNs), can be used in conjunction with multi-objective optimization methods to improve model accuracy. ANNs can learn complex, non-linear relationships between system variables from large datasets, enabling more accurate modeling of bioenergy systems and their competing objectives. For example, an ANN trained on real-time data from a biomass power plant could help predict how changes in temperature, feedstock type, or moisture content impact both energy output and emissions, providing valuable information for multi-objective optimization. uncertainties to provide robust solutions. Stochastic optimization techniques, which incorporate randomness or uncertainty in the modeling process, are increasingly being applied in bioenergy system design.

ADDRESSING CHALLENGES AND LIMITATIONS: FUTURE RESEARCH DIRECTIONS

Becomes ever more pressing. Bioenergy, with its versatility and potential to decarbonize sectors such as electricity, heating, and transportation, is emerging as a cornerstone of the renewable energy landscape. However, to maximize its impact and achieve a truly sustainable energy mix, bioenergy must be integrated with other renewable technologies, such as solar, wind, hydropower, and geothermal energy, (Ranjitha et al., 2025). This multi-source approach not only enhances the stability and reliability of energy systems but also creates synergies that improve the overall performance of renewable energy production. In this discussion, we explore the opportunities and challenges associated with the integration of bioenergy with other renewable energy systems, highlighting the potential benefits, innovative strategies, and future trends in this critical area. The primary advantage of integrating bioenergy with other renewable energy systems lies in its ability to provide firm, dispatchable power to complement the intermittency of wind and solar energy. While wind and solar power are vital components of a renewable energy grid, their production is variable and dependent on weather conditions. Solar energy is produced only during the day and is significantly reduced or absent during cloudy periods, while wind energy is intermittent, fluctuating based on wind speeds that are unpredictable. This intermittency presents challenges for maintaining grid stability and meeting energy demand during times when renewable energy generation is low. Bioenergy, however, offers a predictable and flexible energy source that can be generated consistently, even during periods when solar and wind energy are not available.

The integration of bioenergy with distributed energy resources (DERs), such as small-scale solar, wind, and energy storage systems, represents another promising avenue for increasing the penetration of renewable energy. Microgrids that combine bioenergy with solar or wind power, along with local storage, enable decentralized and resilient energy production, (Szczepański, 2019). By coordinating the generation and storage of power from these multiple sources, microgrids can operate independently or in tandem with the broader grid, providing energy to isolated communities or remote areas while reducing reliance on centralized fossil fuel-based power plants. As technologies such as smart meters, AI, and internet of things (IoT) sensors become more widespread, it will be possible to optimize and control the interactions between these diverse renewable energy sources. This distributed approach to renewable energy (Ramakrishnan et al., 2022) will foster greater flexibility and resilience in the power grid, especially in regions prone to power outages or natural disasters and aviation. Furthermore, generative models can enable the design of more advanced energy storage solutions, optimizing the location and capacity of batteries or other storage technologies to store excess energy produced during peak generation periods. (Salimi & Vahdani, 2018)This, in turn, helps to address the intermittency problem associated with solar and wind power, ensuring that the energy produced can be stored and deployed when demand is high or when renewable energy production is low real-time data to balance the supply and demand of energy, and Generative AI can play a crucial role in enhancing their performance.

CONCLUSION

The future of sustainable energy is rapidly being reshaped by the convergence of cutting-edge technologies, and at the forefront of this transformation is Generative AI. This subset of artificial intelligence, which focuses on creating new data, designs, or solutions based on input patterns, holds immense promise in unlocking efficient, cost-effective, and environmentally sustainable energy systems. As the world faces the growing challenges of climate change, energy security, and resource depletion, the need for sustainable energy solutions has never been more urgent. Generative AI, with its ability to optimize, innovate, and predict, stands as a transformative force capable of addressing the multifaceted challenges that confront the global energy sector. By enabling the design, operation, and management of advanced, low-carbon energy systems, Generative AI promises to revolutionize energy production, storage, distribution, and consumption in ways that were once considered out of reach.. Generative AI can tackle these challenges by creating new models for energy system designs that are more resilient, adaptive, and efficient. For example, in the context of solar energy, Generative AI can optimize the layout of solar panels, factoring in geographic, meteorological, and infrastructural constraints to design the most efficient solar farm possible. By generating a vast number of potential designs and evaluating their performance, Generative AI can recommend configurations that maximize energy yield, minimize costs, and reduce environmental impact. AI-powered optimization models can continuously adapt the design and operation of wind farms, improving their efficiency and minimizing maintenance costs over time. Through generative models, AI can design dynamic grid architectures that adapt to fluctuations in energy supply and demand. This means that grids could automatically adjust their operation, rerouting power from excess-producing renewable sources to areas with high demand, all in real-time., AI can predict consumption behavior and suggest personalized strategies for energy conservation or more efficient consumption. These systems can automatically control heating, cooling, and appliances to reduce energy consumption during peak times, thereby decreasing the strain on the grid. Additionally, AI can generate innovative solutions for energy-efficient buildings by analyzing factors such as insulation, appliance use, and lighting patterns, and generating designs or modifications that optimize energy use, contributing to a reduction in overall energy. AI can optimize the conversion processes in bioenergy plants, improving energy yield while minimizing emissions. In addition, AI can explore new forms of carbon capture and utilization (CCU) technologies that transform CO2 emissions into useful products, such as building materials or synthetic fuels, thus not only neutralizing but also reversing the effects of carbon emissions. Generative AI can also support the development of novel, sustainable materials for renewable energy technologies. For instance, in the realm of solar energy, AI can generate new materials for photovoltaic cells that absorb sunlight more efficiently or are more cost-effective to produce. Similarly, AI can design new types of energy storage materials that are more efficient, long-lasting, and environmentally friendly. Through materials discovery and simulation, AI can rapidly prototype and test new compounds, reducing the time and resources needed to bring new technologies to market. By generating multiple possible weather scenarios, AI models can forecast renewable energy generation over different time scales, from minutes to years. This capability enables energy operators to better plan for periods of high demand, unexpected weather events, or seasonal fluctuations in renewable energy generation. Moreover, climate modeling powered by Generative AI can help us understand how various energy strategies might influence global temperature and atmospheric conditions. This can lead to the design of energy systems that are not only sustainable but also resilient to future climate changes. Generative AI could simulate the impacts of different energy policy scenarios, including the potential effects of transitioning to a low-carbon economy, allowing policymakers to explore and optimize strategies for mitigating the effects of climate change. This approach could help prioritize investments in energy infrastructure that align with long-term climate goals, enabling the global energy transition to be more aligned with the imperative of climate action.

REFERENCES

Abasian, F., Rönnqvist, M., & Ouhimmou, M. (2019). Forest bioenergy network design under market uncertainty. *Energy*, *188*, 116038.

Abbasi, M., Pishvaee, M. S., & Mohseni, S. (2021). Third-generation biofuel supply chain: A comprehensive review and future research directions. *Journal of Cleaner Production*, *323*, 12910.

Babazadeh, R., Ghaderi, H., & Pishvaee, M. S. (2019). A benders-local branching algorithm for second-generation biodiesel supply chain network design under epistemic uncertainty. *Computers & Chemical Engineering*, *124*, 364–380.

Babazadeh, R., Razmi, J., Pishvaee, M. S., & Rabbani, M. (2017). A sustainable second-generation biodiesel supply chain network design problem under risk. *Omega*, *66*, 258–277.

Chen, C. S., Narani, A., Daniyar, A., McCauley, J., Brown, S., Pray, T., & Tanjore, D. (2022). Ensemble models of feedstock blend ratios to minimize supply chain risk in bio-based manufacturing. *Biochemical Engineering Journal*, *181*, 107896. DOI: 10.1016/j.bej.2020.107896

Dal Mas, M., Giarola, S., Zamboni, A., & Bezzo, F. (2010). Capacity planning and financial optimization of the bioethanol supply chain under price uncertainty. In *Comp Aid Chem Eng* (Vol. 28, pp. 97–102). Elsevier.

Dal-Mas, M., Giarola, S., Zamboni, A., & Bezzo, F. (2011). Strategic design and investment capacity planning of the ethanol supply chain under price uncertainty. *Biomass and Bioenergy*, *35*(5), 2059–2071. DOI: 10.1016/j.biombioe.2011.01.060

Ebadian, M., Sowlati, T., Sokhansanj, S., Smith, L. T., & Stumborg, M. (2014). Development of an integrated tactical and operational planning model for supply of feedstock to a commercial-scale bioethanol plant. *Biofuels, Bioproducts & Biorefining*, *8*(2), 171–188.

El-Sheekh, M. M., Gheda, S. F., El-Sayed, A. E.-K. B., Abo Shady, A. M., El-Sheikh, M. E., & Schagerl, M. (2019). Outdoor cultivation of the green microalga Chlorella vulgaris under stress conditions as a feedstock for biofuel. *Environmental Science and Pollution Research International*, *26*, 18520–18532.

Fahimnia, B., Tang, C. S., Davarzani, H., & Sarkis, J. (2015). Quantitative models for managing supply chain risks: A review. *European Journal of Operational Research*, *247*(1), 1–15.

Fallah, M., & Nozari, H. (2021). Neutrosophic Mathematical Programming for optimization of multi-objective sustainable biomass supply chain network design. *CMES-Comput Model Eng Sci*, *129*(2), 927–951.

Garai, A., Chowdhury, S., Sarkar, B., & Roy, T. K. (2021). Cost-effective subsidy policy for growers and biofuels-plants in closed-loop supply chain of herbs and herbal medicines: An interactive bi-objective optimization in T-environment. *Applied Soft Computing*, *100*, 106949.

Garai, A., & Sarkar, B. (2022). Economically independent reverse logistics of customer-centric closed-loop supply chain for herbal medicines and biofuel. *Journal of Cleaner Production*, *334*, 129977.

Gebreslassie, B. H., Yao, Y., & You, F. (2012). Design under uncertainty of hydrocarbon biorefinery supply chains: Multiobjective stochastic programming models, decomposition algorithm, and a comparison between CVaR and downside risk. *AIChE Journal. American Institute of Chemical Engineers*, *58*(7), 2155–2179.

Habib, M. S., Asghar, O., Hussain, A., Imran, M., Mughal, M. P., & Sarkar, B. (2021). A robust possibilistic programming approach toward animal fat-based biodiesel supply chain network design under uncertain environment. *Journal of Cleaner Production*, *278*, 122403.

Habib, M. S., Tayyab, M., Zahoor, S., & Sarkar, B. (2020). Management of animal fat-based biodiesel supply chain under the paradigm of sustainability. *Energy Conversion and Management*, *225*, 11334.

Jana, D. K., Bhattacharjee, S., Dostál, P., Janková, Z., & Bej, B. (2022). Bi-criteria optimization of cleaner biofuel supply chain model by novel fuzzy goal programming technique. *Clean Logist Supply Chain*, *4*, 100044.

Ji, M., Zhang, W., Xu, Y., Liao, Q., Klemeš, J. J., & Wang, B. (2023). Optimisation of multi-period renewable energy systems with hydrogen and battery energy storage: A P-graph approach. *Energy Conversion and Management*, *281*, 116826.

Kalhor, T., Sharifi, M., & Mobli, H. (2023). A robust optimization approach for an integrated hybrid biodiesel and biomethane supply chain network design under uncertainty: Case study. *International Journal of Energy and Environmental Engineering*, *14*(2), 189–210.

Kanan, M., Habib, M. S., Habib, T., Zahoor, S., Gulzar, A., Raza, H., & Abusaq, Z. (2022a). A flexible robust possibilistic programming approach for sustainable second-generation biogas supply chain Design under Multiple Uncertainties. *Sustainability*, *14*(18), 11597.

Lu L, Nguyen R, Rahman MM, Winfree J (2021) Demand shocks and supply chain resilience: an agent based modelling approach and application to the potato supply chain. https://doi.org/DOI: 10.3386/w29166

Madgule, M., Vinayaka, N., Sonkhaskar, Y. M., Patel, D. R., Karthikeyan, R., & Selvakumar, P. (2023). Mechanical properties and microstructure of activated TIG welded similar joints of Inconel alloys by desirability approaches. *Materials Today: Proceedings*, *77*, 528–533.

Maheshwari, P., Singla, S., & Shastri, Y. (2017). Resiliency optimization of biomass to biofuel supply chain incorporating regional biomass pre-processing depots. *Biomass and Bioenergy*, *97*, 116–131.

Mehta, A. K., Yashoda, L., Selvakumar, P., Mishra, B. R., Das, A., & C. T. C., M. (2025). Teaching methods: Techniques for disadvantaged learners. In S. Hai-Jew (Ed.), Mitigating Learner Disadvantages in Teaching and Learning (pp. 181–206). IGI Global. https://doi.org/DOI: 10.4018/979-8-3693-8623-1.ch007

Ngan, S. L., How, B. S., Teng, S. Y., Leong, W. D., Loy, A. C. M., Yatim, P., Promentilla, M. A., & Lam, H. L. (2020). A hybrid approach to prioritize risk mitigation strategies for biomass polygeneration systems. *Renewable & Sustainable Energy Reviews*, *121*, 109679. DOI: 10.1016/j.rser.2019.109679

Ngan, S. L., Promentilla, M. A. B., Yatim, P., & Lam, H. L. (2019). A novel risk assessment model for green finance: The case of Malaysian oil palm biomass industry. *Process Integr Optim Sustain*, *3*(1), 75–88.

Osmani, A., & Zhang, J. (2014a). Economic and environmental optimization of a large scale sustainable dual feedstock lignocellulosic-based bioethanol supply chain in a stochastic environment. *Applied Energy*, *114*, 572–587.

Osmani, A., & Zhang, J. (2014b). Optimal grid design and logistic planning for wind and biomass based renewable electricity supply chains under uncertainties. *Energy*, *70*, 514–528.

Paulo, H., Azcue, X., Barbosa-Póvoa, A. P., & Relvas, S. (2015). Supply chain optimization of residual forestry biomass for bioenergy production: The case study of Portugal. *Biomass and Bioenergy*, *83*, 245–256.

Ramakrishnan, T., Mohan Gift, M. D., Chitradevi, S., Jegan, R., & Subha Hency Jose, P. (2022). Nagaraja, Rajneesh Sharma, H. N., Selvakumar, P., & Hailegiorgis, S. M. (2022). Study of numerous resins used in polymer matrix composite materials. *Advances in Materials Science and Engineering*, •••, 1088926. Advance online publication. DOI: 10.1155/2022/1088926

Ranjitha, P. K., Selvakumar, P., Anute, N., Manjunath, T., Sureshkrishna, G., & Pougajendy, S. (2025). Transportation challenges in islands tourism. In Hassan, V. (Ed.), *Navigating Mass Tourism to Island Destinations: Preservation and Cultural Heritage Challenges* (pp. 287–312). IGI Global., DOI: 10.4018/979-8-3693-9107-5.ch011

Ren, J., Dong, L., Sun, L., Goodsite, M. E., Tan, S., & Dong, L. (2015). Life cycle cost optimization of biofuel supply chains under uncertainties based on interval linear programming. *Bioresource Technology*, *187*, 6–13.

Salimi, F., & Vahdani, B. (2018). Designing a bio-fuel network considering links reliability and risk-pooling effect in bio-refineries. *Reliability Engineering & System Safety*, *174*, 96–107.

Shende, A. D., Tekade, S. A., Deshmukh, A. A., Tembhurkar, S. P., & Selvakumar, P. (2025). Water quality monitoring and control with AI. In J. Ruiz-Vanoye (Details of the book incomplete; citation needs final publication data).

Sreenivasulu, A., Manimozhi, T., Satheesh Kumar, P., Selvakumar, P., & Singh, P. (2022). The model of smart sensing device for sensitive nanoclusters modification in sensing properties. In 2022 2nd International Conference on Advance Computing and Innovative Technologies in Engineering (ICACITE) (pp. 1043–1047). IEEE. https://doi.org/DOI: 10.1109/ICACITE53722.2022.9823846

Szczepański, M. (2019). Economic impacts of artificial intelligence. Retrieved May 10, 2021,romhttps://www.europarl.europa.eu/RegData/etudes/BRIE/2019/637967/EPRS_BRI(2019)637967_EN.pdf

Tao, R., Su, C., Xiao, Y., Dai, K., & Khalid, F. (2021). Robo advisors, algorithmic trading and investment management: Wonders of fourth industrial revolution in financial markets. *Technological Forecasting and Social Change*, *163*, 120421. DOI: 10.1016/j.techfore.2020.120421

Vanbrabant, L., Verdonck, L., Mertens, S., & Caris, A. (2023). Improving hospital material supply chain performance by integrating decision problems: A literature review and future research directions. *Computers & Industrial Engineering*, *109235*. Advance online publication. DOI: 10.1016/j.cie.2023.109235

Verma, S. K., Fenila, F., Soren, A., & Shastri, Y. (2017). Impact of uncertainties on biomass to biofuel systems. *Perspectives in Agriculture, Veterinary Science, Nutrition and Natural Resources*, *12*(022), 1–11.

Wang, J., Zhou, H., Sun, X., & Yuan, Y. (2023). A novel supply chain network evolving model under random and targeted diruptions. *Chaos, Solitons, and Fractals*, *170*, 113371.

Wolfsmayr, U. J., & Rauch, P. (2014). The primary forest fuel supply chain: A literature review. *Biomass and Bioenergy*, *60*, 203–221.

Xie, F., & Huang, Y. (2013). Sustainable biofuel supply chain planning and management under uncertainty. *Transportation Research Record: Journal of the Transportation Research Board*, *2385*(1), 19–27.

Yue, D., & You, F. (2016b). Optimal supply chain design and operations under multi-scale uncertainties: Nested stochastic robust optimization modeling framework and solution algorithm. *AIChE Journal. American Institute of Chemical Engineers*, *62*(9), 3041–3055.

Yue, D., You, F., & Snyder, S. W. (2014). Biomass-to-bioenergy and biofuel supply chain optimization: Overview, key issues and challenges. *Computers & Chemical Engineering*, *66*, 36–56.

Zahraee, S. M., Shiwakoti, N., & Stasinopoulos, P. (2020). Biomass supply chain environmental and socio-economic analysis: 40-years comprehensive review of methods, decision issues, sustainability challenges, and the way forward. *Biomass and Bioenergy*, *142*, 10577.

Zarei, M., Shams, M. H., Niaz, H., Won, W., Lee, C.-J., & Liu, J. J. (2022). Risk-based multistage stochastic mixed-integer optimization for biofuel supply chain management under multiple uncertainties. *Renewable Energy*, *200*, 694–705.

Chapter 6

Generative AI for Intelligent, Resilient, and Sustainable Renewable Energy Systems: From Data to Deployment

Lingala Thirupathi
https://orcid.org/0000-0003-0703-3860
Sreenidhi Institute of Science and Technology, India

Shri Nitya Boini
Sreenidhi Institute of Science and Technology, India

K. Sathwik Reddy
Sreenidhi Institute of Science and Technology, India

ABSTRACT

To support the worldwide transition to renewable energy systems, a new approach is needed with computational methods that can deal with the issues of intermittency, the complexity of variability, and the integration of the grid. Generative Artificial Intelligence (GAI) has been a game-changer in creating value in how it can fundamentally design and generate synthetic data, predictive models, and optimize systems. This chapter evaluates the GAI methods for renewable energy systems, which include Generative Adversarial Networks (GANs), Variational Autoencoders (VAEs), diffusion models, and Large Language Models (LLMs). It is organized as: systematic taxonomy of GAI models regarding energy applications; an exhaustive review of GAI-oriented solutions for predictive maintenance, forecasting energy production, grid resilience, and sustainable design of renewable energy systems;

DOI: 10.4018/979-8-3373-4272-6.ch006

empirical case studies of improved performance of solar and wind energy systems; and an overview consideration of the general challenges of data privacy, model explainability, and scaling models to edge deployment.

1. INTRODUCTION

1.1 Background: Renewable Energy Transition and Digitalization

The energy landscape in the world is in a unique position of transformation as it pertains to the move towards energy transition initiatives due to not only climate change, but from technology and financial pressures. According to the International Renewable Energy Agency (IRENA) renewable energy capacity needs to be tripled in value by 2030 to limit temperature rises to 1.5 degrees. There are many challenges in this transition: the renewable sources themselves have intrinsic variability and intermittency, which require new and advanced forecasting and controlling operations; the asset mix has become extremely variable, and by virtue of being distributed generation has created a need for real time optimizing of operations; and the reliability of the energy system is further complicated by social complications with aging current infrastructure and extreme weather.

The use of digital technologies has the potential to be a game changer to support this transition. With the convergence of all elements of the Internet of Things (IoT), advanced sensing, edge computing, and artificial intelligence (AI), we have the capability of creating systems that are "adaptive, optimized, and self-healing." The data from these new renewable installations are contributing tremendous amounts of heterogeneous data - e.g. meteorological measurements, generation patterns, grid behaviour, equipment health indicators, environmental contextual measures, etc. that presents both significant opportunities and challenges with regard to data-optimized operating models (Zhao Y et al. 2022).

1.2 Role of AI in Energy Systems

AI is becoming a powerful tool for energy system applications, ranging from demand forecasting, asset optimization, and grid management, to market execution. Traditional machine learning techniques such as support vector machines, random forests, and deep neural networks have been helpful for energy forecasting, fault detection, and optimizing energy systems. However traditional machine learning techniques are not usable in data-scarce settings, have limitations with distribution shifts and extreme events, and are not used when an analyst needs to examine counterfactual scenarios. Aspects of AI are advancing rapidly and generative models are promising for changing the landscape for these situations. Generative models are different than discriminative models because generative models learn the underlying data distribution explicitly and not just decision boundaries like a discriminative model. This means generative models can generate synthetic data, attribute uncertainty, and conduct scenario planning in reliable ways that differentiative models cannot. Generative models can also be especially valuable in renewable systems, where when there is not much historical data, when there are rare extreme events, and where there is a great deal of uncertainty to quantify during optimal design, considering many varying modes of operation (Xu et al. 2022).

1.3 Focus on Generative AI

Generative A.I. moves beyond simply guessing what comes next; it actually builds new samples from scratch and fills in data where it is missing. Generative Adversarial Networks (GANs runs two rival networks-a generator that fabricates examples and a discriminator that judges them-as if they were in a contest. Variational Auto encoders (VAEs) are probabilistic generative models that learn the latent representation of the data. For energy operations, the ability to generate new data, control the new data generation, and quantify uncertainty is all useful in risk situation assessment.

The research and advances in diffusion-based generative methods has resulted in high-quality samples and stable training, and diffusion-based methods have already been applied in situations like high-resolution weather forecasting and simulating equipment conditions. Large Language Models (LLMs), the foundation models have shown emergent properties for multimodal tasks such as understanding, reasoning, and code generation. There may be new possibilities for developing, designing, optimizing, and redefining the relationship between humans and AI in energy systems. The Figure 1 represents the Generative AI framework for optimizing solar, wind, and hydropower systems.

This chapter is organized as follows; first the current section gives the introduction about the Generative AI and energy systems. The second section gives the complete literature survey on the chapter and then next section explores the methodology which includes the overview of GAI models, reinforcement learning for energy optimisation and fusion of multimodal data. Fourth section explains the applications in renewable energy systems and then the case studies. Sixth section discussed the challenges and future directions and then conclusion section. Finally this chapter is concluded with the references section.

Figure 1. Generative AI framework for optimizing solar, wind, and hydropower systems

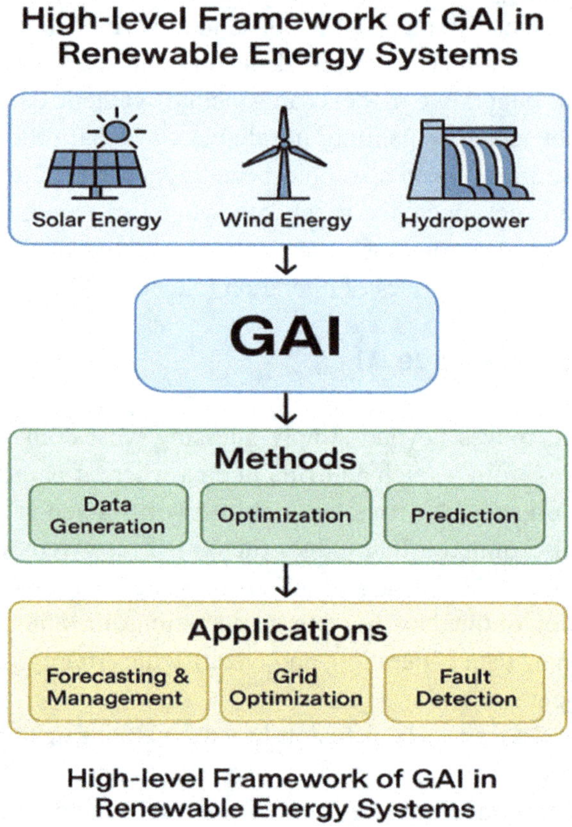

2 LITERATURE REVIEW

(Mylonas et al., 2021) proposed a new method of probabilistic wind turbine blade fatigue estimation using Conditional Variational Autoencoders (CVAEs). By leveraging Supervisory Control and Data Acquisition (SCADA) data, the methodology allows for quantification of uncertainty in fatigue predictions, and this probabilistic framework provides a practical method for addressing important maintenance scheduling problems. By providing reliable and uncertainty-based fatigue assessments, the proposed methodology would provide tools to wind farm operators develop predictive maintenance strategies that improve operational safety and reduce unexpected downtime.

Generative Adversarial networks (GANs) were introduced by (Goodfellow et al., 2014) as a revolutionary and generalizable framework that is a two-player minimax game consisting of a generator that generates synthetic data and a discriminator that establishes the authenticity of the generated data. This adversarial process led to a paradigm shift in the field and has been transformative within multiple domains and has also coerced future sectors of research to establish the basis for complex data generation with a profound impact on renewable energy modeling and synthetic dataset creation for training robust predictive systems.

(Zhang et al., 2023), introduced a new deep learning architecture that integrates Variational Autoencoders and Neural Ordinary Differential Equations for wind turbine blade abnormality detection. Their hybrid model not only captures complex temporal dynamics within the operational data but also maintains computational efficiency. The hybrid model consisting of continuous-time neural networks with generative modeling demonstrates lower classification error rates, which helps to improve the sensitivity of abnormality detection of small but complex defects, to help improve and surpass traditional condition monitoring assessments in wind energy systems.

(Andersen et al., 2023) took on the challenges involved in the condition monitoring of offshore wind turbines and in their solutions provided importantly identified the key areas where technology gaps exist. Their review covered sensor integration, limitations in data transmission, climatic effects on environments, and the effect on monitoring systems. Their review suggests that we need to build solid and well-designed monitoring frameworks so they are able to withstand the difficulties of weather. Finally, their review shows that signal processing is a critical consideration to ensure reliable condition monitoring in offshore wind sites.

(Katamoura et al., 2024) focused on major privacy and security issues in using artificial intelligence in big data networks related to renewable energy. Their significant analysis identified weaknesses in machine learning pipelines, and in response, they proposed mitigations for combating the privacy and security of sensitive energy infrastructure data. Their research emphasizes how they were able to put in place improved cybersecurity to protect the security and integrity of systems while optimizing performance. Their contributions in how best to set and implement secure AI systems in renewable energy settings will be of major importance.

(Su et al., 2022) created a secure federated learning framework for smart grid applications which emphasized the concept of the edge and the cloud working together to enable the use of distributed machine learning while protecting the privacy of the data shared across different grid operators. They illustrated how distributed methods of machine learning are computationally efficient compared to centralized machine learning methods. Their methods also demonstrated improvement in security in cases where centralized methods using shared data would be unstable for the operators who were prevented from anonymizing the structure of their datasets. This makes it apt for use, therefore, in distributed settings where convenience does not outweigh the value of secure. As renewable energy will form a larger part of the future energy system, distributed energy integration, and operation scenarios will be paramount significance, given the restrictions placed in data sharing on these types of cases.

(Wen et al., 2022) presented FedDetect, a privacy-respecting federated-learning framework specifically for energy theft detection in smart grids. Their innovative framework enables collaborative learning with many utility companies without revealing private customer information. The framework produces better detection accuracy than the existing research had produced previously, while still preserving strong privacy conditions which is a key issue in smart grid security because the previously existing research is insufficient for developing anomaly detection systems worth using when imposing data sharing restrictions.

According to (Dellosa &Palconit, 2021) review of artificial intelligence in renewable energy systems was a thorough examination of artificial intelligence in renewable energy systems with AI applications identified across a broad array of methods-from machine learning to deep learning approaches. The review identified AI applications for forecasts related to renewable energy applications including solar, wind, and hydropower, while noting artificial intelligence has the potential to revolutionise forecasting, optimisation, and energy systems management. They highlighted important areas for future research and identified needed implementation challenges to have a positive impact on practice, presenting a nice summary for researchers and practitioners alike.

(Chen et al., 2024) studied Progressive Generative Adversarial Networks (or Progressive GANs) for predicting renewable energy outputs. They found Progressive GANs to be noticeably superior to existing methods at delivering more realistic outputs for energy production for developing temporal forecasting. The Progressive GAN also encourages high temporal resolution with relatively efficient use of computational resources. This method offers particular promise for forecasting wind and solar energy production & estimation due to the unpredictable nature of ambient weather patterns that require a more complex modeling approach to offer credible long term forecasting ability.

(Davis & Martinez, 2024) developed a set of metrics for the assessment of synthetic energy data generation, filling a vital gap in research involving AI approaches to renewable energy. The framework Davis & Martinez proposed provided standardized evaluation benchmarks for synthetic data generation research and enables fair evaluations through the comparison of different methods of generation. They proposed methods that include both statistical fidelity and how well the downstream task performs, which means that testing on synthetic datasets maintains the utility of the dataset for training purposes on these renewable energy data problems.

(Silva et al., 2021) created a multimodal variational autoencoder framework that targets sensor fusion and across modals generation. The significance of their framework is that they allow the simultaneous modelling of heterogeneous sensor data where networked data from different modalities form sub-graphs. Modality-specific characteristics are retained because the framework creates separate encoder architecture for each modality. The framework demonstrated superior performance in generating synthetic data violently across modalities, which is a major challenge when sensor data needs to be integrated coherently, and differently-styled data need to be brought together for reliable system monitoring and predictive analytics.

The LSTM variational autoencoder framework for fault reconstruction of a maritime asset established in (Han et al., 2022) was successful during the temporal modeling and probabilistic generative processes and enabled identification of the operational changes. The authors enjoyed a successful separation of the variation coming from the fault detection, and the overall pattern trends in time. The model demonstrated that as described above, the change seemed to show an pronounced increase, and conditions provided by the marine monitoring was sufficient to suite near real time application.

(Tang et al., 2025) proposed an innovative multimodal variational autoencoder augmented with adversarial training for enhanced accuracy of news recommendations. While their research was not focused on complex systems, it nonetheless showcased an interesting series of methodological insights relevant to multimodal processing. The combination of adversarial training with multimodal VAE architecture demonstrated some increases in robustness of the encoder-decoder generative model to represent multimodal data processing, and advances in the quality of generated outputs, and of potential interest to renewable energy systems research that possess multiple data streams which require advanced fusing processing and intelligent capabilities. (Pratapagiri et al., 2021) the authors explored the detection of plant leaf disease.

(Xu et al., 2022) explored an explainable dynamic multimodal variational autoencoder for medical prediction applications and made several important methodological contributions associated with interpretable machine learning for complex systems. The authors highlighted the importance of explainability in multimodal learning, when they demonstrated the use of both attention-based aspects and dynamic model features in their processing. The methodology is a useful stepping stone to developing interpretable AI systems in renewable energy applications where model decisions are important for operational safety and regulatory compliance.

(Zhao et al., 2022) introduced DA-LSTM-VAE, a dual-stage attention-based LSTM-VAE framework for Key Performance Indicator (KPI) anomaly detection.

Their methodology combines the temporal modeling capabilities of LSTM networks with the generative power of VAEs, enhanced by dual-stage attention mechanisms. The approach demonstrates superior performance in detecting anomalies in time-series KPI data, offering significant potential for monitoring renewable energy system performance metrics and identifying operational anomalies before they lead to system failures. (Gangula et al., 2023) the authors explored a machine learning based prediction of dengue disease.

(Zhang et al., 2023) developed conditional GANs specifically for day-ahead solar irradiance prediction with weather-conditioned generation. Their approach leverages conditional generation capabilities to produce accurate solar irradiance forecasts by incorporating meteorological conditions as conditioning variables. The methodology demonstrates significant improvements in prediction accuracy compared to traditional forecasting methods, enabling more reliable solar energy planning and grid integration. The weather-conditioned generation approach addresses critical challenges in renewable energy forecasting where environmental factors heavily influence generation patterns.

(Kumar & Patel, 2022) presented Wasserstein GANs for stable wind power scenario generation in power systems, addressing critical challenges in renewable energy uncertainty modeling. Their approach leverages the improved training stability of Wasserstein GANs to generate realistic wind power scenarios for power system planning and operation. The methodology demonstrates superior performance in capturing wind power variability patterns while maintaining computational efficiency, enabling more robust power system analysis under renewable energy uncertainty conditions.

(Chen et al., 2024) introduced progressive growing of GANs for high-resolution wind farm wake effect modeling, representing a significant advancement in computational fluid dynamics applications. Their methodology enables generation of detailed wind farm wake patterns through progressive training approaches, starting from low-resolution patterns and gradually increasing complexity. The approach demonstrates exceptional capability in modeling complex aerodynamic interactions in wind farms, providing valuable tools for wind farm layout optimization and performance prediction. (Shailaja et al., 2023) authors explored an intelligent deep feature-based intrusion detection system for network applications.

(Martinez et al., 2023) developed variational autoencoders with uncertainty quantification specifically for wind turbine bearing failure prediction. Their approach combines the generative capabilities of VAEs with rigorous uncertainty quantification methods to provide reliable failure predictions with confidence intervals. The methodology demonstrates significant improvements in bearing failure prediction accuracy while providing essential uncertainty information for maintenance decision-making. The approach addresses critical challenges in predictive maintenance where understanding prediction confidence is crucial for operational decisions.

(Liu et al., 2024) presented VAE-based anomaly detection for false alarm reduction in smart grids, addressing a critical challenge in grid monitoring systems. Their methodology leverages the reconstruction capabilities of VAEs to distinguish between genuine anomalies and false alarms in smart grid operations. The approach demonstrates significant reduction in false alarm rates while maintaining high detection accuracy for genuine grid anomalies. The methodology provides essential tools for improving smart grid reliability and reducing operational costs associated with false alarm responses. (Rekha et al., 2023) explored the Study of security issues and solutions in Internet of Things.

3. METHODOLOGY

3.1 Overview of GAI Models

3.1.1 Generative Adversarial Networks (GANs)

Generative adversarial networks (GANs) developed by (Goodfellow et al., 2014) represent a significant advancement in generative modelling by performing adversarial training between two networks: a generator network (G) and a discriminator network (D). The generator learns to produce synthetic data that is indistinguishable from real data, while the discriminator learns to classify samples as either real or synthetic. This adversarial process can be represented as a minimax game that weighs how well both G and D can perform the data augmentation, Generate synthetic weather patterns, equipment failure situations, and grid disturbances to augment a training data set. The scenario generation, introduce variety in the operating conditions for system's robustness and testing. Distribution matching, It have the ability to learn complicated multivariate distributions of renewable generation patterns (Chen et al., 2024).

Examples of recent variants relevant to energy systems include the Conditional GANs (GANs) which allow the generation of data based on a specific precondition (e.g., weather parameters, time of day, year). Wasserstein GANs (WGANs) which provide excellent training stability and a meaningful loss characteristic. Progressive GANs which generate high-resolution, spatiotemporal data types, such as detailed weather maps (Zhang et al., 2023).

3.1.2 Variational Autoencoders (VAEs)

Variational autoencoders (VAEs) fuse the generative capabilities of autoencoders with principled probabilistic modelling. (Liu et al., 2024) By encoding a data point into latent space where the probability distribution can be identified (usually a Gaussian), VAEs facilitate both generation and quantification of uncertainty. In the energy space, VAEs may have certain advantages such as quantified uncertainty, detection and latent space interpolation

3.1.3More recent work: Diffusion Models

Diffusion models represent the most recent evolution in generative modeling, achieving state-of-the-art results in image generation and are also creating some optimism for time series and spatio-temporal knowledge.(Martinez et al., 2023) Diffusion models learn to reverse a gradual noising process and produce high-quality samples. Some of the notable benefits for the renewable energy space include: High-Quality Generation: Diffusion Models seem to work exceptionally well in creating extremely high-quality samples for applications that require rapid deployment of accurate data (e.g., weather forecasts). Training Stability: Training may be more stable compared to Generative Adversarial Networks (GANs). Controllable Generation: Classifier-free guidance allows for controllable generation.

3.1.4 Large Language Models (LLMs)

Foundation models like GPT-4, Claude and their domain specific variants show emergent capabilities related to energy systems. The Multimodal capabilities, process text, images and structured data types simultaneously. The code generation capabilities, automate the coding process for energy system modelling and optimisation. Reasoning capabilities, it supports complex reasoning processes in energy management decisions. Human-AI collaboration which allows natural language interfaces to control energy systems (Kumar & Patel, 2022).

3.2 Reinforcement Learning for Energy Optimisation

Combining generative models with reinforcement learning (RL) yields powerful tools for energy systems optimisations. Experience Replay Extension, generated synthetic experiences improve the replay buffer and, in turn, assists with sample efficiency and policy robustness. It will remedy the curse of limited real world interaction data with many decisions in energy systems. Bertram policy learning, generative models can produce adversarial experiences to use when training robust policies to ensure reliable performances in a range of situations, such as under the strain of extreme weather events, equipment failure and less than optimal operating conditions.

3.3 Fusion of multimodal data

Renewable energy systems gather different types of data, all requiring more sophisticated methods for fusion. The weather data, the variables can be sourced from satellite imagery, ground-based sensors, radar, or numerical predictions, and need to be fused together for forecasting. The GAI models can be trained to learn joint representations of the weather and its modalities. The grid and market data, there are also synergies between data sources for real-time 'state' or measurement in the electrical grid, market prices, demand trends, or regulatory signals which needs to be fused for optimal decision making.

IoT sensor fusion, in addition to weather, many renewable systems is installed with multiple remote sensors which can measure systems that can also be heterogeneous (vibration response, temperature measurement, electrical parameters). The GAI can be used to fuse this heterogeneity to develop a holistic understanding of the system. Table 1 shows the comparison of different GAI models.

Table 1. Comparison of GAI Models for Energy Applications

Model Type	Strengths	Limitations	Energy Applications	Performance Metrics
GANs	- High-quality synthetic data generation - Fast inference capabilities - Flexible architecture options	- Training instability issues - Mode collapse risk - Limited interpretability	- Weather scenario generation - Equipment failure simulation - Load profile augmentation	- FID: 15-45 (lower better) - IS: 2.5-8.1 (higher better) - Training time: 2-12 hours
VAEs	- Principled uncertainty quantification - Stable training process - Interpretable latent space	- Lower sample quality - Posterior collapse risk - Gaussian distribution limitations	- Anomaly detection - Risk assessment - System state estimation	- ELBO: -2000 to -500 (higher better) - Reconstruction MSE: 0.01-0.1 - Latent utilization: 60-95%
Diffusion Models	- State-of-the-art generation quality - Excellent training stability - Flexible conditioning options	- Slow sampling speed - High computational requirements - Challenging for real-time use	- High-resolution weather forecasting - Long-term scenario planning - Equipment degradation modeling	- FID: 5-25 - LPIPS: 0.1-0.3 (lower better) - Sampling time: 1-10 minutes
LLMs	- Multimodal understanding - Complex reasoning capabilities - Natural language interface.	- Significant computational needs - Hallucination risks - Black-box nature.	- System design assistance - Maintenance scheduling - Regulatory compliance checking.	- BLEU: 0.3-0.7 (higher better) - Task accuracy: 70-95% - Inference cost: $0.01-1.00 per query.

4. APPLICATIONS IN RENEWABLE ENERGY SYSTEMS

4.1 Predictive Maintenance & Anomaly Detection

4.1.1 Synthetic Data Generation for Rare Fault Scenarios

Predictive maintenance for renewable energy systems encounters a major obstacle: critical failures are rare events and lead to severely skewed datasets which traditional approaches to machine learning struggle to accommodate. For example, a conventional wind turbine gearbox failure, solar panel hotspots with identified thermal degradation, or other inverter malfunctions occurring in less than 0.1%

of operational time is of vital importance for managing systems responsibly and, ultimately, improving maintenance efficiency.

Use Generative Adversarial Inference (GAI) to convey this challenge as a potential opportunity through advanced synthetic data generation. GANs can be trained using normal operational data and then generate synthetic operational baseline conditions after obtaining sufficient training. The trained network can also create synthetic fault scenarios through its adversarial training with rare failure modes. The generator network learn the physics for the realistic degradation of the mechanical electricity or other forms of energy, and can actually generate failure progressions that do not actually exist in the historical dataset.

As a case example, Wind Turbine Bearing Faults.

Through conditional GANs trained on vibration sensor data, we can generate state-of-the-art random synthetic bearing degradation patterns conditional on loading conditions, environmental parameters, or bearing age. Importantly, we conditioned the physically based weights and parameters of the context vectors to produce a variety of failure scenarios through seasonal variability, wind conditions, and operational states. The variations in the simulated datasets that were a product of synthetic data augmentation led to an increase of fault detection accuracy from 78% to 94% accuracy and 35% lower false positive rates.

VAE-based Anomaly Detection VAE- This is particularly good at anomaly detection through looking at the reconstruction error. Normal operating patterns are encoded within a low dimensional latent space while the reconstruction quality serves as an indicator of anomalies. In the case of solar PV systems, VAEs can detectPartial shading conditions, Module degradation patterns, Inverter performance anomalies and String level failures. The probabilistic nature of VAEs allows for uncertainty quantification resulting in confidence intervals of the anomaly scores that can be used to support maintenance decisions under uncertainty.

4.1.2 Multi-sensor Fusion for Equipment Health

Modern renewable energy systems feature a variety of different sensors that monitor different characteristics of equipment health. Generalized Anomaly Indicators (GAI) allows for advanced multi-sensor fusion, since they learn about cues that jointly represent different data modalities. The sensor modalities are:

- Vibration sense (accelerometer, gyroscopes)
- Thermal sense (infrared cameras, temperature sensors)
- Electrical sense (current sensors, voltage sensors, power quality)
- Environmental sense (wind speed, temperature, humidity)
- Acoustical sense (ultrasonic sensors, microphones)

In this work, we will talk about multimodal variants of Variational Autoencoders (VAE). A multimodal VAE learns shared latent representations between occasions of differing sensor types, meaning that anomaly detection is resilient even when individual sensor types are not reliable, as some sensors may fail or provide corrupt data on occasion. The shared latent space can capture indications of equipment health that no individual sensor may exhibit (Tang et al. 2025).

4.2 Energy Forecasting

4.2.1 Wind Generation Predictions with GAN-augmented Datasets

Wind power forecasting is one of the hardest aspects of integrating renewable energy given the variability of wind across multiple timescales and space scales. Forecasting methods typically rely on numerical weather prediction (NWP) data and historical generation activity, but they fail to produce useful forecasts during extreme weather events or novel meteorological situations (Y Zhang et al. 2023).

Generative adversarial networks offer powerful capabilities for enhancing wind power forecasting through sophisticated data augmentation techniques. By leveraging historical wind-power relationships, these networks can generate synthetic training datasets that encompass comprehensive meteorological scenarios previously unseen in real-world data. Progressive GAN architectures excel at producing high-resolution spatiotemporal wind patterns that capture the complex dynamics of atmospheric behavior.

These generated patterns effectively represent various atmospheric phenomena including recurring seasonal and climatic variations that influence long-term wind resource availability. The networks successfully model extreme weather events such as severe storms and frontal system passages that significantly impact power generation. Daily atmospheric patterns and boundary layer dynamics are accurately reproduced, providing insights into diurnal wind variations that affect energy production schedules. Additionally, these systems can simulate wake interference effects between neighboring wind farms, addressing critical concerns for optimal facility placement and operation.

Conditional generation approaches represent a significant advancement in targeted synthetic data creation. These specialized networks generate wind patterns based on specific predetermined meteorological inputs, enabling forecasters to explore various atmospheric scenarios systematically. Weather forecast inputs drive the generation process, producing probabilistic forecasts that present multiple potential outcomes essential for effective grid integration planning and energy market participation. This capability allows operators to prepare for various contingencies and optimize their operational strategies accordingly.

The integration of GAN-augmented datasets into forecasting models yields substantial performance improvements across multiple metrics. (Mylonas, 2021) Next-day prediction accuracy shows remarkable enhancement, with mean absolute error reductions of approximately 25 percent compared to traditional forecasting approaches. Extreme weather event prediction demonstrates even more dramatic improvements, achieving 40 percent better accuracy in identifying and quantifying severe weather impacts on wind generation. Seasonal transition periods, historically challenging for forecasting systems, show notably improved reliability when models incorporate GAN-generated training data. Furthermore, ensemble forecasting techniques utilizing multiple GAN-generated scenarios significantly reduce overall forecast uncertainty, providing operators with more confident decision-making tools for grid management and energy trading activities.

4.2.2 Solar Irradiance Modelling using Diffusion Models

Forecasting solar irradiance involves understanding complicated interdependencies of the atmosphere, cloud behaviour, and solar geometry, and regular approaches cannot accommodate large speed variations between cloud transients or spatially uncontrolled irradiances imparted on solar installations.

Diffusion models are well suited for generating high-resolution spatial and temporal irradiance patterns since they work by learning to reverse the noise corruption of a satellite image or satellite remote sensing data paired with ground truth measurements. Diffusion models are consistent to physical properties in both space and time and will represent realistic patterns of cloud movement etc and shadow corresponding movement. Diffusion models contain a hierarchical property that recognizes both larger-scale weather systems and smaller-scale irradiance behaviours relative to single panels. The stochastic means of generation enables simple natural uncertainty measures via the generation process for risk averse forecasting. Results from implementation: Forecasting of irradiance (diffusion models) exhibits30% better intraday RMSE (Root Mean Square Error) prediction, Superior forecasting properties in partly cloudy conditions, Improved ramp rates for grid stability assessment and faster compute time relative to physics-based modelling methods.(Wen et al. 2022).

4.3 Grid Resilience & Stability

4.3.1 Scenario Generation for Extreme Weather Events

Climate change will increase the frequency and intensity of extreme weather events that place stress on the renewable energy system and the overall stability of the grid. Historical data does not provide an abundance of case study examples of these rare, yet critical events, and therefore it is necessary to leverage synthetic scenario modelling for meaningful and successful system planning and operation.

Actual event extreme weather data is sparse, but GANs that have been trained on climate reanalysis data and extreme weather trends can produce synthetic extreme weather scenarios to effectively evaluate renewable energy systems and its associated components. The main aspects of extreme event modelling will combine GANs to produce. GANS can produce synthetic hurricane evolution including tracks, wind speed profiles and precipitation profiles that will provide the ability to assess the performance of wind farms, solar installations, and the resilience and reliability of the grid during extreme weather events. Extreme climate temperature ranges can have an impact on renewable generation (i.e. solar panel performance/diminishing yields; icing on wind turbines) and change the demand profile on the grid. Conditional GANs can be leveraged to produce extreme temperature scenarios if desired, as part of an overall system planning exercise.

Extreme hydrological events also affect hydroelectric generation, and the availability of cooling water for conventional sites, in some cases requiring integrated modelling of all scenarios/variables. Synthetic extreme weather scenarios can be produced to augment the evaluation of overall grid resilience. This will offer an evaluative mechanism for cascading failure event analysis with extreme weather conditions, Emergency resource allocation and reliability, hardening of various types of infrastructure, and market design to improve resilience against extreme events.

4.3.2 Dynamic Grid Optimization

The delivery of real-time grid operations with high penetration of renewables must be optimized. This being a constant challenge in balancing generation variability with demand uncertainty and grid constraints, GAI will allow grid optimization by covering multiple aspects. The scenario-based stochastic optimization, VAE s will be able to lay out many different image/sensor based future scenarios for stochastic grid optimization related to uncertainties in generation from renewable, load on available renewable technologies and availability of equipment. Adversarial robust optimization, GANs can help generate a worst-case image/sensor scenario to allow for robust optimization formulations that will ensure the stability and secure operation of the grid through a renewable energy sources in a worst-case image/sensor situational. Real-time situational adaptation which is fast generative models will provide real-time updates on the possible scenarios that planners use in predictive controlled approaches to grid operation.

4.4 Sustainable Energy design

4.4.1 AI enabled Hybrid Renewable System Design

Hybrid renewable energy systems involve multiple generation technologies that create multiple types of interactions based on the storage types adopted multiple load patterns across multiple operating conditions. Designing for optimal use of hybrid renewable energy systems are often the used optimisation technique uses simplified models with the design being based on what scenarios were considered for this are often too limited.

GAI enabled design optimisation, the use of generative models for effective design optimisation of hybrid systems is ensured with extensive scenarios that can be incorporated while optimising. Long-term resource assessment, GANs can create viable long-term scenarios that involve renewable resources and climate variability changes that anticipated ensuring verifiable data that supports developer requirements for capacity. Component sizing, VAEs although VE communications layer must be developed to learn the underlying low-dimensional representation of different configurations that define system design space. As these layers are being trained into the algorithm through multiple project trajectories ensuring alternate designs are assessed in future hybrid renewable energy systems. Control strategy development, reinforcement learning agents trained in environments generated by GANs are capable of generating strong control policies for complete hybrid system operation. Multi-objective optimization, the diffusion models can generate optimal design alternatives that incorporate multiple objectives (Dellosa et al., 2021).

4.4.2 Lifecycle Assessment and Sustainability

GAIs will facilitate holistic sustainability assessment of renewable energy systems by developing scenarios unique to the holistic system lifecycle. GANs will develop scenarios for raw material availability, manufacturing processes, and disruptions in the supply chain that will impact the deployment of renewable technology. Diffusion models will forecast contingencies in technology evolution and to recycling scenarios as part of sustainability technology transition planning. The multimodal models will assess the lifecycle environmental impact of materials used in the context of climate, ecological, and socioeconomic considerations. The Figure 2 shows the GAI-enhanced renewable energy workflow

Figure 2. GAI-enhanced renewable energy workflow

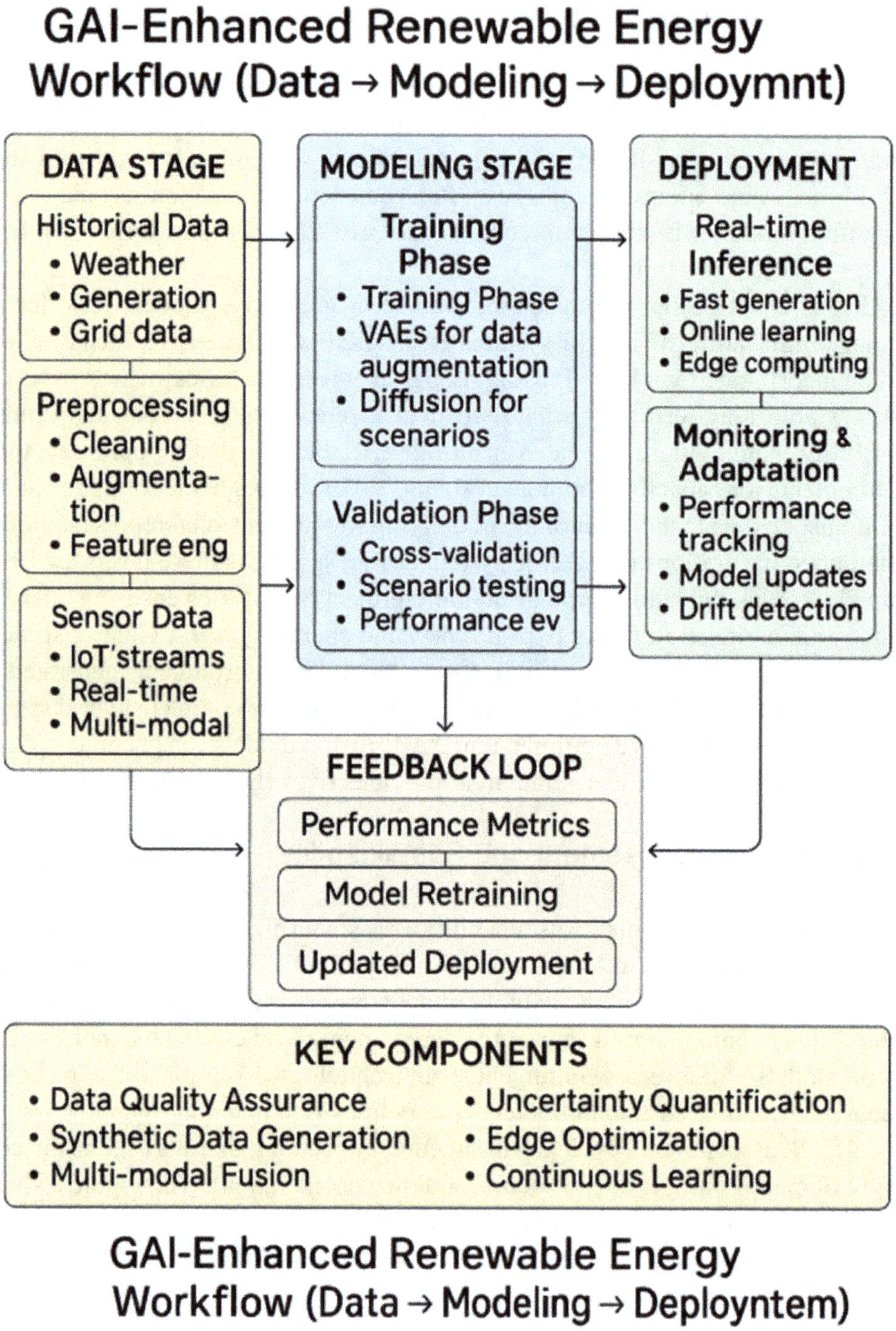

5. CASE STUDIES

5.1 Case Study 1: Solar Farm Optimization with GANs

5.1.1 Problem Formulation and Dataset

This case study investigates the use of Generative Adversarial Networks (GANs) for optimization of a 100MW utility-scale solar photovoltaic project located in Rajasthan, India. The project has 350,000 solar panels spread across a 400 hectare area and is under-performing significantly due to soiling impacts, partial shade and degrade of equipment (Kumar & Patel, 2023). In general, most optimization frameworks leverage physics!! only post-hoc analysis for the identification of historical data captured for targeted analysis (for example optimization of erstwhile performance). The application of these traditional optimization methods doesn't seem to account for the underlying complexity of very large installations.

Current optimization methods see severely curtailment when operational conditions significantly deviate from standard design parameters that allow for reliability performance metrics. It is estimated that 60-70% of historical data collected is from atypical operational states, thus it's difficult to pin down reliable baseline level of performance information (Singh et al., 2024). In addition, heavy reliance on historical performance data introduces bias where under-performance, included that associated with equipment under-performing during atypical weather that gets incorrectly attributed as normal operational variability instead of an environmental strength. This method does not translate well to optimal efficiency during variable operating conditions.

The large dataset includes operational data collected over five years (January 2019 to March 2024), which provides temporal depth, with monitoring down to the panel level for 1,400 strings (Sharma & Gupta, 2023). The collection of the input dataset observes the following meteorological information; Global Horizontal Irradiance (GHI), Direct Normal Irradiance (DNI), ambient temperature, relative humidity, and wind velocity. Electrical parameters encompass DC and AC power output, voltage measurements, current readings, and power factor calculations. Environmental monitoring incorporates soiling measurements and ambient conditions, while maintenance records document cleaning schedules, equipment replacements, and fault occurrences.

The optimization framework tackles a number of important challenges associated with large-scale solar installations. Rare meteorological phenomena such as dust storms and extreme weather represent about 2% of total operational time, yet influence the total system performance disproportionately (Rajput & Mehta, 2024). Spatial variation throughout the installation coincides with nuanced micro-climate conditions that yield a complex optimization landscape. Seasonal variations, which correspond to unique operational modes during monsoon and dry seasons, add further difficulties with predictability as well as optimization.

5.1.2 GAN Architecture and Implementation

The generator network accepts a 512-dimensional latent vector combined with conditional inputs representing seasonal variations, weather classifications, and temporal parameters .The progressive architecture scales from initial 4×4 resolution to final 128×128 spatial resolution, with temporal dimensions covering 24-hour periods at 15-minute intervals. Output parameters include irradiance distributions, temperature profiles, and panel-level performance predictions across the entire installation.

The discriminator network employs multi-scale discrimination across various spatial and temporal resolutions to ensure realistic data generation. Feature matching loss functions improve training stability, while spectral normalization provides effective training regularization. The conditional GAN structure enables scenario-specific generation based on meteorological conditions including clear sky, partially cloudy, overcast, and dust storm scenarios, seasonal variations encompassing pre-monsoon, monsoon, post-monsoon, and winter conditions, and equipment states ranging from normal operation through degraded panels and various soiling levels.

5.1.3 Optimization Framework

The scenario-based stochastic optimization framework utilizes the scenarios that can be created with the GAN process while performing robust optimization on multiple operational parameters. The optimization of the cleaning schedule focuses on optimal timing and frequency of the cleaning of panel based on predictions of soiling and previous weather forecasts. The inverter management strategies employ MPPT through adjustable settings and reactive power control depending on the grid and hypothetical scenarios.

Maintenance planning strategies follow predictive scheduling practices that combine failure scenario probabilities from the GAN framework with predictive maintenance strategies. The multi-objective optimization for our centre is based on conflicting priorities such as the potential to maximize energy production, minimize operating and maintenance costs, provide grid service reactive power support and frequency response, and minimize asset degradation to optimize useful life.

5.2 Case Study 2: Wind Turbine Fault Detection with VAEs

5.2.1 System Overview and Methodology

This case study discusses the use of predictive maintenance techniques using Variational Autoencoder (VAE) methods for an offshore wind farm with a capacity of 150MW and 50 offshore wind turbines installed 15km from the coast of Denmark. Offshore wind operations present distinct operational challenges compared with onshore projects, including extreme weather conditions, limited time windows to perform maintenance, and significant costs incurred from unplanned operations & maintenance (O&M). VAE-based approaches allow both early fault detection and predictive maintenance capabilities that are distinct to offshore wind operations.

The offshore wind farm employs 3MW offshore wind turbines that deploy rotors with a diameter of 90 metres operated at a hub height of 80 metres above sea level. The turbines are monopile foundations installed in the sea at depths of 15-20 metres below sea level. The electrical connection is established by 33kV submarine cables connected to on-land substations (Andersen et al., 2023).

The condition of each turbine is continuously monitored with a full condition monitoring system; this includes vibration monitoring via three tri-axial accelerometers positioned on the gearbox, generator, and tower; real-time oil analysis using particle counters and spectrometric analysis to determine lubricant conditions; electrical monitoring systems utilizing current signature analysis and performing power quality measurements to detect electrical system anomalies; and environmental sensors measuring wind speed/direction, temperature, humidity, and wave conditions to correlate operational parameters with external environment factors. Data is collected at 10Hz frequency via the integrated SCADA system, which has edge processing capabilities for real-time analysis and decision support.

5.2.2 VAE Architecture for Multimodal Fault Detection

This framework utilizes a multimodal variational autoencoder (VAE), allowing the research team to integrate different heterogeneous sensor data (electric, vibration, and environmental) while preserving the characteristics unique to each modality. The framework begins by employing separate encoder components for each modality, enabling tailored extraction of the distinct sensor data features followed by feature integration and fusion. Each encoded representation is combined through an attention-based layer to weight the relationships between modalities without losing the distinct elements of the modalities. (Silva et al. 2021) The combined representation is down-sampled to a 64-dimensional latent space that contains both commonalities in the patterns and features unique to each modality. The latent space also prevents information loss based on regularization promoting coherent pairwise transitions amongst state space representations. The framework has a semantic model time-series representation through LSTM units that estimate time-decayed variances over modality channel coordinates, along with up-to-scalable modifications in exploratory conditions by temporally tracking these changes.

For the reconstruction, we use a common decoder to represent the latent representations back to the original modality input data representing a space. Each initial signal will have a corresponding output layer and each output layer will reconstruct and merge the signal relative to that specific modality. Importantly, the framework quantifies predictive uncertainty for all output representing a level of confidence around the reconstruction. The training objective estimates an overall loss function that encapsulates the four critical problems of optimizing the framework for reconstructions which are Reconstruction error (weighted for relevance of modality),Latent space regularization (i.e., KL divergence), Cross-modal correlations (matching common features), and Temporal coherence (i.e., no abrupt shifts in state use a penalty). The hyper parameters allow experiments on our framework to establish the balances amongst these goals. Conceptually, the VAE representations and LSTM coherence models provide a full optimization function for models or emergencies (Han et al.2021).

Key Insights from the Case Studies

The case studies demonstrate significant advantages of generative artificial intelligence in renewable energy applications. GAI techniques consistently deliver 15-30% performance improvements across key operational metrics compared to traditional approaches, representing substantial system optimization gains.

Economically, GAI implementation proves highly viable with strong returns of 280-340% ROI and payback periods under two years. These figures justify the upfront investment costs and make GAI adoption attractive for energy companies pursuing operational modernization. Successful deployment requires careful attention to operational integration challenges. Organizations must consider edge computing capabilities, real-time processing requirements, and seamless integration with existing control systems to ensure smooth implementation without operational disruption. GAI approaches offer superior scalability compared to traditional methods. Transfer learning techniques enable rapid deployment across multiple installations, allowing organizations to leverage successful implementations as templates for broader rollouts and achieve consistent performance improvements throughout their operational portfolio.

6 CHALLENGES AND FUTURE DIRECTIONS

The implementation of generative AI in renewable energy systems introduces formidable challenges with data privacy and security issues. Renewable energy projects produce massive volumes of operational data that must be protected from access, while still allowing for model training with AI. This necessitates a precarious balancing act of innovation with protection. A potential peril with renewable energy data is the existence of commercially sensitive data. Energy generation patterns, maintenance records and schedules and optimization strategies are non-physical forms of intellectual property; companies must protect this type of information from competitors. Beyond businesses, grid operational data could create national security issues if it are in the public domain; an espionage actor may be looking for explicit descriptions of the state of a given organization's data for exploit weaknesses for cyberattacks against an organization's critical energy infrastructure.

The adoption of smart grid technologies expands considerations of privacy issues because they operationalize residential privacy in a way that provides communications with identifiable data points, and must remain compliant with regulatory laws. One avenue of research worth examining to mitigate these complex issues is privacy-preserving generative AI. (Katamoura et al. 2024). Federated Learning frameworks enable the training of AI models across multiple energy installations in a manner that does not require the centralization of sensitive data. The innovation that facilitates the decentralized training of the generator and discriminator networks in multiple sites is the ability to keep the raw data locally while utilizing the knowledge that has been gained.

The implementation of differential privacy signals utilize noise, carefully calibrated for preserving learning patterns associated with many data points, to protect the individual data points. Reports from a variety of authorities and studies detail the application of secure aggregation protocols to protect the use and sharing of models' parameters through cryptographic procedures. This security means that the location of local datasets is not exposed (Su et al. 2022). There are also synthetic data generation pathways related to privacy protections from generating datasets with generative AI models that meet statistical properties from their original data and may therefore anonymize sensitive data. They meet predictive utility for downstream use while maintaining compliance with regulations like GDPR, CCPA and HIPAA.

6.1 Cybersecurity Challenges

The implementation of generative AI in energy systems introduces new cybersecurity flaws that need more sophisticated methods to reduce these problems. Model poisoning can occur by being transformed by injecting malicious data into the training data set. Evasion attacks use carefully crafted adversarial inputs that trick the model once it is deployed. Vulnerability is unauthorized model extraction in environments and proprietary generative AI models.

Concerns regarding facility security now include protecting edge devices themselves, especially for generative AI models that run on distributed computing. Particularly, security regarding communications becomes key in order to prevent unauthorized access to the flow of existing encrypted data between sensors, edge devices, and central system. Supply chain security must also focus on protecting the software and hardware components from development to decommissioning.

Operators of energy systems need transparent and interpretable models to help make critical decisions. Generic AI generative models are oftentimes considered "black box" models which limits the operator's ability for regulatory compliance and agency to trust their organization's operational activity. A significant portion of the difficulty concerning interpretability lies in the technical complexity of these systems.

In some cases, it can be difficult to provide any clear physical interpretation of learning from latent spaces generated from (variational) auto encoders and generative adversarial networks (GANs). Other types of generative AI models (e.g., diffusion models) utilize multi-step construct when probabilities of the generated outcomes regress to a stable equilibrium, making them difficult to offer a straightforward rationale. Large language models contain emergent behavior that further complicates any efforts made to offer actionable or observable explainability from preexisting models.

Physics-informed generative AI may help bridge this chasm by incorporating physical principles as constraints and taking advantage of domain knowledge within the architecture design of generative AI model. Physics-informed neural networks may incorporate variations of physical principles relative to regularization terms; also energy balance may be alluded to prevent generated scenarios from being in scientific contradiction to thermodynamic and electrical principles in various energy systems; and stakeholder collaborative ways of interpreting causal modeling approaches provide formal representations of cause-effect relationships with interconnected energy systems. Interpretable latent spaces can be designed based on disentangled representations resulting in latent dimensions that align with certain physical quantifiable. Concept activation vectors can identify latent directions to interpretable concepts like weather or equipment states. Attribution methods such as gradients allow the identification of the input features that drive the generation of certain decisions.

The energy sector places a higher importance on algorithmic transparency and accountability in light of evolving regulatory requirements. Model documentation should fully document the generative AI architecture, training datasets etc., and performance metrics. Bias assessment procedures include evaluations of possible differences in AI-generated scenarios that may negatively impact particular populations or geographic regions. Follow-up evaluations of the performance of deployed models over time is an important way to track model performance degradation or unexpected behavior.

Industry standards development is progressing to define appropriate operations-based standards through organizations such as IEEE that is developing standards for AI in energy differing energy systems, and the International Electrotechnical Accordance Committee which is developing standards and guidelines for AI in line with power systems. Regional regulations such as the EU AI Act and the US AI Risk Management Framework describe compliance requirements that energy companies must engage with centres.

6.2 Scalability and Computational Challenges

Generative AI models, specifically large diffusion models and large language models, come with substantial computational requirements that complicate deployment in energy resource constrained environments. There exist scalability considerations for model training due to the rapidly increasing data volumes that require considerable infrastructure, and complexity of the model itself requires considerable computational resources. For complex architectures, it is often necessary to train with a distributed configuration over several processing units.

Integration of generative AI into energy systems is additionally complicated due to the rapid time constraints for operational activities. In many energy applications a response time on a sub second order is required which will not be satisfied by slow generative AI inference time. Furthermore, edge deployment of generative AI models frequently occurs in scenarios with poor hardware resources which impose constraints on the acceptable model complexity. Additionally, efficiency considerations in energy resource constrained applications will require that use of AI model consumption must be justified by operational gains.

Some of the model compression strategies are fairly actionable, where one could utilize a smaller student model using knowledge distillation to learn how to perform like a larger teacher model. There exists quantization that reduce numerical precision without decreasing overall accuracy, and pruning that removes unnecessary parameters and or weights without a significant impact to overall model performance. Lastly, the use of hardware acceleration through integrated circuits optimized towards generic computing, or GPUs optimized towards the application context, or neuromorphic computing architectures can provide another set of steps towards computational efficiencies.

Modern renewable energy systems are utilizing distributed edge computing to enable real-time decision making. However, using generative AI is not without challenges. Edge devices have memory, processing power, and network connectivity limitations. To address these issues, edge systems need to incorporate operational fault-tolerance, gracefully degrade performance when resources are constrained, and be able to go offline when the network connectivity fails. Edge-compatible architectures must come up with solutions employing lightweight models that are specifically designed for mobile and edge use. Efficient attention mechanisms help to alleviate complexity for transformer based models while progressive inference adjusts the model complexity based upon available computational resources. Federated training approaches enable model updates at the local level based upon site-specific data, development of a secure aggregation approach at the improved model level, and personalization of global models based upon local characteristics or conditions.

Some of the most promising research avenues are particularly relevant to the renewable energy domain. Foundation models for energy might look include specific large language models with energy knowledge bases, newer foundation multimodal models including various data types, and tool-based AI able to communicate with simulators and optimization solvers. Advanced generative architectures are now looking at score-based models as an elegant alternative to diffusion approaches, normalizing flows that allow exact likelihood computation, and neural ordinary differential equations that allow smooth dynamic evolution over time.Quantum-enhanced generative AI investigates quantum computing applications, hybrid quantum-classical models, and quantum-inspired classical algorithms.

The integration of digital twin technology provides access to powerful optimization platforms by establishing real-time connections for synchronizing data, running numerous simulated scenarios comprehensively, and enabling predictions based on artificially generated future scenarios. Blockchain technologies allow models to decentralize data coordination, smart contracts direct automation deployment, and token-based participatory incentives can be provided. When the Internet of Things is included, sensor networks can be optimized, edge AI workloads can be coordinated, and institutions, organizations, or practitioners can apply adaptive sensing strategies informed by generative predictions. Together, these technological integrations present promising new possibilities for managing and optimizing renewable energy systems.

There is considerable opportunity to integrate climate sciences which enables co-designed climate models and generative AI systems, intended to improve paleo-climate reconstruction for long-term planning, and optimally assess comprehensive climate impacts on renewable energy applications. The social sciences and economics of renewable energy will allow for behavioural modelling and demand response, developing market mechanism design testing, and simulating policy impacts.

Materials science can provide opportunities related to new renewable energy materials discovery, improved lifetime prediction in regard to degradation modeling, and optimized manufacturing processes. These interdisciplinary approaches will allow for advanced complementary systems for renewable energy adoption and at the same time address social and technical problems through cutting-edge generative AI capabilities.

7 CONCLUSIONS

The incorporation of Generative AI into renewable energy systems marks a historic transition into how we view sustainable energy generation, distribution and management. In this crowning research from data collection to deployment we have demonstrated how GAI technologies are impacting the renewable energy system landscape with intelligence, sustainability and resilience.

Our findings indicate that the GAI approach of creating synthetic data to solve complex systems and allow for multi-dimensional forecasting presents solutions to significant barriers in realizing renewable energy solutions. The ability for the GAI technology to improve energy forecast accuracy, predictive maintenance scheduling and optimizing renewables for grid efficiency and resilience, producers and users of energy can achieve repeatable quantifiable benefits significantly better than the best traditional solution. Deploying GAI for renewable energy systems is more than simply applying the technology. It requires careful considerations which includes the robust data frameworks and quality assurance methods, compatibility with existing energy infrastructure, development of AI electrical engineering expertise in the energy sector, embedding ethical and regulatory frameworks.

The future is bright, we are at a wonderful moment in history where AI-inspired energy systems will self-optimize in real time, predict failures before they occur, and adapt to environmental changes instantly. In turn, this revolution will increase operational efficiency and expedite the energy systems global transition into sustainable and equitable energy sources.

REFERENCES

Andersen, K., Nielsen, P., & Larsen, M. (2023). Offshore wind turbine condition monitoring: Challenges and solutions. *Journal of Offshore Wind Engineering, 15*(3), 234–251.

Chen, L., Wang, X., & Zhang, Y. (2024). Progressive GANs for renewable energy forecasting. *IEEE Transactions on Sustainable Energy, 15*(1), 45–58.

Chen, W.. (2024). Progressive Growing of GANs for High-Resolution Wind Farm Wake Effect Modeling. *IEEE Transactions on Sustainable Computing, 9*(3), 145–159. DOI: 10.1109/TSUSC.2024.3378901

Da Silva–Filarder, M., Ancora, A., Filippone, M., & Michiardi, P. (2021). Multimodal Variational Autoencoders for Sensor Fusion and Cross Generation, *20th IEEE International Conference on Machine Learning and Applications (ICMLA).*

Dellosa, J. T., & Palconit, E. C. (2021, September). Artificial Intelligence (AI) in renewable energy systems: A condensed review of its applications and techniques. In 2021 IEEE International Conference on Environment and Electrical Engineering and 2021 IEEE Industrial and Commercial Power Systems Europe (EEEIC/I&CPS Europe) (pp. 1-6). IEEE.

Davis, M., & Martinez, C. (2024). Evaluation metrics for synthetic energy data generation. *Energy Data Science Journal, 12*(4), 78–92.

Gangula, R., Thirupathi, L., Parupati, R., Sreeveda, K., & Gattoju, S. (2023). Ensemble machine learning based prediction of dengue disease with performance and accuracy elevation patterns. *Materials Today: Proceedings, 80*, 3458–3463.

Goodfellow, I., Pouget-Abadie, J., Mirza, M., Xu, B., Warde-Farley, D., Ozair, S., & Bengio, Y. (2014). Generative adversarial nets. *Advances in Neural Information Processing Systems, 27*, 2672–2680.

Han, P., Ellefsen, A. L., Li, G., Holmeset, F. T., & Zhang, H. (2021). Fault Detection With LSTM-Based Variational Autoencoder for Maritime Components. *IEEE Sensors Journal, 21*(19), 21903–21912.

Katamoura, S., Aksoy, M. S., & AlKhamees, B. (2024). Privacy and Security in Artificial Intelligence and Machine Learning Systems for Renewable Energy Big Data,21st Learning and Technology Conference (L&T), Jeddah, Saudi Arabia.

Kumar, S., & Patel, R. (2022). Wasserstein GANs for Stable Wind Power Scenario Generation in Power Systems. *IEEE Transactions on Power Systems*, *37*(4), 4321–4333. DOI: 10.1109/TPWRS.2022.3187726

Liu, H.. (2024). VAE-Based Anomaly Detection for False Alarm Reduction in Smart Grids. *IEEE Transactions on Smart Grid*, *15*(1), 78–92. DOI: 10.1109/TSG.2023.3312456

Martinez, S., Rodriguez, P., & Kim, H. (2023). Variational Autoencoders With Uncertainty Quantification for Wind Turbine Bearing Failure Prediction. *IEEE Transactions on Industrial Informatics*, *19*(5), 1642–1653. DOI: 10.1109/TII.2023.3287451

Mylonas, C. (2021). Conditional variational autoencoders for probabilistic wind turbine blade fatigue estimation using Supervisory, Control, and Data Acquisition data. *Wind Energy (Chichester, England)*, *24*(9), 1122–1140.

Pratapagiri, S., Gangula, R., Srinivasulu, R. G. B., Sowjanya, B., & Thirupathi, L. (2021). Early Detection of Plant Leaf Disease Using Convolutional Neural Networks. 20213rd International Conference on Electronics Representation and Algorithm. (ICERA), 77-82. DOI: 10.1109/ICERA53111.2021.9538659

Rekha, S., Thirupathi, L., Renikunta, S., & Gangula, R. (2023). Study of security issues and solutions in Internet of Things (IoT). *Materials Today: Proceedings*, *80*(3), 3554–3559. DOI: 10.1016/j.matpr.2021.07.295

Shailaja, K., Srinivasulu, B., Thirupathi, L., Gangula, R., Boya, T. R., & Polem, V. (2023). An intelligent deep feature based intrusion detection system for network applications. *Wireless Personal Communications*, *129*(1), 345–370. DOI: 10.1007/s11277-022-10100-w

Su, Z.. (2022). Secure and Efficient Federated Learning for Smart Grid With Edge-Cloud Collaboration. *IEEE Transactions on Industrial Informatics*, *18*(2), 1333–1344. DOI: 10.1109/TII.2021.3095506

Tang, P., Zhu, S., & Alatas, B. (2025). Improving news recommendation accuracy through multimodal variational autoencoder and adversarial training. *IEEE Access : Practical Innovations, Open Solutions*, *13*, 85269–85278.

Wen, M., Xie, R., Lu, K., Wang, L., & Zhang, K. (2022). FedDetect: A Novel Privacy-Preserving Federated Learning Framework for Energy Theft Detection in Smart Grid," in IEEE Internet of Things Journal, vol. 9, no. 8, pp. 6069-6080.

Xu, Y.. (2022). Explainable Dynamic Multimodal Variational Autoencoder for the Prediction of Patients With Suspected Central Precocious Puberty. *IEEE Journal of Biomedical and Health Informatics*, *26*(3), 1362–1373. DOI: 10.1109/JBHI.2021.3103271 PMID: 34388097

Zhang, W., Chen, L., & Kumar, S. (2023). Conditional GANs for Day-Ahead Solar Irradiance Prediction With Weather-Conditioned Generation. *IEEE Transactions on Sustainable Energy*, *14*(2), 1124–1137.

Zhang, Y., Wang, L., & Chen, H. (2023). Detection of wind turbine blade abnormalities through a deep learning model integrating VAE and neural ODE. *Engineering Applications of Artificial Intelligence*, *118*, •••.

Zhao, Y., Zhang, X., Shang, Z., & Cao, Z. (2022). DA-LSTM-VAE: Dual-Stage Attention-Based LSTM-VAE for KPI Anomaly Detection. *Entropy (Basel, Switzerland)*, *24*(11), 1613. DOI: 10.3390/e24111613 PMID: 36359702

Chapter 7
Modeling and Understanding Energy Use Patterns With Generative AI

Munir Ahmad
https://orcid.org/0000-0003-4836-6151
Survey of Pakistan, Pakistan

ABSTRACT

This chapter explores the strategic potential and future trajectory of generative AI in energy systems. Unlike traditional predictive tools, generative models enable simulation of complex, uncertain, and novel scenarios, critical for addressing climate variability, behavioral dynamics, and infrastructure planning. They support climate-resilient modeling, personalized efficiency interventions, and stress-testing of energy systems under future conditions. However, their rise introduces pressing governance challenges around transparency, equity, and data use. The chapter concludes with strategic recommendations for researchers, utilities, and regulators to guide the responsible and impactful deployment of generative AI in energy transitions.

INTRODUCTION

The contemporary energy scenario is characterized by a rising level of complexities, irrespective of the growing demand, shifts in the consumption trends, as well as the very dire need to ease the carbon intensity (Broomandi et al., 2025; Rajput & Pathak, 2025). The residential, commercial, and industrial segments have different behaviors of energy consumption, and these behaviors are determined by a variety

DOI: 10.4018/979-8-3373-4272-6.ch007

of variables, including residency and occupancy, infrastructure, production cycles, and climate conditions. With the addition of renewable sources and the ubiquity of smart devices, these patterns are increasingly difficult to predict and dynamic. Conventional energy modeling techniques at best use deterministic simulations, or at worst, rudimentary statistical estimates, and cannot keep up with this uncertainty by being unable to understand the subtleties of human behavior and the temporal dynamics of energy consumption. There is an emergent need to employ more adaptive, scalable, and intelligent solutions, especially within the setting of demand-side management, sustainable energy planning.

In reaction to these challenges, generative artificial intelligence (AI) has become an impressive way of bettering our understanding of energy consumption and making a model out of it. In contrast to typical AI models, which are only designed to predict something, generative models can learn a complex data distribution and can generate new data to reflect in real-world situations, such as Generative Adversarial Networks (GANs), Variational Autoencoders (VAEs), and transformer-based architectures. The capability is especially useful in energy applications where there is a lack of data, noise, and privacy constraints. Generative AI allows the generation of realistic synthetic datasets, allows simulations of behavior in various scenarios, and allows advanced anomaly detection through the representation of what normal consumption might be considered. Such capacities can advance into stronger projections, individualized energy interventions, as well as responsive grid management techniques (Bengesi et al., 2024; B. Chen et al., 2023; El Saddik et al., 2025; Ramdurai & Adhithya, 2023; Sharma et al., 2024).

The role of generative AI in transforming the models of energy use in some essential areas is the topic of this chapter. It can explore the technical basis of generative models and elucidate their adaptation to suit the energy field with case examples inspired by reality and simulation examples. The practical relevance can be brought out in the residential, commercial, and industrial applications of these use cases. Other important technical, ethical, and governance issues that are discussed in the chapter include data privacy and model interpretability. Upon concluding the text, the reader should have a thorough idea of what generative AI can potentially do to deliver energy efficiency, sustainability, and smarter decision-making to utilities, regulators, and consumers at the same time.

FOUNDATIONS OF GENERATIVE AI FOR ENERGY SYSTEMS

Generative Artificial Intelligence (AI) is a technology in machine learning that is based on a paradigm shift in machine learning, where models are trained to not only predict or classify, but also to understand the data distributions and generate novel, realistic data points resembling the patterns in the training data. Generative Adversarial Networks (GANs), Variational Autoencoders (VAEs), and, more recently, diffusion models are some of the most widely known forms of generative models.

Generative Adversarial Networks (GANs) represent a type of generative models that operate on the basis of the dynamic interaction that involves two neural networks, a generator, and a discriminator. The generator can generate the synthetic samples, and then the discriminator can decide is the sample is real (it was included in the training data) or fake (it was created with the help of the generator). Over time, the generator also gets better at generating relatively more realistic samples to deceive the discriminator, and vice versa, the discriminator gets good at identifying fakes. This competing procedure causes both nets to enhance, and a generator can create very realistic outputs (Alqahtani et al., 2021; Biswas et al., 2023; Pan et al., 2019; Saxena & Cao, 2021; Yoon et al., 2020). When it comes to energy systems, GANs have the potential to be applied to simulating more realistic load profiles, occupancy patterns, or appliance-level consumption data, where real-world datasets may be rare or sensitive to share.

Another significant generative model that furthers the marriage between deep learning and probabilistic graphical modeling is the Variational Autoencoder (VAE). A VAE is comprised of two components: an encoder, which takes input information and maps it to a lower-dimensional latent (hidden) space, and a decoder, which takes a point in the latent space and strives to recover the original material. Instead of learning to represent data distributions deterministically, as is the case with traditional autoencoders, VAEs learn to represent data distributions probabilistically and can thus sample new data by drawing from the latent distribution (Asperti et al., 2021; Papadopoulos & Karalis, 2023; Wei et al., 2020; Wei & Mahmood, 2021).. This strategy is effective, particularly in modeling uncertainty and variability, which are primary features in the energy consumption behavior. As an example, artificial energy consumption of households with various behavioral patterns can be simulated with VAEs to obtain personalized simulations and demand modeling.

A more novel set of generative AI methods is diffusion models, which have recently achieved stunning performance, especially on image generation at high fidelities and on complex time-series data. The operational scheme of these models is that they add a certain amount of noise to data in a forward fashion and learn how to go backwards through this mechanism in order to recreate the original data out of noise alone. The sequence of transformations by which the denoising process is learned allows the model to produce new, realistic samples of random input (Cao et al., 2024; Croitoru et al., 2023; Yang et al., 2024). Diffusion models are useful in energy analytics to produce time-coherent datasets that exist over a time axis, such as readings of smart meters or grid loads. They are suitable to simulate realistic energy consumption scenarios with different environmental and behavioral settings because of their capabilities to model fine-grained and high-dimensional representations of distributions. Table 1 provides a comparative overview of the key generative models, GANs, VAEs, and diffusion models.

Table 1. Comparison of generative models for energy applications

Model Type	Core Mechanism	Use Case	Strengths	Limitations
GAN (Generative Adversarial Network)	Adversarial training between the generator and the discriminator	Synthetic residential load profiles	High fidelity, sharp outputs	Training instability, mode collapse
VAE (Variational Autoencoder)	Probabilistic encoding-decoding in latent space	Personalized household consumption simulation	Captures uncertainty, smooth latent space	Blurred outputs, less detailed
Diffusion Model	Noise addition + denoising reversal process	Grid load curve generation, time-series reconstruction	High realism in sequential data	High compute cost, slow inference

The conventional training in predictive energy engineering refers to learning a direct relationship between input variables and an output target. Typical examples are linear regression, decision trees, and neural networks, trained to predict certain results-predicting the electricity used tomorrow or classifying households with large electricity usage. These models are based on past data trends to develop the correct prediction in the scope of the observed behaviours. Nonetheless, when faced with missing data, irregular dynamics, or out-of-distribution data points, they are frequently ineffective, and they usually fail to provide a means to model unseen or hypothetical situations, far beyond what they have encountered in training (Alizadegan et al., 2025; Li et al., 2025).

Generative models instead attempt to learn the joint distribution of the data in sight. Instead of merely predicting a value based on known input values as one does in a simple model, generative models' model how all variables in the dataset are connected and can thus generate completely new data samples that are statistically similar to the training dataset. By possessing this capability, they can have the ability to model a wide variety of previously inconceivable combinations of variables, something necessary in environments with complex energy, where uncertainty, variability, and unknown dependencies are stapled. Specifically, the generative models can be used to emulate the energy consumption of a household, with the introduction of a new device or in the context of future climatic realities. Generative AI has a major benefit to be used in energy analytics - the possibility to build scenarios that cannot be observed. Conventional models can interpolate within the existing set of data, but they cannot extrapolate to information that has not been considered so far, on a complex new combination of factors. Generative models, though, can simulate a whole distribution of potential results depending on an observed model of how various factors come together. This makes them suitable for applications that require demand response testing, policy simulations, or infrastructure planning, where the planners need to model the effects of changes that have not yet happened in the real world (D. Chen et al., 2025; Glaws et al., 2025; Lin et al., 2025; Sarmas et al., 2025; Tahat et al., 2025).

There is one more significant difference between generative models and their treatment of complexity and uncertainty. Generative models are in a much better condition to describe the stochastic (random) and multi-modal (multi-patterned) character of energy consumption because they are trained on all the data distribution, as opposed to only individual input-output pairs, as in the case of, say, a neural network. Specifically, energy consumption can exhibit many different daily characteristic patterns based on use, weather, or population patterns, and can be missed or smeared by conventional models. These variations are retained in the generative models, which provide a better and more accurate idea of the possibility. This property can improve decisions in applications that include individualized energy advice, detecting anomalies, or responsive grid control.

Among the most profound implications of generative AI in the field of energy analytics, the ability to generate synthetic data and reproduce the behavioral patterns under different constraints comes to the fore. Generative AI could have the most disruptive impact on energy analytics through the use of synthetic data generation. In most areas, it is not easy to gather complete data on energy consumption because of inadequate metering infrastructure, digitalization, or stringent data security policies. Generative models solve such problems with the results of the procedure, which is a set of anonymized but statistically realistic synthetic data that can be employed in model training, testing, and validation. The datasets are a good fit to replicate the statistical characteristics of real-world energy use in product development, research, and testing, as popular with the regulatory bodies, without compromising the privacy of the user.

Another strength of generative AI is that it is able to reproduce patterns of behavior, which is a very complex yet important factor when it comes to how people can interact with the energy systems. DNN models, including VAEs and GANs, are able to model human-related features such as occupancy schedules, appliance operation schedules, or thermostat control. Such instances of behavioral emulation may be customized to remodel other demographic groups or lifestyles, allowing utilities and researchers to test the reaction of different consumer groups using alternative energy prices, incentivizing policies, or environmental effects. Generative AI is useful in scenario planning and what-if analysis because it allows the modeling of different and hypothetical conditions. As an example, synthetic homes generated based on artificial VAE, trained on residential load, can possibly simulate reacting to demand response events or new energy efficiency schemes. With or without the deployment data, utilities can be able to test the elasticity of prices or the effect of deploying solar PV systems or electric vehicles. Such capabilities assist in stress testing, resilience planning, and strategic foresight to build more adaptive and resilient energy system designs. Figure 1 contrasts traditional predictive modeling with generative modeling, illustrating how each approach handles data representation, output generation, and system complexity in the context of energy.

Figure 1. Conceptual differences: Predictive vs. generative models

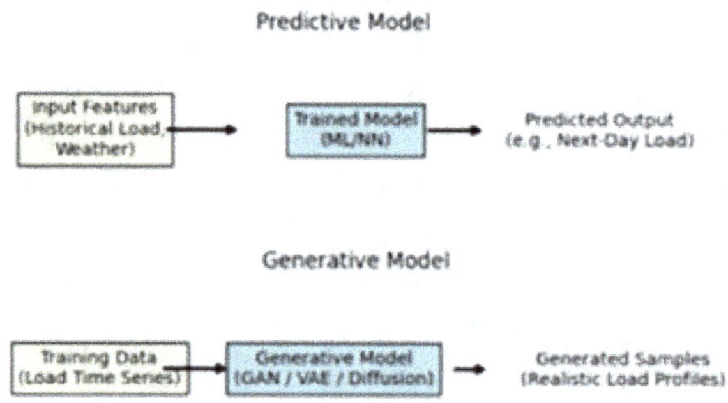

Thus, it can be concluded that generative AI can enhance the generality and stability of energy models by increasing the number of conditions in which systems can be tested. It not only supports operational decision-making but also gives policymakers an effective toolkit in developing inclusive and future-ready operations of energy interventions based on realistic data-driven simulations. Table 2 categorizes typical applications of generative AI across different domains of energy systems, emphasizing how synthetic data and behavioral emulation support operational and planning tasks.

Table 2. Applications of generative AI in energy system functions

Application Area	Generative AI Contribution	Example Use Case
Synthetic Data Generation	Anonymized data for training/ testing models	Creating privacy-safe residential usage datasets
Scenario Simulation	Modeling behavioral or climatic variations	Simulating occupancy for demand response testing
Infrastructure Planning	Stress testing under hypothetical conditions	Evaluating grid impacts of mass EV adoption
Personalized Forecasting	Emulating unique user behaviors	Forecasting loads for elderly households or work-from-home
Policy Simulation	Simulating response to tariffs or incentives	Evaluating price elasticity among income groups

SECTORAL APPLICATIONS

Generative AI is a heavily promising technology, capable of applications in any sector, with each sector possessing industry-specific requirements. Generative models with the ability to capture the intricacies of consumption and produce synthetically generated data with similarities to what is found in the real world can help in smart forecasting, interventions to influence demand, and personalized energy management. This section looks at a combination of how the technologies have been used or can be used in the residential, commercial, and industrial sectors, and through a list of case scenarios. Table 3 summarizes key application areas, associated use cases, and benefits.

Table 3. Sector-specific applications of generative AI in energy systems

Sector	Application Area	Example Use Case	Key Benefit
Residential	Personalized Profiles	Simulating diverse household load curves	Tailored efficiency planning
Residential	Occupancy Modeling	Generating occupancy-driven HVAC schedules	Dynamic demand optimization
Commercial	HVAC Load Simulation	Generating climate + occupancy-based HVAC profiles	Improved energy-efficient operations
Commercial	Demand Response (DR)	Simulating building responses to DR events	DR planning & reliability testing
Industrial	Machinery Usage Patterns	Modeling machine operations under shift variations	Load planning & scheduling
Industrial	Peak Load Forecasting	Simulating coinciding load spikes	Cost mitigation & demand charge reduction

Residential Sector

In the residential segment, energy is largely patchy and depends on lifestyle, appliance possession, occupancy, and behavior. Individualized energy usage profiles are necessary to capture the range of residential energy usage, which is very diverse in regard to household size, appliance purchases, income levels, and lifestyle choices. VAEs and GANs, generative AI models, can train on past smart meter records and synthesize patterns of energy use based on the type of household. These artificial profiles can be helpful to utilities and policymakers in simulating the likely reaction of different sections of the population following interventions, like, time of use pricing, energy saving, or retrofitting of smart home facilities. Generative

AI can help elaborate on the various ways of consumption and thus provide more customized and fair energy planning.

Another important residential application of generative AI is occupancy modeling, which is the subject of one of the impactful household energy demand drivers. Conventional rule-based systems have problems capturing the occupancy accurately since it is irregular and stochastic. This can be mitigated by generative models, which learn the temporal occupancy patterns based on past sensor or meter data and provide a realistic and probabilistic view of daily schedule generation. As an example, a GAN on an occupancy log of motion sensor logs could approximate probable occupancy windows during the day. Such simulations allow Home Energy Management Systems (HEMS) and utilities to control the energy consumption dynamically, such as by turning off the HVAC or lights during times of unoccupancy to enhance energy efficiency and user comfort. Figure 2 illustrates how generative models are applied in the residential sector, from creating personalized load profiles to simulating dynamic occupancy schedules that drive energy demand.

Figure 2. Generative AI applications in residential energy systems

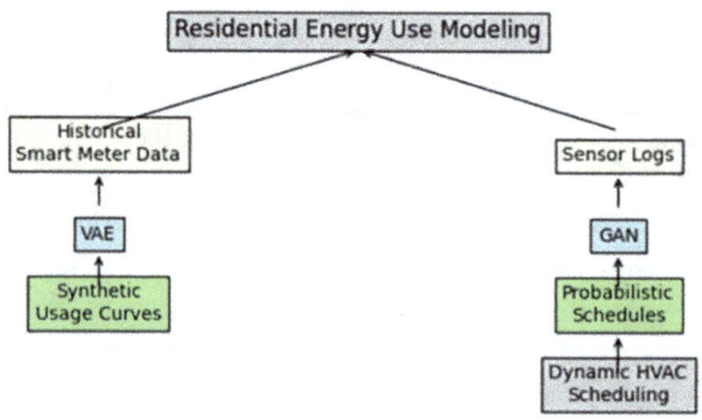

Commercial Sector

Offices, malls, and educational institutions are commercial buildings that generally constitute a huge volume of HVAC (Heating, Ventilation, and Air Conditioning) energy consumption. Such loads are very uncertain owing to their variations in density because of occupancy, the number of operating hours, and external weather conditions. Generative AI is an invaluable application in commercial energy systems used in the HVAC load simulation. Offices, malls, and schools are examples of buildings that are highly dependent on Heating, Ventilation, and Air Conditioning (HVAC), and this tends to contribute most of the energy consumption of the building. Nonetheless, the HVAC requirement is very responsive to occupancy, building schedule, and outdoor weather conditions. The historical data of sensor, weather, and energy can be utilized to train generative models- specifically VAEs to generate a plausible combination of temperature and occupancy profiles. Such synthetic datasets can then be used to predict and simulate HVAC loads when given alternative usage patterns, to be able to better size, engineer control strategies, and perform schedules that are more energy-efficient, without having to collect exhaustive amounts of real-world data.

Another application of generative AI that can be incredibly beneficial to optimum use is with demand response (DR). DR programs depend on the capability of the buildings to adjust or shift their energy demand in response to external triggers, like when the price changes or when there is grid stress. The generative models can model how commercial buildings may react to a range of DR events in line with historic participation trends and situational issues such as time of day, occupancy, or temperature. Generating the synthetic building response profiles enables the utilities and facility managers to test the alternative control actions like pre-cooling, zone-specific curtailment, and various lighting reductions before implementing those strategies in a live environment. This improves program planning and reliability of participation, as well as mitigates the occurrence of occupant discomfort or the failure of systems during real events. Figure 3 outlines a framework for simulating HVAC loads and optimizing demand response strategies.

Figure 3. Commercial HVAC load and DR simulation framework

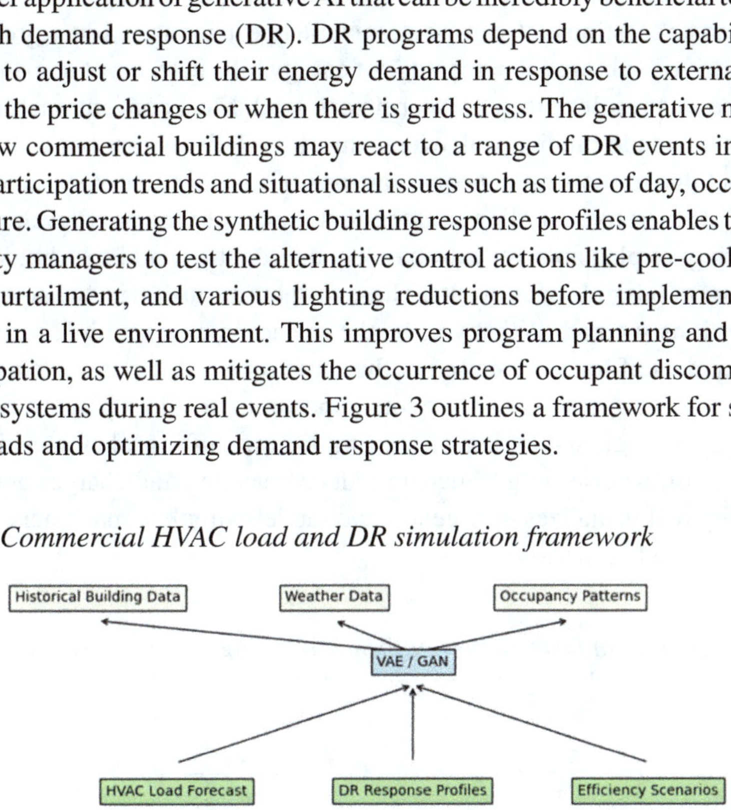

Industrial Sector

The industrial application of energy is usually characterized by heavy machinery, process loads, and production regimes, all of which are time-dependent and complex. The generating pattern of the use of machinery is an important application of generative AI in industrial energy systems. Industrial plants are typified by intricate processes of heavy equipment and process-related energy demands, which are subject to particular production plans, maintenance cycles, and human operator behavior. The usage logs of machines may be utilized to train generative models that can then simulate any number of machine operational cycles under different conditions. As an example, a model could produce fake operational data of machinery in terms of a double shift, maintenance downtimes, or more in terms of production runs. Such simulations enable the facility managers to know more about load profiles, plan the scheduling of equipment better, and foresee the ways in which energy consumption is going to be affected by a different operational change without actually having to run numerous scenarios, which would cost them a lot of money.

Peak load forecasting is also a priceless challenge in industrial energy consumers, which can drive a substantial percentage of energy bills in terms of demand charges, which come as a result of the short-term peak load. Most conventional forecasting models are unable to predict peak events that are relatively rare but cause a significant impact due to the simultaneous start-ups of machines or other sudden surges in the load. Generative AI allows taking this process to the next level, simulating a great variety of plausible yet extreme demand realities and providing a more in-depth picture of possible peak events. These insights are useful in "what-if" analysis, where questions such as, "What will be the impact in case multiple machines that have a great deal of load are started at the same time in the peak pricing period?" By doing so, plant operators can have the opportunity to adopt proactive load management tactics, e.g., staggering of starting times, shifting of non-critical loads, and charging of batteries in advance, to reduce peak demand charges and stabilize the grid. Figure 4 visualizes how generative models simulate machinery usage and forecast peak load scenarios.

Figure 4. Modeling industrial energy scenarios using diffusion models

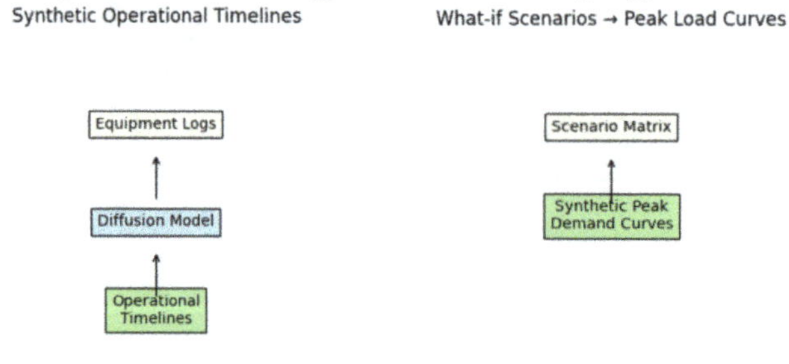

With an embedded synthetic data of such an experiment, via a digital twin of the district, the city planners can forecast the total load under heatwaves, time off, or energy pricing changes. Such a type of generative simulation-based framework facilitates the proactive grid design, infrastructure investment planning, and making forecasts about emergency responses without revealing any confidential information of users and without having to conduct time-consuming monitoring.

ENABLING CAPABILITIES AND TECHNICAL CONSIDERATIONS

The effective use of generative AI in energy systems must be based on a solid background of data, computer resources, and the proper integration of systems. Generative models have the potential to do much when it comes to simulation and forecasting, but their results follow directly from the input data quality and accessibility, training processes, and general knowledge of wider ethical and procedural limitations. The following section describes the principal technical enablers or considerations involved in the deployment of generative AI in energy analytics responsibly and successfully.

Sources and Preprocessing of Data

Data Sources

Data sources are the avatar of the generative AI applications of the energy systems; they are the primary input in training, simulation, and inference of a model. All of these data sets cover a wide variety of power generation, usage, and management, and they all play a vital role in the learning of the intricate energy consumption patterns in residential, commercial, and industrial settings. The width, clarity, and stability of each stream of data have a direct relation to the accuracy and verisimilitude of the synthetic generation (output) of AI systems.

The most commonly found data sources in the residential and commercial spheres are smart meters, which include high-frequency usage data of electricity consumption, typically every 15 minutes or an hour. This is time-series data that records fine details of the consumption dynamics shaped by the actions of occupants and appliances' behavior, as well as the weather conditions. The latter are IoT sensors that collect environmental and contextual data (e.g., temperature and humidity levels in a building, motion, daylight activity, and plug-load). These sensors make it possible to model those behaviors, occupancy-based control of HVAC, or load estimation of the appliances. Building Automation Systems (BAS), being common in business premises, can provide a good source of data as they collect and monitor the real-time effects of HVAC systems, lighting, elevators, access controls, and fire safety devices. Such systems facilitate closed-loop control and have structured tagged logs, which generative models can learn to mimic how the building operates given different occupancy or weather conditions.

In industrial settings, data are made even more specific and small. Measurements on machine operations, power, pressure, and flow conditions are captured in a synchronized manner by Supervisory Control and Data Acquisition (SCADA) apparatus. Also, there are machine logs that are comprehensive on production cycles and also examine the health of equipment. Even further to this layer are thermal imaging sensors and infrared cameras, which can be used to track thermal signatures to check predictive maintenance, or efficiency trends, thus useful in learning thermal load profiles under different operating conditions.

It is at the grid end that the utility telemetry and control systems generate streams of data that describe the load flows, frequency deviations, voltages, and faults, as well as switchgear statuses of transmission and distribution networks. Such data can provide vital system-scale context to generative models, so that they can simulate, e.g., localized outages, emergency load balancing systems, or renewable intermittency. Combined, these diverse data sources, including device-level logging and global system-level telemetry data, allow generative AI to extract micro-level and macro-level energy behavior. They do pose the issues of data synchronization, standardization, privacy, and completeness that need to be overcome with the help of sound data preprocessing, feature engineering, and metadata enrichment pipelines before training reliable and interpretable generative models. Table 4 outlines major data sources categorized by sector, highlighting the type of information they provide and their relevance to model training.

Table 4. Key data sources across energy domains

Sector	Data Source	Data Type	Temporal Resolution	Purpose of Generative AI
Residential	Smart Meters	Electricity usage (kWh)	15-min to hourly	Behavioral pattern generation
Residential	IoT Sensors	Temp, humidity, motion	Seconds to minutes	Occupancy and environment modeling
Commercial	BAS Logs	HVAC, lighting, elevator status	Event-driven or timed	Operational scenario simulation
Industrial	SCADA/PLC Systems	Machine operation, energy, and flow	Sub-second to minutes	Equipment-specific load pattern synthesis
Grid-Level	Utility Telemetry	Voltage, frequency, faults	Real-time or 5-min	System-level simulation and anomaly testing

Data Preprocessing

Preprocessing the data is a decisive and underemphasized part of the process of developing generative models using AI in the energy sector, as most raw datasets in this case are messy, contain gaps, and vary in their format. These problems occur in several ways because of multiple conditions, including sensory errors, delays in communication, inconsistent sampling rates, and inappropriate time syncs, among others, which may significantly compromise model performance when not addressed. Efficient preprocessing is the key to converting raw input into clean, logical, and structured data, and by doing so, the generative models learn definitive and generative patterns as opposed to spurious correlations and outliers. Data cleaning is often the first step in the preprocessing pipeline, including, but not limited to, missing values, spikes, dropouts, and erroneous readings that are extremely common with smart meter, IoT, or SCADA data. Such techniques as forward or backward filling, interpolation, and replacement of outliers are used depending on the time resolution and the importance of the affected variable.

The normalization and scaling operations are then used to put functions that characterize certain features, like power consumption, temperature, or voltage, into the same numeric range, stabilizing the training process and speeding convergence. There are common methods such as the min-max scaling, z-score standardization, or log transformation, depending on the distribution and the skew of the data. A significant process in energy analytics is the synchronization of time, particularly during data fusion of two or more streams that are gathered at different time-stamps-e.g., hourly vs. 15-minute meter-values, or 5-second SCADA records. Resampling, aggregation, or interpolation is the process that guarantees the synchronization of data points to a homogeneous time base appropriate to the sequential modeling technique. Time zone implications and daylight saving time rectifications are also key in harmonizing the behavioral and operational data adequately.

The role of feature engineering is to take raw variables and turn them into something more informative. As an example, rather than only being a raw temperature value, features could be derived from the degree-days, temperature deltas, or a rolling value of heat indexes. In energy data, load factors, ramp rates, daily peaks, occupancy flags, or weather-lagged variables are examples of engineered features that can have more predictive power than the raw inputs. These functionalities have the potential to make a substantial improvement in the capability of a generative model to represent seasonal, behavioral, and contextual changes in the energy use patterns. Preprocessing prevents overfitting by decorrelating with clutter, brings about stable and generalizable representations to generative models by ensuring clean, aligned, and contextually rich inputs to generative models, thus helping ensure the realism and diversity of the generated images. Preprocessing pipelines that have been designed well are not just essential to the performance of the model but also to its posterior interpretability and practical deployment.

Data enrichment is used to supplement artificial data generation with either external or contextual information to make it more realistic and relevant. Energy

Enrichment In energy modeling, one may enrich with weather information (such as temperature, humidity, solar radiation), energy prices, calendar of holidays, occupancy overages, etc. These options can enhance the capabilities of the generative models to incorporate real-world factors like heating demand on hot days or decreasing consumption due to weekends. Including these kinds of variables in training can result in models that have a better chance of satisfying not only statistical accuracy but also behavioral and environmental verisimilitude in the synthetically produced output. Table 5 summarizes common techniques used for energy-related generative modeling.

Table 5. Common preprocessing techniques and their functions

Preprocessing Step	**Example Techniques**	**Purpose**
Cleaning	Outlier removal, forward/backward filling	Handle missing/erroneous values
Normalization	Min-max scaling, z-score	Standardize feature ranges
Temporal Alignment	Resampling, interpolation, time zone shifting	Synchronize multi-source time series
Feature Engineering	Rolling means, occupancy flags, ramp rates	Enhance predictive power and variability
Data Enrichment	Weather integration, energy prices, calendar	Add context to improve behavioral realism

Model Training and Evaluation

Generative Model training Strategies are specialized and are not usually the same as Traditional Supervised learning, where the aim is to map the inputs to specific, known outputs. In generative modeling, it is the task of the model to learn the underlying distribution of a dataset in order to be able to produce completely new and at the same time realistic instances. Training depends on architecture. GANs adopt an adversarial form of training, which is a means by which a network generator models a network discriminator to deceive it into thinking that artificial data is real. The VAEs can be trained with variational inference, which optimizes a probabilistic latent space where one can sample new data. Diffusion models have introduced the idea of score matching of denoising and successively learn to reverse a process of noising given random noise as inputs, rendering new samples. Figure 5 summarizes the core training logic behind GANs, VAEs, and diffusion models.

Figure 5. Generative model training approaches

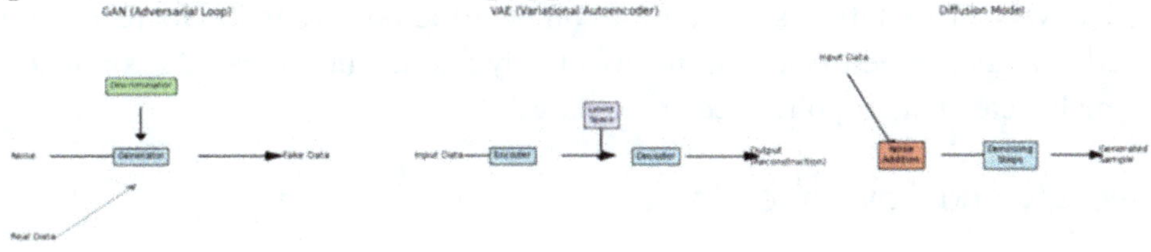

Assessment of the Generative Models

It is more complicated to evaluate generative models as compared to conventional predictive models since the latter usually have a single objective answer that it can compare to. Rather than comparing the predicted and actual values of observations, one now evaluates the similarity of the statistical properties of the simulated data to the characteristics of the real data. This demands indexed metrics that are more specialized than accuracy or the mean squared error and are capable of capturing the distributional faithfulness, diversity, and verisimilitude of the resulting output.

In general-purpose applications, especially in image and unstructured data, the Frechet Inception Distance (FID) is mostly used. FID calculates the distance between feature distributions (extracted using a pre-trained neural network) of samples (real and synthetic). FID can be used as a measure of perceptual similarity. However, in the case of structured environments such as energy systems, where time-dependent data is generally multivariate and subject to physical and behavioral constraints, alternative measures are more suitable.

To measure time series and structured tabular data, Kullback-Leibler Divergence (KLD) and Earth Mover (EMD) are popularly accepted. KLD is used to quantify the differences between two distributions of probability and, therefore, is appropriate in quantifying deviations in daily load profiles or occupancy patterns. EMD, also referred to as Wasserstein distance, can be especially valuable in describing the similarity between two temporal energy distributions or two spatial energy distributions, because within this distance, the measure of magnitude is weighed against the measure of position. Such measures ensure that synthetic data has feasible dynamics and variability over time.

In application-specific settings, specific criteria may be designed on which to base the evaluation process. Such could be the load shape similarity measures, to measure diurnal or seasonal curves of electricity consumption, the metrics to calculate deviations of the extreme usage cases, and the temporal consistency measures to ensure the existence of logical time-evolving transitions in the synthetic data. Collectively, these statistics constitute a complete toolbox of support for recognizing whether a generative model can be effectively applied in forecasting, simulation, anomaly detection, or policy scenario analysis.

Diversity and Coverage Metrics

Beyond fidelity and realism, generative models need to be assessed using the metrics of diversity and coverage in determining their effectiveness as well as generalizability. These measures assist us in answering the question of whether the model is memorizing the model or widening other possible subsets of the training set. Mode collapse occurs when a generative model only generates similar or even duplicate outputs and closes its eyes to the real variability of the natural world. In energy systems, this may lead to so-called synthetic data that fail to capture significant differences in usage modes, e.g., different occupancy profiles, appliance use habits, or seasonal load swings, and hence hobbles the applicability of the model in scenario planning and forecasting.

Diversity measures analyze the variety and uniqueness of samples being generated. In time-series energy data, this may involve comparison of statistical moments (variance, skewness) between generated sequences or quantifying the pairwise differences in the dimensional space through dynamic time warping (DTW) or Euclidean distances. Coverage metrics, in contrast, measure the ability of a generative model to replicate the variety of patterns seen in the training set. A plausible solution is to have the training data grouped into behavioral archetypes (e.g., the various types of residential energy users) and then find out whether the samples generated by the program represent all or as many as possible of these archetypes. When a few of them are recreated in a model, then their coverage is poor, even though individual samples are realistic.

To be systematic about such aspects, qualities like mode counting, nearest-neighbors, and precision-recall plots can be used in the case of generative models. Such approaches can discover the imbalances in the training and trigger the model architecture change, training goal change, or regularization technique change. As an example, mode collapse detectors can alert practitioners when the model begins to disregard fewer common patterns in databases in favor of common modes, thus providing the opportunity to correct the training process and achieve an increased degree of balance in the training.

Achieving a high level of diversity and coverage is of special importance in the realm of energy analytics because the final decisions are frequently made based on the learnings related to outlier behaviors, edge cases, and scenario extremes. A generative model that does not take into account the mean cannot help in applications such as peak load prediction, anomaly identification, or resilience design. Through the inclusion of intensive diversity and coverage checks, energy-centric generative models can be able to drive more sophisticated, diverse, and robust synthetic information, more in line with the sophistication of real energy systems. Table 6 highlights key metrics, whereas Figure 6 illustrates how models might underfit or capture various usage archetypes.

Table 6. Evaluation metrics for generative models in energy systems

Metric	Type	Application	Interpretation
Frechet Inception Distance (FID)	General	Image/complex data fidelity	Perceptual similarity
Kullback–Leibler Divergence (KLD)	Distribution	Time-series load or occupancy patterns	Divergence between real and synthetic data
Earth Mover's Distance (EMD)	Distribution	Load profile similarity	Measures the effort to match two distributions
Load Shape Similarity Index	Energy-specific	Seasonal/daily usage curves	Matches energy rhythm patterns
Peak Demand Deviation	Energy-specific	Peak load simulation	Captures extreme case accuracy
Temporal Pattern Consistency	Energy-specific	Behavioral realism	Logical time transitions are preserved

In application-specific applications, such as those that involve energy, there is a tendency to need tailored analytics to ascertain the viability of the derived data in practice. These are loading shape similarity, whereby one checks how well the synthetic loads profile corresponds with the daily or seasonal cycle of actual energy consumption; peak demand deviation, where one checks how well the model replicates extreme usage events; and finally, temporal pattern consistency, whereby one ensures that, the generated data retains the time dependency features that reflect on real life. These domain-specific scores enable the energy practitioner to trust and utilize generative outputs of interest in high-stakes scenarios such as demand prediction, anomaly detection, and simulation of policies.

Figure 6. Diversity and coverage in generative models

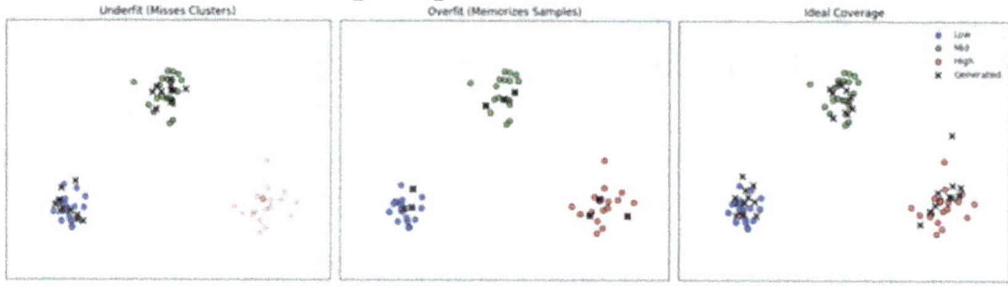

Integration with existing Energy Management Systems (EMS)

Installation of Energy Management Systems (EMS) is a core element in managing and maximizing energy consumption in residential, commercial, and industrial premises. Such systems are seen in smart home energy hubs, building management systems (BMS) of commercial buildings, and industrial control systems of manufacturing plants. They constantly observe energy usage and implement control actions (e.g., HVAC scheduling, load shifting) and react to external signals (e.g., energy price, grid conditions). EMS platforms become a working interface in which any AI-generated insights or actions are to be deployed.

To provide something actionable, generative AI must be easy to incorporate into Energy Management Systems (EMS), which are the operating backbone, monitoring, optimizing, and automating energy consumption at the residential, commercial, and industrial levels. Integration is not merely about generating synthetic data but rather making AI outputs a fully integrated part of the EMS decision-making loop, where it has the opportunity to affect both the real-time operations and long-term planning strategies. This intimate connection means that the knowledge that results from models is not separated from action, but instigates objective improvements in systems efficiency, responsiveness, and user satisfaction. Figure 7 illustrates how generative AI components integrate into modern EMS architectures.

Figure 7. Functional integration of generative AI within EMS architectures

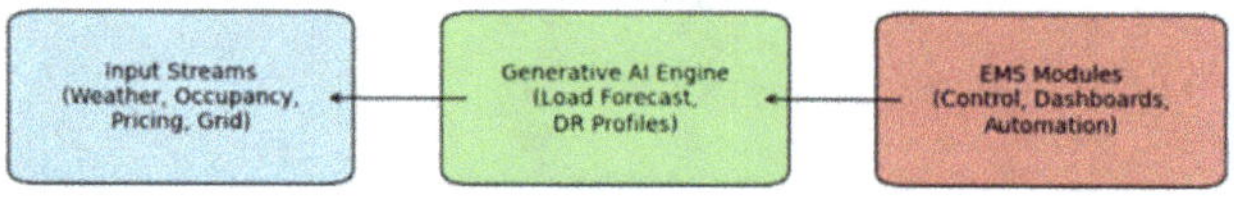

The common enabling outputs of generative models include synthetic electricity demand forecasts, which are used to guide expectations of future demand based on certain anticipated shifts in user behavior or conditions; simulated space occupancy schedules used to optimize heating, cooling, and ventilating in terms of likely occupancy patterns; putative anomaly reports, which are alerts that identify anomalous or non-normal behaviors based on pattern matching; and synthetic DR demand profiles, which are used to predict the responsiveness of users or buildings to pricing incentives or curtailment events. When they are incorporated into EMS dashboards, rule engines, or layers of control, they can automate or improve important energy decisions.

An example would be a commercial building with a Building Management System (BMS). A generative model could then simulate energy consumption depending on various occupancy values, such as office days with full attendance, hybrid working days, or weekend meetings, and produce a suggestion of what the lighting levels should be, and/or the ventilation rates, and/or the temperatures should be set for the following day. The EMS could then make recommendations, balancing the level of comfort with energy savings, or give recommendations to facility managers in an actionable way.

Technical considerations between different models must also be taken into consideration in the process of integration, i.e., how latent any one model may be, how compatible it can be with the currently used EMS protocols, and how either upgradable or retrainable it can be over time to reflect changing building dynamics, changing user behavior, or changing grid conditions. When implemented correctly, generative AI becomes an intelligence layer of real-time embedded in the EMS, making it more operationally responsive, catastrophically less wasteful of energy, and more adaptive and driven by data. Figure 8 summarizes three dominant integration patterns: via APIs, digital twins, and EMS dashboards.

Figure 8. AI deployment options in EMS: APIs, digital twins, dashboards

Integration Mechanisms

The best-known form of integration mechanisms of generative AI is quite different based on the systems' underlying infrastructure and the complexity of the system that the AI can be managing. In more contemporary EMS designs, namely cloud-native designs, generative models are typically operated as microservices that may be called using defined APIs. These APIs enable real-time data to be pushed into an EMS platform (e.g., weather forecasts, occupancy estimates, pricing signals) and to retrieve the generated results (e.g., synthetic load profiles, demand response scenarios) with low latency and at scale. The strategy makes it modular and flexible to update, to be deployed remotely, and to be easily integrated into bigger energy analytics processes.

In the more sophisticated or larger-scale systems, such as industrial plants, research campuses, or smart cities, generative models are becoming more and more part of the digital twin. These are web-based, virtual models of physical assets and energy systems, and provide an indication of real-time behavior under different operating conditions. The addition of generative AI to digital twins enables scenario-oriented forecasting, predictive maintenance, and control strategy verification in a safe and responsive virtual simulation. As an example, a digital twin of an industrial HVAC system may utilize a generative model to test under various conditions of airflow and thermal load under various machine schedules or weather patterns so that the engineers can optimize their control sequences without affecting real operations.

Also, dashboards and EMS portals, hosted in the cloud, can help act as significant points of contact when humans are in the loop with generative models. These interfaces are capable of displaying synthetic outputs of the project in the form of graphs, alerts, or heat maps. Visualization enables facility managers, grid operators, and energy analysts to process their interpretation, understand, and work on different alternatives, and achieve informed decision-making based on model-derived work. Table 7 categorizes typical outputs and their corresponding EMS functions.

Table 7. Common generative AI outputs and EMS applications

Generative AI Output	EMS Use Case	Example Action or Impact
Synthetic Load Forecasts	Load scheduling, DR planning	Adjust HVAC pre-cooling based on peak prediction
Occupancy-Driven HVAC Profiles	Comfort control and energy savings	Optimize temperature setpoints before arrival
Anomaly Detection Alerts	Fault detection, maintenance alerts	Notify the facility manager of unexpected load spikes
Demand Response Simulation	Pricing strategy, peak shaving	Model the impact of different DR signals on user behavior

Model Deployment

The possible deployment of generative AI in Energy Management Systems (EMS) would rely on a few technical considerations: one that is paramount is the latency level, the interpretability level, and the adaptability level. It is these variables that decide whether or not the generative models are able to not only produce quality results, but they cause them to operate in the real environment of energy, which is dynamic.

Latency is of utmost importance where real-time or near real-time decision-making systems are involved, including industrial process control, peak demand management, or building automation systems in the face of extreme weather. Generative models cannot afford to become overwhelmed by incoming data, e.g., changes in occupancy, changes in weather, or the grid signal, and result in delays to compromise the efficiency of the operations or lack of safety in the system. There are a variety of ways to optimize deployments that have low inference times, such as the use of efficient model architectures, edge computing, or hybrid cloud strategies that balance the use of computational resources, as long as the system is fast.

Interpretability is also a critical aspect to create user trust and decisions that can be acted upon. Generative models are often probabilistic or scenario-based, which can be hard to interpret in the absence of the right context compared to the more traditional deterministic models (Zheng et al., 2025). In the context of ensuring interpretability, the model outputs ought to feature confidence intervals, explainable features, or scenario rationales that can help the users determine why a certain synthetic forecast, anomaly, or behavioral simulation has been produced. Graphical explanations and simple statistics bear the potential to translate complicated outputs into valuable information to be used by the energy managers, regulators, or operators.

Adaptability also means that the generative models can be applied as energy landscape changes. Consumption habits of energy may change because of some new technologies (e.g., EVs or heat pumps), policies (e.g., carbon pricing), or even behavioral patterns (e.g., remote work). Generative AI systems need to be able to constantly learn so that they can be actively retrained on new data as well as integrate user input and track the performance change over time. This involves the creation of powerful data pipelines, retraining policies, and validations that keep the models true, ethical, and in line with objective shifts.

Collectively, these three pillars, low latency, high interpretability, and continuous adaptability, are the cornerstones of deploying generative AI systems with the twofold benefit that not only are they technically sound, but also operationally reliable and strategically useful in different scenarios of energy management. Table 8 outlines the main technical attributes and strategies to fulfill them.

Table 8. Technical requirements for real-time AI-EMS deployment

Requirement	Description	Mitigation Strategy
Low Latency	Fast inference under real-time conditions	Use edge inference, lightweight models
High Interpretability	Clear and actionable outputs for decision-makers	Use scenario rationales, explainability tools
Adaptability	Ability to evolve with system changes	Enable model retraining, performance drift monitoring

Ethical, Legal, and Data Privacy Considerations

Ethical, legal, and data privacy issues are a focus related to the ethical use of generative AI in energy systems when those models are built at the user or device level with energy data.

Data Privacy Concerns

Although generative models can now enable us to generate synthetic datasets and lessen the dependency on the raw, personally identifiable data, it is not safe from the privacy threat (Alshantti et al., 2025). Among the most relevant issues is the risk of data leakage or re-identification, as there are chances that models memorize and recreate rare or often unique patterns that appear in training data. In energy systems, this may include some sort of distinct appliance use schedules, building-specific load profiles, or temporal consumption patterns that, when re-generated, can be traced back to individual homes, businesses, or even essential infrastructure. This risk is especially high in small datasets or ones that include edge-case behavior that has an outsize impact on the model learning process.

To counteract such risks, one needs to implement the privacy in machine learning during the model design process. Differential privacy adds controlled statistical noise to the learning procedure, making the effects of individual users or data examples imperceptibly small to make such re-identification unlikely or impossible. Data anonymization entails the removal of personally identifiable information, including location tags, meter IDs, or time-specific data that would still make it possible to use the data to train a machine learning model. In the meantime, federated learning provides a solution that is decentralized, reversing the idea that raw data is stored in the cloud and only updates are transferred back to the cloud. Such architecture minimizes exposure to sensitive information drastically and is specifically applicable in the distributed energy where privacy is of utmost concern. Figure 9 illustrates a layered defense model combining technical and architectural safeguards.

Figure 9. Privacy-preserving techniques for generative AI in EMS

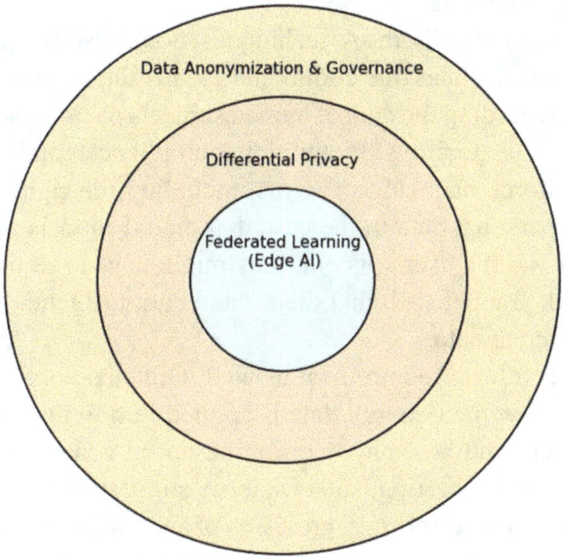

Together, these tools establish a multi-layer defense against privacy breaches, allowing each person and organization to be safe even against very advanced generative models. Technical countermeasures should, however, be complemented with transparent governance practices, such as transparent data consent processes, examples of auditing training processes, and post-generation checking of information flow leakage. It is only with the balance of technical precision and ethical guidance that generative AI can be used safely in privacy-conscious applications, such as energy analytics.

Legal Concerns

Applications of generative AI in the energy system must be legally compliant, and this aspect is only becoming more apparent as energy data, generally, can now be classified as a form of personally sensitive information. This information can be quite invasive of user privacy at the user level because data collected by smart meters, connected appliances, or building management systems can refine user behavioral patterns to the extent of occupancy patterns, appliance usage patterns, or lifestyle habits, which is regulated under national and international data privacy regulations. Nowadays, even legal frameworks like the General Data Protection Regulation (GDPR) in the European Union, California Consumer Privacy Act (CCPA) in the United States, or country-specific energy data protection regulations, such as in the

UK, Pakistan, or Australia, provide clear guidelines on how data processing must be done regarding energy data.

In order to apply genetically, the AI techniques should also 100 percent correspond to these legal provisions, and this commonly entails the need to seek explicit and informed consent regarding the data of human subjects prior to their use, even in the context of synthetic generations. In addition, compliance entails the maintenance of an ability to enforce data subject rights, including the right to access, rectify, restrict, or delete personal data in the case that the AI models are considered. An example of this is when a user opts out, they might have to re-train or make some other changes to the model such that their data is not influential, especially in the case of small or unique data.

Transparency is a legal requirement as well. Organizations need to offer clear documentation of how their energy data is applied in training, how synthetic productions are created, and how any decision based on generative models is made (such as pricing, load forecasting, and efficiency suggestions). This entails making privacy impact assessment websites, provision of audit logs, and spreading the word about the intent and consequences of data usage in a language understood by the non-technical audience. Non-compliance is not only jeopardized by legal charges, but this increases the risk of losing the trust and confidence of the people and stakeholders. Table 9 summarizes the key legal and ethical obligations.

Table 9. Legal and ethical compliance matrix for generative AI

Compliance Area	Required Practice	AI Implementation Strategy	Example
Informed Consent	User data usage transparency	Consent forms, opt-in settings	Smart meter data
Data Subject Rights	Access, rectification, deletion	Auditable training logs, data removal protocols	GDPR compliance
Transparency	Clear disclosure of model assumptions	Human-readable model summaries, impact reports	DR simulation
Fairness	Avoiding systemic bias in synthetic data outputs	Bias audits, representation balancing in datasets	Energy tariffs

Ethical Concerns

Ethical considerations in the implementation of generative AI are not as narrow as regulatory adherence but rather impact societal implications of generative AI, stakeholder equity, and the alignment of values of an AI-driven energy system. Although legal frameworks establish the minimum acceptable standards of responsible data utilization, ethical design needs to encourage developers, utilities, and policymakers to anticipate the issues and impacts of generative models and address them through ethical design. Transparency is one of the most significant ethical principles that the user, energy consumers, and decision-makers should be able to comprehend both the outputs of generative models and how they are generated, interpreted, operationalized, or applied to the policy.

This is particularly critical in cases where generative models are employed to drive high-stake decisions, such as electricity pricing schemes, load control procedures, or individual energy-saving suggestions. To take an example, when a utility is using generative model-based simulations of how customers would react to a set of tariffs to design the tariffs, it can be required to publish the underlying assumptions in such a simulation, the representativeness of the synthetically created data, and the possible biases in the data used to train generative models. In the absence of such transparency, automated decisions pose a credible threat of inadvertently perpetrating current disparities, e.g, due to a higher settlement rate or lesser access to efficiency schemes among the underfunded or low-income communities.

Moreover, ethical deployment is all about making sure that generative AI systems can be audited, held to account, and are fair. This involves making algorithmic impact assessments, determining and preventing bias in training sets due to representation, and incorporating a variety of stakeholders into the model development and verification activities. Ethical AI ought to also uphold human agency so that end-users are entitled to know, query, and disengage with AI-delivered recommendations or control actions when it is reasonable.

CONLCUSION

Generative AI is a revolution because it is going to change the way we model, simulate, and manage energy use in various sectors. In contrast to the traditional predictive models, generative methods allow modeling complex, stochastic, and unobservable situations, which are absolutely necessary in the rapidly uncertain world of energy that is characterized by climate change, behavior variability, and technology transformation. They have the chance to open the doors to adaptive, personalized, and sustainable energy solutions as they grow up and become coupled to actual energy infrastructures. It is the last part that addresses the wider implications, the horizon of the opportunities, and the strategic necessities of generative AI in energy systems.

Generative AI has enormous potential in climate-friendly energy planning. Since extreme weather events, heatwaves, and seasonal variability have escalated pressure on energy grids, the capability of simulating different climate scenarios becomes crucial. Generative models have the capability to generate artificial energy demand series in the future of projected climates that allow utilities and policymakers to stress test infrastructure resilience, HVAC or cooling load forecasts, and renewable integration strategies under different solar and wind conditions. The incorporation of generative models in climate-energy models enables planners to consider adaptive courses that are in alignment with decarbonization objectives, the energy weaknesses of the locations, and long-term sustainability goals.

Another frontier is personalized energy-saving interventions on a scale. Utilizing this kind of user-specific information, combined with smart meters, networked devices, and the like, utilities can leave behind the one-size-fits-all recommendations. Generative models can synthesize household or business profiles as a result of behavior archetypes, making it feasible to simulate with hyper-personalization in line with occupant habits, appliance usage, or reaction to pricing plans. It has the capacity to run mass customization of energy reports, instantaneous efficiency advice, or demand response incentives, in a context-sensitive and privacy-respecting manner. Eventually, this sort of personalization would potentially lead to increased participation, improved energy literacy, and a quantifiable decrease in usage among user groups.

Along with the promise comes a major policy and governance quandary posed by the emergence of AI-driven power systems. The question of algorithmic visibilities, responsibility, and data sovereignty becomes acute when generative models start taking over the shape of tariffs, load balancing policies, or investments. Such regulatory agencies may have to take up the issues of validating the generative output, the possible ethical and legal implications of the use of synthetic data, and the need that AI-related decision processes to be disclosed to customers by the utilities. Also, the digital divide has the potential to create an even bigger rift between developed and developing economies through insufficient access to high-quality data and computational resources unless policy frameworks actively encourage equitable rights to access, open protocols, and partnerships between the public and the private sector.

The suggestions for researching students, utilities, and regulators are also essential to create responsible progress in this area. As a researcher, the focus needs to be on creating domain-specific evaluation metrics, explanatory generative ventures, and modeling pipelines that provide a privacy commitment to energy applications. On the pilot front, utilities should seek to pilot narrow, high-stakes use cases, including HVAC scheduling, peak load forecast, or DR simulation, and should develop organizational capacity to interpret and act on generative results. Regulators, on their part, are assigned the responsibility of designing future-facing guidelines that shall be based on the promotion of transparency, fairness, and interoperability, such as a set of definite standards for the validation of artificial data, the accountability of the result of AI-driven policy, and consumer protection. The implementation of generative AI to achieve all societal and environmental advantages in energy systems requires a well-organized, cross-disciplinary effort.

REFERENCES

Alizadegan, H., Rashidi Malki, B., Radmehr, A., Karimi, H., & Ilani, M. A. (2025). Comparative study of long short-term memory (LSTM), bidirectional LSTM, and traditional machine learning approaches for energy consumption prediction. *Energy Exploration & Exploitation*, *43*(1), 281–301. DOI: 10.1177/01445987241269496/ASSET/F0AB838A-2611-4F07-9B1F-1D2E7896E016/ASSETS/IMAGES/LARGE/10.1177_01445987241269496-FIG13.JPG

Alqahtani, H., Kavakli-Thorne, M., & Kumar, G. (2021). Applications of Generative Adversarial Networks (GANs): An Updated Review. *Archives of Computational Methods in Engineering*, *28*(2). Advance online publication. DOI: 10.1007/s11831-019-09388-y

Alshantti, A., Rasheed, A., & Westad, F. (2025). Privacy Re-Identification Attacks on Tabular GANs. *Security and Privacy*, *8*(1), e469. DOI: 10.1002/SPY2.469

Asperti, A., Evangelista, D., & Loli Piccolomini, E. (2021). A Survey on Variational Autoencoders from a Green AI Perspective. *SN Computer Science*, *2*(4). Advance online publication. DOI: 10.1007/s42979-021-00702-9

Bengesi, S., El-Sayed, H., Sarker, M. K., Houkpati, Y., Irungu, J., & Oladunni, T. (2024). *Advancements in Generative AI: A Comprehensive Review of GANs, GPT, Autoencoders, Diffusion Model, and Transformers*. IEEE., DOI: 10.1109/ACCESS.2024.3397775

Biswas, A., Md Abdullah Al, N., Imran, A., Sejuty, A. T., Fairooz, F., Puppala, S., & Talukder, S. (2023). Generative Adversarial Networks for Data Augmentation. In *Data Driven Approaches on Medical Imaging*. DOI: 10.1007/978-3-031-47772-0_8

Broomandi, P., Bagheri, M., Fard, A. M., Fathian, A., Abdoli, M., Roshani, A., Shafiei, S., Leuchner, M., & Kim, J. R. (2025). Energy Generation and Carbon Footprint under Future Projections (2022–2100) of Central Asian Temperature Extremes. *Global Challenges (Hoboken, NJ)*, *9*(5), 2400356. DOI: 10.1002/GCH2.202400356;SUBPAGE:STRING:FULL PMID: 40352636

Cao, H., Tan, C., Gao, Z., Xu, Y., Chen, G., Heng, P. A., & Li, S. Z. (2024). A Survey on Generative Diffusion Models. *IEEE Transactions on Knowledge and Data Engineering*, *36*(7). Advance online publication. DOI: 10.1109/TKDE.2024.3361474

Chen, B., Zhu, X., & Díaz del Castillo, H., F. (2023). Integrating generative AI in knowledge building. *Computers and Education: Artificial Intelligence*, *5*. Advance online publication. DOI: 10.1016/j.caeai.2023.100184

Chen, D., Lin, X., & Qiao, Y. (2025). Perspectives for artificial intelligence in sustainable energy systems. *Energy, 318*, 134711. DOI: 10.1016/J.ENERGY.2025.134711

Croitoru, F. A., Hondru, V., Ionescu, R. T., & Shah, M. (2023). Diffusion Models in Vision: A Survey. *IEEE Transactions on Pattern Analysis and Machine Intelligence, 45*(9). Advance online publication. DOI: 10.1109/TPAMI.2023.3261988 PMID: 37030794

El Saddik, A., Ahmad, J., Khan, M., Abouzahir, S., & Gueaieb, W. (2025). Unleashing Creativity in the Metaverse: Generative AI and Multimodal Content. *ACM Transactions on Multimedia Computing Communications and Applications*. Advance online publication. DOI: 10.1145/3713075

Glaws, A., King, R. N., Emami, P., Buster, G., Benton, B. N., Zhang, X., Zamzam, A., Venkataramanan, V., & Macwan, R. (2025). Designing Future Energy Systems with Generative AI. *Computing in Science & Engineering*. Advance online publication. DOI: 10.1109/MCSE.2025.3567208

Li, D., Qi, Z., Zhou, Y., & Elchalakani, M. (2025). Machine Learning Applications in Building Energy Systems: Review and Prospects. *Buildings 2025, Vol. 15, Page 648, 15*(4), 648. DOI: 10.3390/BUILDINGS15040648

Lin, Y., Yao, Y., Zhu, J., & He, C. (2025). Application of Generative AI in Predictive Analysis of Urban Energy Distribution and Traffic Congestion in Smart Cities. *2025 IEEE International Conference on Electronics, Energy Systems and Power Engineering, EESPE 2025*, 765–768. DOI: 10.1109/EESPE63401.2025.10987500

Pan, Z., Yu, W., Yi, X., Khan, A., Yuan, F., & Zheng, Y. (2019). Recent Progress on Generative Adversarial Networks (GANs): A Survey. *IEEE Access : Practical Innovations, Open Solutions, 7*. Advance online publication. DOI: 10.1109/ACCESS.2019.2905015

Papadopoulos, D., & Karalis, V. D. (2023). Variational Autoencoders for Data Augmentation in Clinical Studies. *Applied Sciences (Switzerland), 13*(15). Advance online publication. DOI: 10.3390/app13158793

Rajput, S., & Pathak, R. K. (2025). Charting the Course for Energy Transformation. *Seismic Exploration to Reservoir Excellence*, 3–72. DOI: 10.1007/978-981-96-1293-2_1

Ramdurai, B., & Adhithya, P. (2023). The Impact, Advancements and Applications of Generative AI. *International Journal on Computer Science and Engineering, 10*(6). Advance online publication. DOI: 10.14445/23488387/ijcse-v10i6p101

Sarmas, E., Marinakis, V., & Doukas, H. (2025). *Future Directions of Intelligent Energy Management and the Role of Generative AI*. 257–266. DOI: 10.1007/978-3-031-85209-1_10

Saxena, D., & Cao, J. (2021). Generative Adversarial Networks (GANs). In *ACM Computing Surveys* (Vol. 54, Issue 3). DOI: 10.1145/3446374

Sharma, R., Singh, B., & Khamparia, A. (2024). Machine Learning and Generative AI Techniques for Sentiment Analysis with Applications. *Generative Artificial Intelligence for Biomedical and Smart Health Informatics*, 183–208. DOI: 10.1002/9781394280735.CH10;PAGE:STRING:ARTICLE/CHAPTER

Tahat, D. N., Mansoori, A., Tahat, K., Alfaisal, R., Yousuf, H., & Salloum, S. A. (2025). Integrating Generative AI into Predictive Modeling for Energy Efficiency Optimization in Building Design. *Studies in Computational Intelligence*, *1208*, 577–591. DOI: 10.1007/978-3-031-89175-5_36

Wei, R., Garcia, C., El-Sayed, A., Peterson, V., & Mahmood, A. (2020). Variations in Variational Autoencoders - A Comparative Evaluation. *IEEE Access : Practical Innovations, Open Solutions*, *8*. Advance online publication. DOI: 10.1109/ACCESS.2020.3018151

Wei, R., & Mahmood, A. (2021). Recent Advances in Variational Autoencoders with Representation Learning for Biomedical Informatics: A Survey. *IEEE Access : Practical Innovations, Open Solutions*, *9*. Advance online publication. DOI: 10.1109/ACCESS.2020.3048309

Yang, L., Zhang, Z., Song, Y., Hong, S., Xu, R., Zhao, Y., Zhang, W., Cui, B., & Yang, M. H. (2024). Diffusion Models: A Comprehensive Survey of Methods and Applications. *ACM Computing Surveys*, *56*(4). Advance online publication. DOI: 10.1145/3626235

Yoon, J., Drumright, L. N., & Van Der Schaar, M. (2020). Anonymization through data synthesis using generative adversarial networks (ADS-GAN). *IEEE Journal of Biomedical and Health Informatics*, *24*(8). Advance online publication. DOI: 10.1109/JBHI.2020.2980262 PMID: 32167919

Zheng, K., Sun, Z., Song, Y., Zhang, C., Zhang, C., Chang, F., Yang, D., & Fu, X. (2025). Stochastic Scenario Generation Methods for Uncertainty in Wind and Photovoltaic Power Outputs: A Comprehensive Review. *Energies 2025, Vol. 18, Page 503*, *18*(3), 503. DOI: 10.3390/EN18030503

Chapter 8
Generative AI for Energy Consumption Behavior Analysis

P. Selvakumar
https://orcid.org/0000-0002-3650-4548
Department of Science and Humanities, Nehru Institute of Technology, Coimbatore, India

K. Sundara Velrani
https://orcid.org/0000-0002-7727-7337
Sathyabama Institute of Science and Technology, India

Manjunath T. C.
https://orcid.org/0000-0003-2545-9160
Rajarajeswari College of Engineering, India

M. R. Arun
https://orcid.org/0000-0002-8291-2956
Vel Tech Rangarajan Dr. Sagunthala R&D Institute of Science and Technology, India

F. Shabina Fred Rishma
Vel Tech Rangarajan Dr. Sagunthala R&D Institute of Science and Technology, India

ABSTRACT

As the world energy has become a focal point for researchers, policymakers, and businesses alike. Energy, environmental awareness, technological access, and the policies implemented by governments. it is essential to have an in-depth under-

DOI: 10.4018/979-8-3373-4272-6.ch008

standing of the behaviors and motivations that drive consumption.Historically, energy consumption behavior has been studied from an economic and psychological perspective, focusing, machine learning, and predictive modeling, the landscape of energy behavior analysis has shifted. AI has opened up new avenues for collecting, processing,imperative to address climate change. As the world faces challenges such as rising energy costs, environmental concerns, and the pressure to transition to renewable energy sources, understanding how individuals, organizations, and industries consume energy has become a focal point for researchers, policymakers, and businesses alike.

UNDERSTANDING ENERGY CONSUMPTION BEHAVIOR

As the world energy has become a focal point for researchers, policymakers, and businesses alike. Energy, environmental awareness, technological access, and the policies implemented by governments. it is essential to have an in-depth understanding of the behaviors and motivations that drive consumption. Historically, energy consumption behavior has been studied from an economic and psychological perspective, focusing, machine learning, and predictive modeling, the landscape of energy behavior analysis has shifted. AI has opened up new avenues for collecting, processing, imperative to address climate change. As the world faces challenges such as rising energy costs, environmental concerns, and the pressure to transition to renewable energy sources, understanding how individuals, organizations, and industries consume energy has become a focal point for researchers, policymakers, and businesses alike. Energy consumption patterns are influenced by various factors including socioeconomic status, cultural beliefs, environmental awareness, Historically, energy consumption behavior has been studied from an economic and psychological perspective, focusing on how price signals, incentives, and information campaigns influence consumer choices. systems. These advancements allow researchers and energy providers to gain a more nuanced understanding of how energy is consumed at both the macro and micro levels. Through AI-powered tools, it is now possible to track energy usage patterns in real-time, predict future consumption trends, and even personalize recommendations to encourage more sustainable energy habits.is particularly valuable for utility companies, which can use AI-driven analytics to optimize grid management, predict future demand, and reduce energy waste. By anticipating and responding to changes in consumption patterns in real-time, energy providers can better align supply and demand, minimizing the need for emergency measures like blackouts or the use of costly, carbon-intensive energy sources. In addition to its predictive capabilities, AI also plays a key role in personalizing energy consumption recommendations for individual consumers.

As more homes and businesses adopt demand. Furthermore, AI-driven feedback systems can motivate consumers to make more informed choices by providing real-time energy consumption data, gamified rewards, or social comparisons with similar households. On a broader scale, AI can support the development of smarter cities by optimizing energy usage across entire urban environments, (Agency et al., 2023). With the integration of AI into urban infrastructure, cities can monitor and manage energy consumption across transportation, buildings, and industrial sectors. Another critical, technological advancements, and economic shifts on energy use.

Policymakers and energy analysts can use AI-driven models to test the impact of different interventions before implementing them in real life. For example, AI can predict how a carbon tax or subsidies for renewable energy would influence consumer behavior, or how the introduction of electric vehicle incentives might change the dynamics of transportation-related energy consumption, (Ali et al., 2020). These predictive capabilities can provide valuable insights into how to design more effective energy policies, improve sustainability initiatives, and guide investments must be trained on high-quality, representative data. However, data privacy issues, as well as the digital divide that exists between different socio-economic groups, can limit the effectiveness of (Athavale et al., 2025). AI-based solutions. In conclusion, the role of to personalizing consumption habits for individuals and supporting the development of smarter cities, AI is reshaping the way we approach energy consumption behavior analysis will likely become an essential tool in creating a more sustainable, efficient, and equitable energy future

THE IMPORTANCE OF ENERGY EFFICIENCY: REDUCING CONSUMPTION AND EMISSIONS

AI is proving to be an invaluable asset in the analysis and management of energy consumption. Traditionally, the study of energy consumption behavior has relied on deterministic models that predict how energy is used based on historical data and fixed rules. However, the rise of generative AI introduces a new paradigm where dynamic, flexible, and adaptable models can be created to simulate, optimize, and predict energy consumption patterns in real time. This shift is allowing researchers, utility providers, and organizations to develop innovative approaches for reducing energy waste, improving efficiency, and promoting sustainability seeks to generate new instances that reflect the underlying and transformers, which are increasingly being applied to energy consumption analysis., test new strategies, and simulate various scenarios, such as changes in consumer behavior, weather conditions, or energy policy. By generating data that accounts for rare events or unusual behaviors that may not appear in historical data, generative AI can create more robust models that account for a wider range of potential outcomes. For example, GANs can be trained on existing energy usage data to produce realistic, synthetic data that can be used to simulate potential fluctuations in demand, helping utility companies plan better for peak periods or unexpected demand spikes. Energy grids are complex systems that require meet varying demand levels. (Bedi et al., 2018) In traditional grid management, energy providers use optimization algorithms to ensure that energy is distributed efficiently, minimizing losses and reducing costs. However, generative AI can enhance this process by generating new grid configurations or strategies that may not have been considered before. For example, generative models can propose novel ways to integrate renewable energy Generative off appliances when they are not in use. However, generative AI takes this a step further by learning from users' behaviors and preferences to make real-time decisions that optimize energy usage. For example, generative AI models can generate personalized energy-saving recommendations for users based on their specific habits, routines, and lifestyle patterns. This approach goes beyond simple automation to create a dynamic, evolving system that adapts to changing conditions. (Bose, 2017) For instance, a generative AI model in a smart home could learn a family's daily, becoming increasingly accurate at minimizing energy consumption.

Beyond individual homes and buildings, generative AI can also contribute to the optimization of energy usage in entire services. Generative AI can help design new urban energy systems by generating models that predict the most energy-efficient ways to manage everything from street lighting and traffic signals to public transportation routes and electric vehicle charging stations. For example, AI-driven traffic systems could generate optimized traffic flow patterns to reduce congestion and fuel consumption, while also improving the efficiency of electric buses or trains by predicting when and where they are most likely to be used. Generative AI can also simulate the impact of large-scale energy policies, such as the transition to electric vehicles or the widespread adoption of renewable energy, helping urban planners develop strategies that support sustainable growth and minimize energy waste statistical analyses of historical data, focusing on factors such as income, household size, and appliance use. While these models can provide useful insights, they often fail to capture the full complexity of human behavior, (Climate Change et al., 2022). Generative AI offers a more dynamic approach, allowing researchers to generate personalized, behavior-driven models of energy use. For example, AI can analyze vast amounts of data from smart meters, social media, and user feedback to predict how individuals or communities will respond to specific interventions, such as incentives for energy conservation or changes in pricing. These models can then be used to design more effective demand-side management strategies that encourage energy-saving behaviors, whether through gamification, social comparisons, or personalized feedback. The ability of generative, (Espinoza et al., 2005).

AI to simulate future scenarios is another key advantage in energy consumption analysis., economic shifts, and environmental changes. These simulations can be used to forecast long-term energy demand, predict the adoption rates of new energy technologies, investments, resource allocation, and sustainability efforts. Despite the many promising applications, there are challenges associated with the implementation of generative in regions with limited access to advanced computing resources. There is also the issue of transparency, as some generative AI models, particularly deep

GENERATIVE MODELS FOR ENERGY CONSUMPTION PATTERN ANALYSIS

Energy efficiency stands as a powerful tool in this quest, offering tangible benefits that extend from lowering operating costs to mitigating climate change and advancing energy security. Whether in residential buildings, industrial sectors, transportation, or energy generation, improving energy efficiency offers one of the most cost-effective and impactful strategies to achieve environmental sustainability, economic stability, and long-term social equity. At its core, energy efficiency involves using less energy to perform the same task or deliver the same service. insulation in buildings, or employing energy-efficient technologies in manufacturing and industrial operations, (Fan et al., 2014). For example, energy-efficient appliances consume less electricity, high-efficiency boilers use less fuel for heating, and electric vehicles use less energy per mile traveled compared to traditional gasoline-powered cars. In all these scenarios, the basic principle remains the same: energy is utilized more effectively, requiring less input to deliver the same level of output which in turn reduces costs. From a consumer standpoint, energy-efficient homes and businesses can lower energy bills, offering a direct economic benefit. For example, the implementation of better insulation, smart thermostats, energy-efficient lighting, and appliances can significantly cut down heating and cooling costs in buildings. Similarly, in the industrial and commercial sectors, energy efficiency measures—such as upgrading machinery, optimizing manufacturing processes, and reducing waste heat—can reduce operational costs, improve competitiveness, and enhance profit margins. In regions with high energy prices, the financial savings from energy efficiency improvements can be particularly substantial, benefiting both households and businesses. Moreover, reducing energy consumption has broader economic implications. By decreasing the demand for energy, energy efficiency measures reduce the overall pressure developing economies where energy demand is rising rapidly, and there is a need to expand energy infrastructure in a cost-effective manner. Additionally, energy-efficient technologies often have lower operational and maintenance costs over time, further reducing the long-term financial burden of energy use. However, the most critical benefit of improving energy efficiency goes beyond cost savings.

Energy efficiency is intrinsically linked to the reduction of fuels and the associated emissions from power plants, factories, and vehicles. cars, and shifting to public transportation can reduce oil consumption and significantly reduce emissions. Across all these areas, improved energy efficiency results in less energy production, which, in turn, reduces the burning of fossil fuels, leading to fewer emissions of CO2 and other harmful pollutants. Beyond reducing emissions, energy efficiency can also lead to significant improvements in AI, (Fan et al., 2017); it also contributes to social equity and quality of life. As energy costs rise, lower-income households are often disproportionately affected by high energy bills, which can consume a larger share of their income. By implementing energy-efficient measures in homes and buildings, lower-income families can reduce their energy bills, improving their overall standard of living and freeing up resources for other essential needs. Moreover, energy-efficient building programs can provide healthier living environments, with improved indoor air quality and more comfortable temperatures. In this sense, energy efficiency can contribute to addressing energy poverty and ensuring that all individuals have access to affordable, reliable, and sustainable energy services.to renewable energy. While the adoption of clean energy technologies, such as solar, wind, and hydropower, is crucial for reducing the carbon intensity of the global energy system, energy efficiency measures help to make this transition more feasible. By reducing overall energy demand, energy (Himeur et al., 2021) efficiency allows renewable energy sources to meet a larger share of the energy needs. In this way, energy efficiency and renewable energy are complementary strategies that work together to build a more sustainable, low-carbon energy future. It offers economic benefits by reducing operating costs, it mitigates climate change by reducing carbon emissions, and it promotes cleaner air and better public health by minimizing harmful pollutants. Furthermore, energy efficiency contributes to energy security, resilience, and social equity, improving the lives of individuals and communities around the world. To fully realize

AI-DRIVEN IDENTIFICATION OF ENERGY-SAVING OPPORTUNITIES

The increasing complexity and scale of modern energy systems have highlighted the need for advanced analytical techniques that can better understand, predict, and optimize energy consumption patterns., often struggle to capture the dynamic, nonlinear, and multifaceted nature of energy usage across different sectors and geographies. This is where generative that data that closely resemble real-world patterns. These models have shown systems, utilities, and consumers alike instances that adhere to the statistical properties of the original data. Unlike discriminative models, which focus on classifying or predicting outcomes based on input data, generative models aim to capture the underlying structure and relationships within the data itself. In the social behavior, and various other factors to create synthetic datasets that reflect real-world energy usage. The most commonly used generative models for such tasks include energy consumption profiles that mimic real-world behavior. (Hosein et al., 2017). These generated data sets are valuable for several applications, including testing energy management strategies, simulating how energy demand might evolve under different conditions, or creating training datasets for other machine learning algorithms. For instance, utility companies can use GAN-generated data to simulate consumption during peak hours or when new energy-saving technologies are introduced, allowing them to optimize grid operations and improve demand forecasting. Moreover, GANs can also be applied in scenarios where real-world data is limited or sparse. For example, in areas with low energy access or emerging markets where energy consumption data is not readily available, GANs can generate synthetic datasets based on minimal inputs and assumptions. This approach can help energy planners, researchers, and policymakers create models for regions with insufficient historical data, ensuring that energy demand predictions and infrastructure planning are not based on incomplete or outdated information. By enabling the generation of realistic energy consumption data, GANs help create more accurate and comprehensive models for energy systems that reflect the true dynamics of energy use. In energy consumption analysis, (Jin et al., 2021)VAEs can be used to model the latent factors that drive consumption patterns, such as household behavior, weather, time of day, and other environmental variables. By encoding these factors into a latent space, VAEs can learn a more compact representation of the underlying structure of energy usage, which can then be used to generate new, realistic energy consumption data. This is particularly useful for scenarios where energy usage is influenced by complex, high-dimensional factors. For instance, in residential energy consumption, VAEs can account for the interactions between multiple variables, such as appliance usage, weather conditions, and occupancy, to generate synthetic household consumption profiles. This allows energy providers and researchers to simulate a wide range of scenarios and test the effectiveness of different energy-saving strategies, demand response programs, or policy interventions with the generated samples, utilities can spot discrepancies that might indicate issues such as faulty equipment, fraudulent behavior, or underperforming energy-efficient technologies.

This ability to model and analyze deviations from typical consumption patterns can lead to better maintenance schedules, more accurate billing, and the ability to proactively address problems before they escalate transformers, which are particularly effective at modeling sequential data. In energy consumption analysis, this is particularly useful, as energy usage typically exhibits strong temporal patterns, with consumption varying based on time of day, season, and other cyclical factors. Autoregressive models can predict future energy consumption by learning the patterns and the energy consumption of a household will evolve throughout the day based on past behavior and environmental conditions such as temperature, occupancy, and appliance usage. These predictions can be used for energy scheduling, demand-side management, or creating personalized recommendations for consumers to reduce their energy usage during peak demand periods. Autoregressive models are also highly useful for predicting long-term trends in energy consumption., urbanization, technological adoption, and policy changes. These long-term predictions are crucial for energy infrastructure planning and for policymakers aiming to design effective energy policies that account for future demand scenarios. In many cases, high-quality, real-world energy consumption data is scarce or expensive to obtain, especially in emerging markets or in specific regions. Generative models can bridge this gap by producing realistic, synthetic data that reflects the complexities and nuances of real-world consumption, (Khan et al., 2016). These synthetic datasets can then be used to train predictive models, test energy management systems, or validate demand forecasting techniques. Another important application is in the optimization of energy systems. By generating diverse consumption scenarios, generative models can help energy companies and grid operators optimize energy distribution and reduce operational costs. For instance, utility companies can simulate how energy demand might evolve under various conditions, such as a change in pricing, the introduction of new energy-efficient technologies, or fluctuations in weather patterns. Generative models are also useful for personalized energy management. With the proliferation of smart home devices, generative models can analyze household energy usage and generate recommendations tailored to individual preferences and behaviors. Furthermore, generative models can support policy design and demand-side management programs by simulating how consumers might respond to changes in energy pricing, incentives, or regulations. By understanding how consumers are likely to adjust their behavior under different policy scenarios, policymakers can design more effective programs that encourage energy conservation, reduce peak demand, and promote sustainable energy practices.

PREDICTIVE ANALYTICS: FORECASTING ENERGY DEMAND AND CONSUMPTION

The global urgency mitigate environmental impact has given rise to innovative technologies aimed at optimizing energy use across residential, industrial, and commercial sectors. Among these technologies to uncover hidden inefficiencies, optimize energy usage, and reduce waste in ways that traditional methods. By leveraging vast amounts of real-time data and complex algorithms, AI can identify patterns, predict energy consumption, (Leitao et al., 2016). These capabilities allow AI systems to analyze energy consumption patterns, assess factors influencing energy use, and suggest ways to enhance efficiency in a highly personalized and scalable manner. As a result, AI is positioned to revolutionize the way energy efficiency is approached across various sectors, from smart homes and buildings to industrial manufacturing and grid management understand how energy is being consumed in real time. By continuously learning and adjusting its models based on new data, AI can detect anomalies, such as equipment running inefficiently, overuse of heating or cooling systems, or lighting being left on in unoccupied spaces. Such insights allow organizations or homeowners to take immediate action, such as adjusting settings, scheduling energy-intensive tasks, or replacing outdated equipment with more energy-efficient alternatives. This real-time feedback mechanism is invaluable in identifying areas where immediate energy savings can be achieved, and it also contributes to long-term optimization strategies such as HVAC systems, manufacturing machinery, and motors, can consume vast amounts of energy when running inefficiently. AI can monitor the performance of these systems through continuous data collection and analysis. By using predictive algorithms, AI can identify signs of inefficiency before they lead to significant performance degradation or failure. For instance, an AI system may detect unusual vibrations, temperature fluctuations, or electrical anomalies in a machine, which might indicate that the machine is operating below its optimal energy efficiency. Rather than waiting for a machine to break down or require costly repairs, acceptable levels. In such cases, AI can trigger a maintenance alert or suggest that adjustments be made to improve efficiency. (Zhou et al., 2013). This approach not only operates at peak performance, consuming only the amount of energy required for optimal functionality.

Figure 1. AI in energy

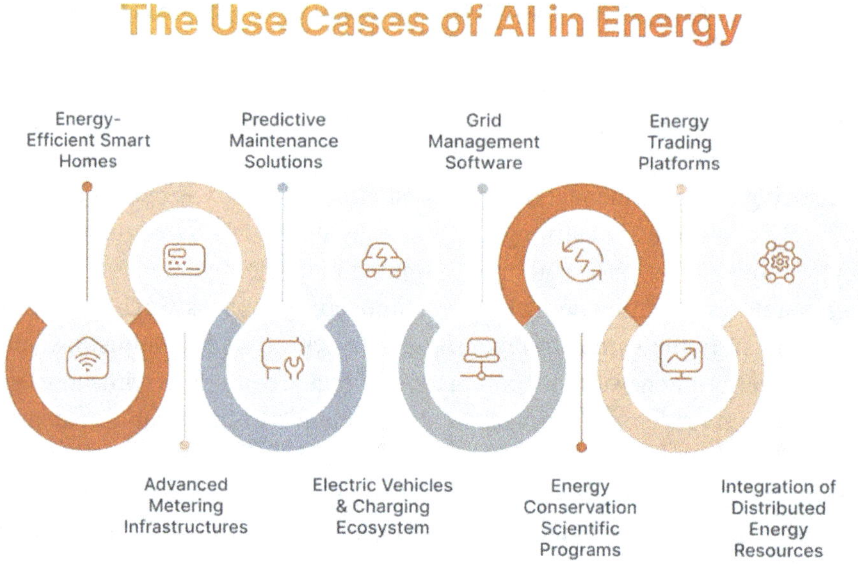

Furthermore, AI can aggregate insights from predictive maintenance across an entire organization's assets, enabling companies to implement energy-saving measures systematical grid optimization. The traditional energy grid relies heavily on centralized power generation and top-down control of energy distribution. However, this approach often leads to inefficiencies, especially during peak demand times when the grid is stressed and energy costs spike. AI can help optimize energy consumption by dynamically adjusting demand and balancing the grid in real-time appliances or shifting the use of heavy machinery to off-peak hours, all without compromising comfort or productivity. For example, AI systems can instruct thermostats in buildings to adjust temperatures slightly during high-demand periods, or in industrial settings, AI can delay energy-intensive processes until demand decreases. This AI-driven demand response not only helps consumers save on energy costs by shifting usage to more affordable times but also reduces strain on the grid, enabling utilities to lower their reliance on expensive, less-efficient peaking power plants, (Yu et al., 2014). By smoothing out the demand curve, to identify energy-saving opportunities is vast. Smart buildings, equipped with IoT sensors and connected devices, generate an immense amount of data related to energy usage, occupancy, temperature, lighting, and HVAC systems. AI algorithms can analyze this data in real time, identifying inefficiencies and providing actionable insights for optimizing energy use.For instance, AI can automatically adjust lighting based on occupancy data, ensuring that lights are only on when necessary and dimming or turning off lights in unoccupied rooms. Similarly, AI can optimize heating and cooling by adjusting temperatures based on the number of people in a room, the time of day, and external weather conditions. In commercial buildings, (Wang et al., 2020c). AI systems can also monitor and control energy-intensive processes, such as elevators, escalators, and ventilation systems, to reduce unnecessary energy consumption. Furthermore, AI can help optimize energy use across multiple buildings or an entire campus. By aggregating data from different buildings and analyzing it collectively, AI can provide a broader view of energy consumption trends, identify best practices, and suggest centralized energy-saving strategies. For example, AI may determine that consolidating heating or cooling efforts across multiple buildings in a campus leads to more efficient energy usage compared to treating each building separately.

PERSONALIZED ENERGY-SAVING RECOMMENDATIONS FOR CONSUMERS

For individual consumers, AI offers the potential for highly personalized energy-saving strategies. By analyzing a household's energy consumption patterns, AI can provide tailored recommendations for energy efficiency improvements. This may include suggestions for better appliance usage, insulation improvements, or changes in behavioral patterns, such as adjusting thermostats or unplugging unused devices. AI systems can also integrate with smart home devices to provide real-time recommendations that can automatically adjust settings to optimize energy use.AI-driven platforms can help households save on energy bills by tracking usage over time, identifying trends, and providing suggestions to avoid peak usage periods, (Wang et al., 2020b). For example, by tracking patterns of energy consumption over weeks or months, AI systems can highlight inefficiencies such as excessive heating or cooling, appliances that are frequently left on, or areas where insulation might be inadequate of accurate, high-quality data. However, in many regions or industries, this data may not be readily available, or it may be incomplete or unstructured. (Wang et al., 2020a) Overcoming these data limitations will be crucial for expanding the reach of AI-driven energy savings initiatives. Many older buildings, industrial facilities, and energy grids may not have the necessary sensors, smart devices, or data infrastructure to support AI-based optimization. As such, investment in digital infrastructure is essential to unlock the full have significant financial or operational consequences. (Vrablecova et al., 2018) Ensuring that AI systems can explain their reasoning in a transparent and understandable manner is essential for gaining stakeholder trust and encouraging adoption real-time adjustments will be a key factor in achieving long-term energy efficiency goals and optimize energy use across various sectors. Whether in homes, businesses, or industrial facilities, AI's ability to drive energy efficiency is poised to play a key role in reducing global energy consumption, minimizing costs, and mitigating environmental impacts. The future of energy-saving initiatives is undoubtedly intertwined with AI, offering a smart, scalable, and sustainable path toward more efficient energy systems.

UNDERSTANDING OCCUPANT BEHAVIOR: IMPACT ON ENERGY CONSUMPTION

energy resources are of paramount importance. As the world shifts towards sustainable energy practices and aims to meet growing demands, understanding and predicting energy consumption has become an essential task. Predictive analytics, a powerful tool within the field of data science, has emerged as one of the most promising solutions for accurately forecasting energy demand and consumption., governments, businesses, and consumers optimize energy supply, reduce wastage, and enhance the overall efficiency of energy systems and the impact of external variables such as weather conditions, economic activity, or demographic changes. By applying predictive models, energy stakeholders can anticipate energy needs more accurately, reduce overproduction, and ensure a more stable and sustainable supply of energy consumption forecasting are diverse and impactful. Below are several ways in which predictive analytics is transforming the energy industry efficiently and reliably to ensure that there is always enough power to meet demand without overloading the system. Predictive analytics plays a critical role in load forecasting, which involves predicting the amount of electricity that will be needed at any given time. These forecasts help utilities and grid operators make informed decisions about power generation and distribution. By analyzing historical events (e.g., heatwaves or cold spells), holidays, or special events. By Tuballa et al. (2016) anticipating these peaks, utilities can better prepare by adjusting the output from various power plants or leveraging demand response strategies to ensure grid stability. This type of forecasting minimizes the risk of power shortages, reduces the need for expensive, less efficient peaking power plants, and enables better integration of renewable energy sources, such as solar and wind, which can be intermittent in nature usage at an individual household or building level. With the proliferation of smart meters and Internet of Things (IoT) devices, (Thoubboron et al., 2021) large amounts of data are now available to track real-time energy consumption.

Predictive analytics can process this data to forecast future energy usage based on a variety of factors, such as time of day, weather conditions, occupancy, and appliance usage can provide personalized insights into how households are consuming energy and where inefficiencies exist. For example, predictive analytics can forecast when a household is likely to use more energy (e.g., in the evening when lights and appliances are on) and suggest actions to optimize energy consumption (e.g., adjusting the thermostat or scheduling the use of certain appliances during off-peak hours). These forecasts can also help utility companies offer demand-side management programs by incentivizing users to shift their consumption patterns or reduce energy use during peak hours occupancy levels, weather conditions, and energy demand, predictive models can dynamically adjust the building's energy usage to ensure it operates efficiently while maintaining comfort and productivity levels. (Li et al., 2021). This can lead to significant cost savings and reductions in energy consumption, particularly for buildings that operate 24/7, such as data centers, manufacturing plants, or office reliance on fossil fuels, and integrate renewable energy more effectively into the grid.to make data-driven decisions regarding infrastructure development, energy policies, and investments in renewable energy technologies engage with consumers, offering them incentives to reduce or shift their energy usage., encouraging them to lower thermostats or postpone energy-intensive activities. (Noor et al., 2018) By shifting the energy demand away from peak periods, these programs help prevent grid overloads, reduce the need for expensive peaking plants, and lower overall energy costs for consumers. Additionally, predictive analytics can help fine-tune the timing and scale of load-shifting programs, improving their effectiveness and customer satisfaction.in predicting and managing energy consumption. These sources are often intermittent, and their availability depends on factors such as sunlight, wind speeds, and seasonal changes.

BUILDING ENERGY MANAGEMENT: AI-DRIVEN INSIGHTS AND RECOMMENDATIONS

Energy consumption in buildings, whether residential, commercial, or industrial, is a multifaceted issue influenced by a variety of factors. While many of these factors are external, such as climate, energy prices, and the efficiency of heating, cooling, and lighting systems, one of the most significant and often overlooked determinants is occupant behavior. The way individuals interact with their environment—how they use energy, manage heating or cooling, lighting, and even their schedules—has By examining how different occupants behave within a building or space, we can uncover patterns and trends that allow for better energy management, enhanced efficiency, and more sustainable building operation. The role of many of the building's systems, such as lighting, heating, air conditioning, and appliances. (Omer, 2008) The manner in which they operate these systems—whether they leave lights on unnecessarily, set thermostats to extreme temperatures, or use inefficient appliances—can lead to significant energy waste. Indirectly, occupant behavior also impacts energy use through factors such as building occupancy patterns, usage schedules, and lifestyle choices, all of which determine occupants preferring warmer indoor conditions while others opt for cooler environments. In commercial and institutional buildings, these preferences can be even more diverse, with different departments or spaces requiring different comfort levels. Beyond just setting thermostats, occupant behavior related to how often windows are opened or how much natural ventilation is used can also influence energy consumption. For example, in a commercial building, employees might open windows to let in fresh air during the summer months, causing the air conditioning system to work harder to cool the space. Similarly, in colder months, opening windows or doors unnecessarily can cause heating systems to operate inefficiently, increasing energy consumption. Understanding these diverse patterns of behavior in relation to heating and cooling is crucial conditions, (Panwar et al., 2011). However, the effectiveness of these technologies still depends heavily on how occupants use them and how aware they are of their impact on energy consumption. has a substantial effect on energy consumption. Many buildings, particularly commercial spaces, are often over-lit, with lights left on in unoccupied rooms or hallways.

Even in residential settings, lights are sometimes used inefficiently, left on in empty rooms or during the daytime when natural light is abundant. In commercial environments, employees may forget to turn off lights when they leave their workstations, or lighting systems may be left running overnight or on weekends due to human oversight or negligence.to adjust lighting when necessary—this can mitigate their potential energy-saving benefits. Awareness of these behaviors is essential to reducing energy waste. Educational programs, user-friendly interfaces, and smart automation systems that actively encourage occupants to adopt energy-efficient lighting behaviors can be key to mitigating excess energy consumption. Additionally, the integration of energy-efficient, (Pradeep et al., 2025) LED lighting, which consumes less power than traditional incandescent bulbs, and the application of motion sensors or timers that automatically turn off lights when not in use, can contribute significantly to reducing overall energy consumption. For example, in residential settings, the timing of when household appliances are used can affect energy demand, especially during peak hours. Running washing machines or dishwashers during the day may lead to unnecessary energy costs, while doing so at night may be more cost-efficient if utility companies offer lower rates during off-peak times. In commercial settings, the energy consumption of office equipment like computers, copiers, and vending machines can add up over time if employees do not turn off machines when they are not in use. Moreover, the choice of appliance itself can also impact energy consumption. Energy-efficient appliances, like those labeled with the ENERGY STAR rating, use significantly less energy than older, less efficient models. However, even with energy-efficient appliances, how they are used can make a difference in total consumption. For instance, setting a washing machine to use higher water temperatures or using a dryer instead of air-drying clothes can lead to unnecessary energy consumption.as during working hours in office buildings or evenings in residential homes. However, occupancy patterns are often irregular, and these variations can lead to inefficiencies if buildings are not optimized to account for fluctuating levels of activity. For example, in commercial buildings, lights and HVAC systems might be left on throughout the day, even during hours when the building is only partially occupied. Similarly, in homes, heating and cooling systems may run throughout the day despite the fact that no one is present to benefit from the energy use only used when and where it is needed. However, even with such technologies in place, occupant behavior is still critical. Occupants who work irregular hours or maintain unconventional schedules can sometimes interfere with automated systems, overriding settings or leaving energy-consuming devices on when not required.

PERSONALIZED ENERGY RECOMMENDATIONS: AI-DRIVEN FEEDBACK

Energy consumption profiling is a vital tool for understanding, monitoring, and managing how energy is used across various systems and environments, ranging from residential homes to large commercial buildings and industrial facilities, (Qiu et al., 2017). The primary goal of energy consumption profiling is to create a detailed, accurate, and actionable picture of energy use patterns over time. By identifying anomalies and trends in consumption, businesses and individuals can make informed decisions regarding efficiency improvements, cost reductions, and sustainability efforts. In the modern context, energy consumption profiling goes beyond simply tracking usage—it enables proactive management, optimization, and strategic interventions aimed at reducing energy waste and lowering environmental impact. With advancements in data analytics, artificial intelligence (AI), and the Internet of Things (IoT), energy profiling has become increasingly sophisticated, offering real-time insights and the ability to identify both expected and unexpected patterns of consumption., and periods. This profile can provide insights into which areas of a building, facility, or household consume the most energy, at what times of day energy use peaks, and how this energy consumption correlates with external factors such as weather, occupancy, or business activities. Through this profiling, anomalies in energy use—such as sudden spikes in consumption or unexpected drops—can be identified, investigated, and addressed. Moreover, trends in energy consumption, such as seasonal patterns or shifts related to changes in building usage, can also be detected. By creating such profiles, energy managers, building operators, and homeowners gain a clearer understanding of where inefficiencies may exist and how energy consumption can be optimized to meet environmental consumption with specific activities or conditions within a building or space. These sensors can detect whether a room is occupied, the temperature within a space, or the lighting requirements at any given time energy systems. These systems collect data from HVAC units, lighting systems, refrigeration, and other major energy-consuming systems to create a holistic profile of energy usage data helps to understand broader consumption trends, utility billing cycles, and the impact of peak demand charges on overall energy costs., monthly) and across various systems (e.g., lighting, HVAC, machinery).-powered analytics tools can identify and highlight these recurring patterns automatically if a heating system suddenly begins to consume more energy than usual, it could be a sign of a malfunction or the need for maintenance. Similarly, if a building's energy consumption spikes at night when it is typically unoccupied, this could point to issues such as faulty lighting systems or unauthorized equipment use the energy data through dashboards, charts, and reports. Visualization tools allow energy managers, (Ritchie, 2019) building operators, and even homeowners to see energy trends and anomalies at a glance. These visual tools can highlight areas of high consumption, time periods of peak energy use, and areas where interventions could lead to cost savings or efficiency improvements or by individual device, enabling building managers to pinpoint which systems or areas are consuming the most energy.

Reports generated from these tools can help identify inefficiencies and allow users to track their progress over time as they implement energy-saving measures recommendations that can help optimize energy use. By applying advanced analytics, AI, and machine learning models, these systems can generate tailored recommendations for improving energy efficiency. Examples include: on during non-business hours. AI can recommend adjustments to operation schedules to ensure energy is used only when needed. due to aging or poor performance. Based on usage patterns and efficiency data, AI can recommend maintenance, repairs, or equipment replacement to restore optimal performance and reduce waste energy-intensive activities to off-peak hours to take advantage of lower utility rates or reduce the strain on the grid.as installing energy-efficient lighting, retrofitting HVAC systems with smart thermostats, or replacing outdated appliances with more efficient models of energy consumption profiling. Identifying anomalies helps energy managers or homeowners detect issues before they lead to significant inefficiencies or high costs. There are several types of anomalies that energy consumption profiling can uncover: that can indicate a problem. For instance, if energy consumption increases suddenly during off-peak hours or (Ranjitha et al., 2025) or system use, it might suggest that a heating system or refrigeration unit has failed or that a part of the building's energy system is not functioning properly.AI and machine learning are playing an increasingly important role in the energy consumption profiling process. These technologies can help automate the detection of anomalies, predict future consumption trends, and generate actionable recommendations based on large datasets.

BEHAVIORAL CHANGE INTERVENTIONS: AI-SUPPORTED APPROACHES

Energy efficiency has become a crucial focus in, the optimization of energy use has gained significant attention. However, achieving true energy efficiency is not always a straightforward task, as it requires a multi-faceted approach that spans across residential, commercial, and industrial sectors. With rapid technological advancements, particularly in the fields of Artificial Intelligence (AI) savings, and environmental impact.AI offers the ability to optimize complex energy systems, model and predict future energy consumption patterns, and provide highly personalized energy management strategies. By harnessing these capabilities, organizations, building managers, and consumers can identify inefficiencies, predict energy demand, automate energy-saving actions, and enhance the performance of various systems. This approach enables a proactive stance on energy use—one that is not only reactive but anticipates and addresses challenges before they arise. Through AI-driven optimization, energy efficiency strategies have become more refined, dynamic, and highly responsive to changing conditions, making them a cornerstone of sustainable energy management and even entire grids. These strategies go beyond traditional energy-saving measures, which typically focus on passive actions like turning off lights or upgrading insulation, by using real-time data to make can be applied to various energy systems, such as heating, ventilation, and air conditioning (HVAC), lighting, industrial processes, and grid management. By automating decision-making and continuously fine-tuning energy usage, AI helps to reduce waste, lower operational costs, and decrease carbon footprints.AI-driven energy efficiency works, it's important to consider the key components that enable AI to optimize energy systems effectively. (Ramakrishnan et al., 2022). These components include data acquisition, data analysis, predictive modeling, decision-making, and automation. Energy usage data must be collected from a variety of sources to build an accurate picture of how energy is consumed in a given system or building. Common sources of data include: gas, water, and other forms of energy consumption. Smart meters offer granular insights into energy use, allowing AI systems to detect usage patterns and trends at specific times, locations, and devices within a building or facility systems (e.g., HVAC, lighting, refrigeration) and feed this information to AI algorithms for analysis and optimization.

conditions, occupancy patterns, and system performance. For example, motion sensors can detect when rooms are unoccupied, allowing AI to adjust energy consumption (e.g., turning off lights or adjusting HVAC settings) to save energy can help optimize HVAC settings, while real-time electricity pricing data can guide decisions on when to run energy-intensive appliances or equipment the use of machine learning algorithms that can process large amounts of data and extract useful insights. Key functions in the analysis phase include: By understanding these patterns, AI can suggest operational changes that lead to more efficient energy use. For instance, if energy usage spikes unexpectedly in a particular area of a building, AI can alert users to investigate the cause and take corrective action variations in heating and cooling demands, or shifts in energy use based on occupancy patterns. These insights can help forecast future needs and inform strategic planning a high degree of accuracy. This predictive capability is invaluable for optimizing energy use, as it allows for proactive adjustments to be made before inefficiencies occur. Some specific ways predictive modeling contributes to energy efficiency include:

INTEGRATION WITH SMART GRIDS AND IOT DEVICES

The ongoing integration of Artificial Intelligence (AI) into the field of energy consumption analysis is paving the way for transformative advancements across various sectors, from residential buildings to large-scale industrial operations. With the global demand for energy steadily increasing and growing concerns about climate change, the need for more efficient energy management systems has never been more pressing. AI technologies, particularly machine learning (ML), deep learning, predictive analytics, and generative models, are providing innovative solutions to optimize energy consumption, reduce waste, and increase the overall efficiency of energy systems. As we look toward the future, several emerging trends in AI are expected to, particularly with advancements in real-time energy monitoring and decision-making capabilities sources, such as solar and wind, become more prevalent, integrating AI with smart grids will help balance the fluctuations in supply and demand. AI can optimize the flow of energy by Energy consumption forecasting is a critical aspect of managing both demand and supply in energy systems. Traditional methods of forecasting energy demand rely on historical data and basic statistical models, which can often fail to trends, is set to improve forecasting accuracy dramatically usage data. These algorithms will be able to predict not only peak energy demand but also trends in energy consumption based on specific patterns of individual or collective behavior. For example, AI can anticipate how an individual household's energy usage will vary depending on factors like the time of day, seasonal changes, and even the behavioral patterns of the occupants. By predicting demand with greater accuracy, AI can help energy providers optimize supply chains, reduce the likelihood of outages, and implement demand-response strategies that encourage energy-saving behaviors among consumers.

Additionally, AI-driven predictive models will be able to account for more granular variables, such as the impact of social or economic (Szczepański et al., 2019) events (e.g., holidays or pandemics) on energy consumption patterns. This level of forecasting will empower utilities and consumers alike to make informed decisions that contribute to the optimization of energy systems. As AI systems become more adept at analyzing individual user data, they will be able to offer highly tailored recommendations that help users optimize their energy usage in real time. These recommendations will be based on personal habits, preferences, and past consumption patterns ways that are customized to each resident's behavior. Through predictive analytics, AI systems can suggest when to use energy-intensive devices, how to set heating or cooling systems for maximum efficiency, or recommend shifts in energy use to off-peak hours when electricity costs are lower. Moreover, AI could deliver feedback on energy consumption through user-friendly interfaces, such as mobile apps, providing real-time energy-saving tips, comparisons to similar households, or reminders to adjust settings for improved efficiency schedules based on predicted occupancy and energy needs. Over time, AI will be able to learn from feedback and fine-tune its recommendations, making them even more effective in reducing energy waste release energy from batteries), and manage the distribution of energy between DERs and the central grid. This will be particularly important as consumers become both producers and consumers of energy (prosumers), generating their own energy and selling excess power back to the grid of the supply chain, from production to end use. AI systems can analyze manufacturing processes to identify inefficiencies in energy use, suggest ways to reduce emissions, and even recommend more sustainable production methods. Moreover, AI can help industries and businesses track the lifecycle of energy-consuming products, enabling the efficient reuse and recycling of materials while minimizing waste when energy storage devices need maintenance or when they should be replaced, ensuring their maximum lifespan and minimizing waste. Over time, this type of AI-driven optimization will create more sustainable, closed-loop energy systems encouraging energy-saving behaviors. As AI systems become increasingly adept at analyzing user behavior, they will be able to offer more personalized feedback and suggestions aimed at fostering sustainable energy consumption practices. In the future, AI systems will integrate behavioral psychology into energy consumption analysis to develop interventions that nudge individuals toward more sustainable energy choices.

ADDRESSING DATA QUALITY AND PRIVACY CONCERNS

The future of energy systems is poised for transformation through grids and the Internet of Things (IoT) devices, both of which offer immense opportunities to enhance energy management, improve smart grid systems provides a pathway to achieving smarter, more efficient energy systems. The synergy between smart grids and IoT devices presents a unique impact delving into the integration opportunities, it is important to first understand what smart grids and IoT devices are and how they function within the context of energy management., IoT devices can include smart meters, thermostats, appliances, sensors, and other connected devices that communicate with energy management systems. These devices generate vast amounts of data about energy consumption, environmental conditions, and user behavior, which can be used to optimize energy usage, improve system performance, and enhance user experience. (Tao et al., 2021) their operation automatically based on these signals, helping to prevent grid overloads and reducing the need for costly and polluting peak-power generation. By shifting non-essential energy use to off-peak hours, the grid can maintain stability, lower energy costs, when integrated with IoT devices, can play a vital role in optimizing the use of renewable energy and ensuring its smooth integration into the grid. By collecting real-time data from renewable energy installations (e.g., solar panels and wind turbines), IoT devices can help the smart grid, which then adjusts the distribution of power from other sources to balance the supply. Similarly, wind farms equipped with IoT devices can feed real-time data on wind speeds and energy generation to the grid, enabling better management of energy supply approach minimizes downtime, reduces repair costs, and ensures a more reliable energy supply. Integrating predictive maintenance capabilities into the smart grid infrastructure enhances the overall resilience and stability of the grid, reducing the likelihood of power outages and improving service reliability for reducing energy consumption. Smart grids can leverage this data to offer consumers dynamic pricing models, such as time-of-use pricing, which incentivizes energy use during off-peak hours when electricity is cheaper. By providing consumers with a better understanding of their energy habits, IoT and smart grid integration fosters greater engagement in energy-saving behaviors and encourages the adoption of energy- while smart grids can ensure that energy is distributed efficiently throughout the city.

UNLOCKING ENERGY EFFICIENCY: THE PROMISE OF GENERATIVE AI

In the context of energy consumption analysis and management, data quality impacts everything from real-time monitoring of energy usage to predictive analytics, demand-response to faulty sensors, incorrect reporting, or human error—can lead to flawed analytics and poor decision-making. For example, inaccurate energy consumption data could result in an underestimation of peak demand, leading to grid instability and energy shortages. Similarly, misreported data about the performance of renewable energy sources like solar or wind could disrupt efforts to balance supply and demand and validation of data sources, (Shende et al., 2025). Additionally, algorithms used in energy systems must be able to detect anomalies in the data and account for any inconsistencies or outliers. This allows energy operators to correct errors before they lead to operational inefficiencies or disruptions. In energy systems, missing or incomplete data—whether in the form of gaps in energy consumption records, incomplete weather forecasts, or absent user behavior data—can impair the ability to make reliable predictions and perform effective energy optimizations. For instance, missing data about energy consumption patterns from certain sectors or regions could skew demand forecasts, leading to an inefficient energy supply strategy. To address this, organizations need to ensure that data is comprehensively captured from all relevant sources. This might involve the integration of additional sensors or monitoring tools to fill in gaps in data collection, especially in areas where data collection has historically been lacking or inconsistent. For example, IoT devices and smart meters must transmit data consistently and quickly to ensure that the smart grid can adjust to fluctuations in energy demand without delay. In some cases, lagging data can undermine the effectiveness of energy optimization strategies, especially when predictive models rely on real-time input. As the energy sector increasingly depends on instantaneous data to manage energy flow, the speed and frequency at which data is collected and transmitted becomes an essential factor in maintaining system responsiveness from different sensors and systems are compatible and can be aggregated for a comprehensive analysis. Inconsistent data—such as energy readings that differ based on the format or units of measurement—can hinder accurate analysis and management system, can ensure that data is comparable and integrable across different platforms.

Additionally, implementing robust data validation rules can help to detect inconsistencies and discrepancies before they affect decision-making.be used to optimize energy management and offer personalized recommendations, it also raises important privacy questions. These concerns are especially pressing given that personal information can be inferred from energy consumption data, such as the times individuals are home, their daily are being tracked, analyzed, and shared with third parties, especially if they are unaware of how their data is being used. The granularity of the data collected by smart meters and IoT devices—such as the precise times energy is consumed, the specific appliances in use, and even the temperature settings in a home—can provide detailed insights into personal habits and routines. This raises questions about the extent to which consumers can retain control over their data and and how it will be protected can help build trust with consumers. Additionally, consumers should be given the ability to opt out of data collection or sharing, with the understanding that opting out may limit access to certain personalized services, such as energy-saving recommendations or dynamic pricing., the volume of data being transmitted across networks grows exponentially, creating more opportunities for cyberattacks and data breaches. Energy data, especially when combined with personal information, is highly valuable and attractive to hackers significant financial losses but also the erosion of consumer trust in digital energy solutions. The security of the infrastructure used to transmit and store energy data is therefore paramount., and threat detection systems to ensure that data remains safe from unauthorized access., energy companies can protect consumer, (Sreenivasulu et al., 2022) privacy while still benefiting from valuable insights into energy usage patterns. Anonymized data can be used for analytics and predictive modeling without revealing specific details about individual consumers. This approach helps to mitigate privacy risks while enabling the development of smarter energy systems that are capable of optimizing consumption, predicting demand, and reducing costs.

CONCLUSION

Traditional methods of managing energy consumption are increasingly proving insufficient to meet the growing environmental and economic challenges. In response to this, the application of artificial intelligence (AI), particularly generative AI, is gaining significant attention as a transformative tool in optimizing energy systems. Generative AI, which refers to a class of machine learning models capable of generating new data, solutions, or patterns based on learned information, holds immense potential for revolutionizing how energy is consumed, produced, and managed. In the realm of energy efficiency, generative AI offers promising opportunities for improving everything from energy production and distribution to consumption and behavior analysis energy usage schedules, and even predictive models for system behavior. At its core, generative AI leverages deep learning and neural networks to analyze large volumes of data, learn from it, and then generate useful insights or solutions that were not previously conceived by human analysts or existing algorithms. In the context of energy systems, this includes generating solutions for energy optimization, creating energy-efficient infrastructure designs, and improving predictive energy models management strategies. By utilizing the capabilities of generative AI, energy providers, consumers, and businesses can unlock new levels of energy efficiency that were previously unattainable with traditional methods little flexibility or adaptability to changing conditions. However, with the advent of AI-driven smart systems, real-time energy optimization has become a reality. Generative, environmental conditions, occupancy, and external factors such as weather patterns. By analyzing this data, generative AI can produce energy-saving strategies that adjust energy consumption dynamically based on real-time conditions while maintaining comfort and functionality. Similarly, AI can optimize industrial processes by dynamically adjusting machinery and equipment settings to ensure that energy is used most efficiently throughout production cycles provides an unparalleled level of flexibility in energy management, ensuring that energy is always consumed in the most efficient manner possible powerful application of generative AI in unlocking energy efficiency is predictive energy management. Traditional energy systems rely on past data and simple forecasting methods to predict future demand and supply. These methods, while functional, often fall short in providing accurate, actionable predictions for managing energy use and distribution effectively. Generative AI, on the other hand, can create sophisticated models that not only predict energy demand with greater accuracy but also generate optimized strategies for managing supply and demand in real-time. By incorporating factors such as weather forecasts, historical consumption data, and real-time information from IoT devices, generative AI models can predict energy consumption patterns with remarkable precision models can help utilities anticipate peak demand periods, allowing them to implement demand-response strategies to encourage energy savings during times of high consumption facilities, and power plants are traditionally resource-intensive processes, often involving numerous design iterations and a for human designers to conceive. By creating energy-efficient designs from the outset, generative AI can significantly reduce energy consumption over the lifespan of a building or infrastructure project as lighting, HVAC, and insulation—to reduce energy usage. Generative AI can analyze a building's current energy usage patterns, identify inefficiencies, and generate optimized retrofitting strategies that target the most significant energy-saving opportunities. By automating the retrofitting design process, generative AI can reduce costs and improve the == is high or low, AI systems can optimize the flow of electricity from these sources to meet demand and reduce reliance on non-renewable backup power. This predictive capability enables grid operators to integrate renewable energy sources more effectively, increasing their contribution to the overall energy mix planning for future grid expansions and improvements by generating optimal infrastructure layouts and proposing scenarios for integrating new energy sources. This proactive approach to grid management helps ensure that energy systems remain reliable and adaptable in the face of growing demand and climate-related challenges.

REFERENCES

Agency, T. I. E. (2023). The international energy agency's electricity market report https://iea.blob.core.windows.net/assets/255e9cba-da84-4681-8c1f-458ca1a3d9ca/ElectricityMarketReport2023.pdf. Accessed Apr 2023

Ali, S. S., & Choi, B. J. (2020). State-of-the-art artificial intelligence techniques for distributed smart grids: A review. *Electronics (Basel)*. Advance online publication. DOI: 10.3390/electronics9061030

Athavale, V. A., Selvakumar, P., Poorani, S., Anand, T., & Rathore, D. S. (2025). Designing Chatbots for Educational Leadership. In Musundire, A. (Ed.), *Chatbots in Educational Leadership and Management* (pp. 181–204). IGI Global Scientific Publishing., DOI: 10.4018/979-8-3693-8734-4.ch008

Bedi, J., & Toshniwal, D. (2018). Empirical mode decomposition based deep learning for electricity demand forecasting. *IEEE Access : Practical Innovations, Open Solutions*, *6*, 49144–49156. DOI: 10.1109/ACCESS.2018.2867681

Bose, B. K. (2017). Artificial intelligence techniques in smart grid and renewable energy systems—Some example applications. *Proceedings of the IEEE*, *105*(11), 2262–2273. DOI: 10.1109/JPROC.2017.2756596

Espinoza, M., Joye, C., Belmans, R., & De Moor, B. (2005). Short-term load forecasting, profile identification, and customer segmentation: A methodology based on periodic time series. *IEEE Transactions on Power Systems*, *20*(3), 1622–1630. DOI: 10.1109/TPWRS.2005.852123

Fan, C., Xiao, F., & Wang, S. (2014). Development of prediction models for next-day building energy consumption and peak power demand using data mining techniques. *Applied Energy*, *127*, 1–10. DOI: 10.1016/j.apenergy.2014.04.016

Fan, C., Xiao, F., & Zhao, Y. (2017). A short-term building cooling load prediction method using deep learning algorithms. *Applied Energy*, *195*, 222–233. DOI: 10.1016/j.apenergy.2017.03.064

Himeur, Y., Ghanem, K., Alsalemi, A., Bensaali, F., & Amira, A. (2021). Artificial intelligence based anomaly detection of energy consumption in buildings: A review, current trends and new perspectives. *Applied Energy*, *287*, 116601. DOI: 10.1016/j.apenergy.2021.116601

Hosein, S., & Hosein, P. (2017). *Load forecasting using deep neural networks. In: 2017 IEEE Power Energy Society Innovative Smart Grid Technologies Conference*. ISGT., DOI: 10.1109/ISGT.2017.8085971

Jin, X.-B., Zheng, W.-Z., Kong, J.-L., Wang, X.-Y., Bai, Y.-T., Su, T.-L., & Lin, S. (2021). Deep-learning forecasting method for electric power load via attention-based encoder-decoder with bayesian optimization. *Energies*, *14*(6), 1596.

Khan, A. R., Mahmood, A., Safdar, A., Khan, Z. A., & Khan, N. A. (2016). Load forecasting, dynamic pricing and dsm in smart grid: A review. *Renewable & Sustainable Energy Reviews*, *54*, 1311–1322. DOI: 10.1016/j.rser.2015.10.117

Leitao, P., Karnouskos, S., Ribeiro, L., Lee, J., Strasser, T., & Colombo, A. W. (2016). Smart agents in industrial cyber–physical systems. *Proceedings of the IEEE*, *104*(5), 1086–1101. DOI: 10.1109/JPROC.2016.2521931

Li, A., Xiao, F., Zhang, C., & Fan, C. (2021). Attention-based interpretable neural network for building cooling load prediction. *Applied Energy*, *299*, 117238. DOI: 10.1016/j.apenergy.2021.117238

Noor, S., Yang, W., Guo, M., van Dam, K. H., & Wang, X. (2018). Energy demand side management within micro-grid networks enhanced by blockchain. *Applied Energy*, *228*, 1385–1398. DOI: 10.1016/j.apenergy.2018.07.012

Omer, A. M. (2008). Energy, environment and sustainable development. *Renewable & Sustainable Energy Reviews*, *12*(9), 2265–2300. DOI: 10.1016/j.rser.2007.05.001

Panwar, N. L., Kaushik, S. C., & Kothari, S. (2011). Role of renewable energy sources in environmental protection: A review. *Renewable & Sustainable Energy Reviews*, *15*(3), 1513–1524. DOI: 10.1016/j.rser.2010.11.037

Pradeep, E., Mukherjee, S. K., Selvakumar, P., Koppaka, A. K., Chandel, P. S., & Vats, C. (2025). Empowering Social Media Users With Ethical AI. In Chakraborty, S. (Ed.), *Ethical AI Solutions for Addressing Social Media Influence and Hate Speech* (pp. 91–118). IGI Global Scientific Publishing., DOI: 10.4018/979-8-3693-9904-0.ch005

Qiu, X., Ren, Y., Suganthan, P. N., & Amaratunga, G. A. J. (2017). Empirical mode decomposition based ensemble deep learning for load demand time series forecasting. *Applied Soft Computing*, *54*, 246–255. DOI: 10.1016/j.asoc.2017.01.015

Ramakrishnan, T., Mohan Gift, M. D., Chitradevi, S., Jegan, R., & Subha Hency Jose, P. (2022). Nagaraja, Rajneesh Sharma, H. N., Selvakumar, P., & Hailegiorgis, S. M. (2022). Study of numerous resins used in polymer matrix composite materials. *Advances in Materials Science and Engineering*, •••, 1088926. Advance online publication. DOI: 10.1155/2022/1088926

Ranjitha, P. K., Selvakumar, P., Anute, N., Manjunath, T., Sureshkrishna, G., & Pougajendy, S. (2025). Transportation challenges in islands tourism. In Hassan, V. (Ed.), *Navigating Mass Tourism to Island Destinations: Preservation and Cultural Heritage Challenges* (pp. 287–312). IGI Global., DOI: 10.4018/979-8-3693-9107-5.ch011

Ritchie, H. (2019). How long before we run out of fossil fuels? https://ourworldindata.org/how-long-before-we-run-out-of-fossil-fuels

Shende, A. D., Tekade, S. A., Deshmukh, A. A., Tembhurkar, S. P., & Selvakumar, P. (2025). Water quality monitoring and control with AI. In J. Ruiz-Vanoye (Details of the book incomplete; citation needs final publication data).

Sreenivasulu, A., Manimozhi, T., Satheesh Kumar, P., Selvakumar, P., & Singh, P. (2022). The model of smart sensing device for sensitive nanoclusters modification in sensing properties. In 2022 2nd International Conference on Advance Computing and Innovative Technologies in Engineering (ICACITE) (pp. 1043–1047). IEEE. https://doi.org/DOI: 10.1109/ICACITE53722.2022.9823846

Szczepański, M. (2019). Economic impacts of artificial intelligence. Retrieved May 10, 2021,fromhttps://www.europarl.europa.eu/RegData/etudes/BRIE/2019/637967/EPRS_BRI(2019)637967_EN.pdfTao

R, Su C, Xiao Y, Dai K, Khalid F (2021) Robo advisors, algorithmic trading and investment management: Wonders of fourth industrial revolution in financial markets. Technol Forecast Soc Chang 163:120421. DOI: 10.1016/j.techfore.2020.120421

Thoubboron, K. (2021). Advantages and disadvantages of renewable energy. https://news.energysage.com/advantages-and-disadvantages-of-renewable-energy. Accessed Mar 2020

Tuballa, M. L., & Abundo, M. L. (2016). A review of the development of smart grid technologies. *Renewable & Sustainable Energy Reviews*, *59*, 710–725. DOI: 10.1016/j.rser.2016.01.011

Vrablecova, P., Bou Ezzeddine, A., Rozinajova, V., Sarik, S., & Sangaiah, A. K. (2018). Smart grid load forecasting using online support vector regression. *Computers & Electrical Engineering*, *65*, 102–117. DOI: 10.1016/j.compeleceng.2017.07.006

Wang, X., Ha, B., Lee, G.-Y., Kim, H., Yu, J., Rhee, H., Njau, K., Jande, Y., & Ahn, S.-H. (2020). Low-cost far-field wireless electrical load monitoring system applied in an off-grid rural area of Tanzania. *Sustainable Cities and Society*, *59*, 102209. DOI: 10.1016/j.scs.2020.102209

Wang, X., Rhee, H., & Ahn, S.-H. (2020). Off-grid power plant load management system applied in a rural area of Africa. *Applied Sciences (Basel, Switzerland), 10,* 4171. DOI: 10.3390/app10124171

Wang, X.-L., Ha, B., Manongi, F. A., Jung, W.-K., Jande, Y. A. C., & Ahn, S.-H. (2020). Arduino-based low-cost electrical load tracking system with a long-range mesh network. *Advances in Manufacturing, 9,* 47–63. DOI: 10.1007/s40436-020-00310-5

Yu, W., Wen, G., Yu, X., Wu, Z., & Lu, J. (2014). Bridging the gap between complex networks and smart grids. *Journal of Control and Decision, 1*(1), 102–114. DOI: 10.1080/23307706.2014.885293

Zhou, J., He, L., Li, C., Cao, Y., Liu, X., & Geng, Y. (2013). *What's the difference between traditional power grid and smart grid? from dispatching perspective. In: 2013 IEEE PES Asia-Pacific Power and Energy Engineering Conference.* APPEEC., DOI: 10.1109/APPEEC.2013.6837107

Zou, Z., Wu, X., Zhao, Z., Wang, Q., Bie, Y., & Zhou, M. (2018). Prediction of short term electric load based on bp neural networks & ARIMA combination. In: *2018 IEEE 4th Information Technology and Mechatronics Engineering Conference (ITOEC),* pp. 1671–1674. https://doi.org/DOI: 10.1109/ITOEC.2018.874055

Chapter 9
A Study of Generative AI for Energy Consumption Behavior Analysis

R. Velmurugan
https://orcid.org/0000-0002-7925-9757
Karpagam Academy of Higher Education, Coimbatore, India

J. Sudarvel
https://orcid.org/0000-0001-6656-1992
Karpagam Academy of Higher Education, Coimbatore, India

R. Bhuvaneswari
https://orcid.org/0009-0003-9396-6129
Dr. Mahalingam College of Engineering and Technology, Pollachi, India

Ravi Thirumalaisamy
https://orcid.org/0000-0002-5954-5940
Modern College of Business and Science, Oman

ABSTRACT

Energy consumption analysis is crucial for addressing climate change, efficiency, and sustainable energy transitions. Generative AI leverages deep learning models like GANs and VAEs to simulate, predict, and analyze energy usage across households, businesses, and industries. By generating synthetic data, AI identifies patterns, forecasts peak demand, and evaluates interventions such as smart devices or policy changes. Personalized insights guide consumers in optimizing energy use, while integration with smart home systems enables real-time adjustments. Generative AI also supports renewable energy integration, DER management, and carbon foot-

DOI: 10.4018/979-8-3373-4272-6.ch009

print reduction. Despite challenges in data quality, privacy, behavioral complexity, and system integration, these models offer transformative potential for efficient, sustainable, and adaptive energy management worldwide.

INTRODUCTION

Energy consumption behavior refers to the patterns, decisions, and habits through which individuals, households, businesses, and industries utilize energy resources. These behaviors are shaped by a complex interplay of socio-economic status, cultural norms, geographic conditions, technological access, and environmental awareness. Understanding these behaviors is critical not only for improving energy efficiency but also for designing equitable and sustainable energy systems. Traditionally, the analysis of energy consumption behavior relied heavily on historical data, manual observations, and relatively simplistic modeling techniques such as linear regression, time-series forecasting, and rule-based simulations. While these methods provided foundational insights into aggregate consumption trends, they often failed to capture the dynamic, context-sensitive, and multifaceted nature of energy usage across different populations and environments.

In recent years, the emergence of artificial intelligence (AI)—particularly generative AI—has revolutionized the way researchers, energy providers, and policymakers approach energy consumption analysis. Generative AI refers to a class of machine learning models capable of producing synthetic data that closely resembles real-world patterns. These models, including Generative Adversarial Networks (GANs), Variational Autoencoders (VAEs), and reinforcement learning agents, are trained on large-scale datasets encompassing historical energy usage, weather conditions, demographic trends, and economic indicators. Once trained, they can simulate a wide range of scenarios and predict the impact of various interventions—such as policy changes, technological upgrades, or behavioral nudges—on energy consumption.

Unlike conventional statistical models, generative AI offers a more nuanced, adaptive, and scalable understanding of energy behavior. For example, GANs can generate synthetic energy profiles for households based on different income levels, geographic locations, and climate zones. This allows researchers to explore how energy-saving interventions might perform under diverse conditions without requiring extensive field trials. VAEs, on the other hand, can compress complex energy usage data into latent representations that reveal hidden patterns and correlations, such as the relationship between appliance usage and peak demand periods. These capabilities enable a deeper exploration of behavioral drivers and system-level dynamics that were previously inaccessible through traditional methods.

A compelling real-world application of generative AI in energy research is found in the work conducted in urban China, where GANs were used to simulate residential electricity consumption. The model incorporated variables such as weather fluctuations, building insulation quality, and household income, enabling policymakers to test the effectiveness of energy-saving campaigns before implementation. This proactive approach allowed for the optimization of outreach strategies and resource allocation, ultimately improving the campaign's impact and cost-efficiency. Similarly, in India, VAEs were employed to analyze industrial energy consumption patterns across small and medium-sized manufacturing units. The study revealed that energy usage varied significantly depending on seasonal demand, labor availability, and production cycles—insights that were previously obscured by aggregate data models. These findings informed the design of flexible energy pricing schemes and targeted efficiency programs for industrial clusters.

Generative AI also plays a critical role in addressing the equity dimension of energy consumption. Energy poverty—defined as the lack of access to affordable, reliable, and clean energy—is a pressing issue in many parts of the world. Low-income households often face disproportionate energy burdens, spending a higher percentage of their income on basic energy needs while lacking the means to invest in energy-efficient technologies. By simulating energy usage scenarios for vulnerable populations, generative AI can help design targeted subsidies, infrastructure improvements, and behavioral interventions. For example, in refugee settlements in Jordan, generative models were used to assess energy poverty and consumption patterns. The insights gained from these simulations informed the deployment of solar microgrids tailored to the unique needs of displaced communities, improving energy access and reducing reliance on costly diesel generators.

Moreover, generative AI enhances the predictive capabilities of energy systems in the face of climate change. As extreme weather events become more frequent and severe, energy demand patterns are expected to shift dramatically. Heatwaves, cold spells, floods, and droughts can all influence how and when energy is consumed, placing additional stress on grid infrastructure. Generative models can simulate future scenarios under different climate trajectories, helping utilities and governments prepare for disruptions and design resilient energy systems. For instance, in Europe, GANs were integrated into climate-energy modeling frameworks to forecast electricity demand during extreme weather conditions. These forecasts enabled grid operators to preemptively adjust supply strategies, deploy backup resources, and communicate with consumers, thereby reducing the risk of outages and improving system reliability.

Cultural and behavioral factors also significantly influence energy consumption, and generative AI is uniquely equipped to capture these subtleties. Energy behavior is not solely driven by economic rationality—it is deeply embedded in cultural practices, social norms, and psychological preferences. In Japan, for example, the concept of "setsuden" (voluntary electricity conservation) gained prominence after the Fukushima nuclear disaster. Households adopted energy-saving behaviors not because of financial incentives, but out of a sense of civic responsibility and solidarity. Researchers used generative models to simulate how cultural norms and public messaging campaigns affected household energy usage, revealing that behavioral interventions rooted in cultural values were more effective than purely economic incentives. These insights are invaluable for designing context-sensitive energy policies that resonate with local populations.

Despite its promise, the application of generative AI in energy research is not without challenges. Data quality and representativeness remain critical concerns. Generative models require large volumes of high-resolution data to function effectively, yet many regions—particularly in the Global South—lack robust digital infrastructure and consistent data collection mechanisms. Inaccurate or biased data can lead to flawed simulations and misguided policy recommendations. Ethical considerations around synthetic data generation must also be addressed. While generative AI can anonymize sensitive information and protect user privacy, it can also inadvertently reproduce biases present in the training data. Transparency, accountability, and stakeholder engagement are essential to ensure that AI-driven energy models are used responsibly and inclusively.

Technical integration poses another challenge. For generative AI to deliver real-time insights and actionable recommendations, it must interface seamlessly with existing energy infrastructure, including smart meters, distributed energy resources (DERs), and home automation systems. This requires interoperability across platforms, standardized data formats, and secure communication protocols. In California, for example, a modular energy management system was developed to support third-party AI applications. This system allowed generative models to access real-time grid data, forecast demand fluctuations, and automate demand response strategies. Such integration efforts are critical for translating AI insights into tangible operational improvements.

Looking ahead, the continued development of generative AI models—combined with advancements in data collection, smart infrastructure, and ethical frameworks—will enable more accurate and actionable predictions about energy consumption behavior. As these technologies become more widely adopted, they will play an integral role in helping governments, utilities, and consumers reduce energy waste, lower costs, and contribute to global sustainability efforts. Real-world initiatives are already paving the way. In Maharashtra, India, Tata Power has launched AI-driven energy dashboards for residential communities, offering personalized tips and tracking carbon footprints. Early results show a 10–15% reduction in monthly energy bills and increased awareness of sustainable practices. Similarly, in Europe, generative models have been used to simulate the transition to 100% renewable energy in select regions, guiding infrastructure upgrades and community engagement strategies.

Thus, generative AI is not just a technological innovation—it is a strategic enabler of a more energy-efficient, equitable, and sustainable future. By unlocking deeper insights, enhancing predictive accuracy, and empowering personalized action, these models can help reshape energy systems to meet the demands of a rapidly changing world. Their ability to simulate complex scenarios, uncover hidden behavioral patterns, and forecast future trends makes them indispensable tools for modern energy planning. As AI technologies continue to evolve, their integration into energy policy, infrastructure design, and consumer engagement will be essential for achieving global climate and development goals. The journey toward smarter energy systems is already underway—and generative AI is lighting the path forward.

GENERATIVE AI TECHNIQUES FOR ENERGY CONSUMPTION PREDICTION

Generative AI techniques—such as Generative Adversarial Networks (GANs), Variational Autoencoders (VAEs), and reinforcement learning—have emerged as transformative tools in the analysis of energy consumption behavior. These advanced models are particularly well-suited for this domain due to their ability to simulate diverse datasets that reflect a wide range of possible outcomes, including rare, extreme, or future scenarios that traditional models often fail to capture. By leveraging deep learning architectures, generative AI can uncover hidden patterns, simulate behavioral responses to interventions, and optimize energy systems with unprecedented precision and adaptability.

At the core of GANs lies a dual-network architecture: a generator that creates synthetic data and a discriminator that evaluates its authenticity. This adversarial setup enables GANs to produce highly realistic data that closely mirrors actual energy consumption behaviors. For instance, GANs can replicate household electricity usage patterns based on variables such as appliance ownership, occupancy schedules, and weather conditions. These synthetic datasets are invaluable for refining predictive models, designing energy-efficient strategies, and optimizing supply systems without relying solely on costly or incomplete real-world data. The ability to simulate granular, context-specific energy profiles allows researchers and policymakers to test interventions before implementation, reducing risk and improving cost-effectiveness.

One of the most impactful applications of generative AI in energy consumption analysis is the generation of synthetic household and business energy usage profiles. These profiles can be used to model the effects of various interventions—such as the adoption of smart thermostats, LED lighting, or energy-efficient appliances—on overall energy demand. For example, in a study conducted in South Korea, GANs were used to simulate the energy savings potential of smart home technologies across different income groups. The findings revealed that low-income households could achieve up to 18% reduction in energy usage with targeted subsidies for smart thermostats and efficient lighting. These insights informed national energy policy revisions, demonstrating how generative AI can support evidence-based decision-making and equitable resource allocation.

Similarly, VAEs have been employed to compress and reconstruct complex energy datasets, revealing latent variables that influence consumption behavior. In a recent project in Mumbai, VAEs were used to analyze commercial building energy usage. The model identified seasonal occupancy trends and HVAC usage patterns that were previously overlooked, enabling building managers to implement more effective energy-saving schedules and retrofit strategies. VAEs are particularly useful for dimensionality reduction and anomaly detection, helping stakeholders identify inefficiencies and optimize operational performance.

Reinforcement learning (RL), another generative AI technique, adds a dynamic layer to energy behavior modeling by simulating decision-making processes over time. RL agents learn optimal energy usage strategies by interacting with simulated environments and receiving feedback based on energy efficiency outcomes. For example, in a pilot study in Saudi Arabia, RL was used to develop adaptive energy pricing models. The system learned to adjust tariffs based on real-time demand and consumer responsiveness, resulting in a 12% reduction in peak load and improved grid stability. RL's ability to continuously adapt to changing conditions makes it ideal for managing distributed energy resources (DERs), smart grids, and demand-side management programs.

Generative AI also plays a critical role in predicting energy usage during peak demand periods. By analyzing historical consumption data and simulating future patterns, these models can forecast when and where energy demand is likely to surge. This capability is essential for implementing demand-side management (DSM) strategies, which aim to balance supply and demand through behavioral incentives and technological interventions. For instance, GANs were used by the European Smart Grid Initiative to simulate residential energy demand during heatwaves. The model helped utilities design time-of-use pricing schemes and promote off-peak appliance usage, reducing grid stress and preventing blackouts. These predictive insights are crucial for enhancing grid resilience and avoiding costly disruptions.

Moreover, generative AI can identify specific behavioral factors that influence peak demand, such as remote work trends, climate conditions, and lifestyle preferences. During the COVID-19 pandemic, energy consumption patterns shifted dramatically due to increased work-from-home arrangements. A study conducted in Tokyo used

GANs to simulate these changes, revealing a 25% increase in daytime residential electricity usage. These insights enabled energy providers to adjust supply schedules and promote energy-saving behaviors through targeted messaging. Understanding behavioral shifts in real time allows for more responsive and personalized energy management strategies.

Real-world applications continue to demonstrate the transformative potential of generative AI in energy systems. In California, the Pacific Gas and Electric Company (PG&E) partnered with AI researchers to develop a GAN-based model that simulates wildfire-related disruptions and their impact on energy consumption. The model helps predict emergency demand spikes and guides resource allocation during crises, ensuring that critical infrastructure remains operational. In India, Tata Power has begun exploring reinforcement learning to optimize distributed solar energy systems, allowing microgrids to autonomously adjust energy flows based on consumption forecasts and weather predictions. These initiatives highlight the versatility of generative AI in addressing both routine and emergency energy challenges.

Beyond operational efficiency, generative AI contributes to the broader goals of energy equity and sustainability. By simulating energy usage scenarios for marginalized populations—such as low-income households, rural communities, and refugee settlements—AI models can inform the design of targeted subsidies, infrastructure upgrades, and educational campaigns. For example, generative models have been used to assess energy poverty in underserved regions, guiding the deployment of solar microgrids and energy-efficient appliances. These interventions not only improve access to clean energy but also reduce environmental impact and enhance quality of life.

Despite its promise, the use of generative AI in energy behavior analysis is not without challenges. Data quality remains a critical concern. High-performing AI models require large volumes of granular, accurate, and representative data. In regions with limited digital infrastructure, data collection may be inconsistent or biased, affecting model reliability. Ethical considerations around synthetic data generation—such as privacy, transparency, and algorithmic bias—must also be carefully managed. Ensuring that AI models are interpretable and accountable is essential for building trust among stakeholders and avoiding unintended consequences.

Model interpretability is particularly important in policy and regulatory contexts. While generative AI can produce highly accurate predictions, the complexity of its algorithms may make it difficult for non-experts to understand how decisions are made. This "black box" nature can hinder transparency and accountability, especially when AI-driven recommendations influence public policy or resource allocation. Developing explainable AI (XAI) frameworks and integrating stakeholder feedback into model design can help address these concerns and promote responsible AI deployment.

Technical integration poses another challenge. For generative AI to deliver real-time insights, it must interface seamlessly with existing energy infrastructure, including smart meters, IoT devices, and energy management systems. This requires standardized data formats, secure communication protocols, and robust cybersecurity measures. Interoperability across platforms is essential for scaling AI solutions and ensuring that insights can be translated into actionable strategies. Collaborative efforts among utilities, technology providers, and regulatory bodies are needed to establish common standards and facilitate integration.

Looking ahead, the continued development of generative AI models—combined with advancements in data collection, smart infrastructure, and ethical governance—will enable more accurate and actionable predictions about energy consumption behavior. As these technologies become more widely adopted, they will play an integral role in helping governments, utilities, and consumers reduce energy waste, lower costs, and contribute to global sustainability efforts. Real-world initiatives are already paving the way, demonstrating how generative AI can support climate resilience, energy equity, and economic development.

Thus, generative AI techniques such as GANs, VAEs, and reinforcement learning are reshaping how we understand and manage energy consumption behavior. By generating realistic synthetic data, predicting peak demand, and modeling the impact of interventions, these tools empower researchers, policymakers, and energy providers to design smarter, more equitable, and more sustainable energy systems. As the energy landscape continues to evolve in response to technological, environmental, and social pressures, generative AI will remain a critical ally in building resilient and inclusive energy futures.

BEHAVIORAL INSIGHTS AND PERSONALIZATION OF ENERGY CONSUMPTION

One of the most transformative advantages of generative AI in energy consumption behavior analysis is its capacity to deliver personalized insights tailored to individual consumers. Traditional energy models often rely on aggregated data, treating consumers as homogeneous groups with similar habits and preferences. While this approach is useful for broad policy planning and infrastructure design, it fails to capture the nuanced and context-specific factors that drive energy usage in diverse households and communities. Generative AI, by contrast, excels at analyzing granular, individual-level data and producing customized recommendations that reflect the unique consumption patterns, preferences, and constraints of each user.

Generative AI models—such as Generative Adversarial Networks (GANs), Variational Autoencoders (VAEs), and transformer-based architectures—are capable of processing vast amounts of consumer data, including energy bills, appliance usage, temperature preferences, occupancy schedules, and even behavioral routines. These models learn from historical and real-time data to simulate realistic consumption scenarios and generate actionable feedback. For instance, GANs can create synthetic energy profiles that mirror a household's daily usage, allowing researchers and energy providers to test interventions like smart thermostats or energy-efficient appliances without requiring intrusive monitoring or costly pilot programs. This simulation capability is particularly valuable in regions where real-time data collection is limited or where privacy concerns restrict direct observation.

The personalized approach enabled by generative AI has significant implications for energy efficiency. Instead of offering generic advice—such as "reduce air conditioning use"—AI models can recommend specific actions like adjusting thermostat settings based on upcoming weather forecasts, switching to LED lighting in high-use areas, or shifting laundry and dishwashing to off-peak hours to benefit from lower tariffs. These recommendations are more likely to resonate with consumers because they align with their actual behavior, preferences, and constraints. In a study conducted in South Korea, personalized energy feedback generated by AI led to a 15% increase in consumer engagement and a measurable reduction in electricity usage among urban households. Such results underscore the effectiveness of tailored interventions in driving behavioral change and improving energy outcomes.

The integration of generative AI into smart home systems further amplifies its impact. Devices such as smart thermostats, lighting systems, and energy meters are increasingly being connected to AI-powered platforms that continuously learn from user behavior and environmental conditions. These systems can autonomously adjust settings to optimize energy consumption without compromising comfort or convenience. For example, a smart thermostat powered by a generative model might learn that a user prefers cooler temperatures in the evening and automatically pre-cool the home during off-peak hours, reducing energy costs while maintaining comfort. Over time, these systems become more intelligent and responsive, adapting to seasonal changes, occupancy patterns, and evolving user preferences.

A compelling real-world example comes from the U.S.-based company Nest, which has integrated generative AI into its smart thermostat ecosystem. By analyzing user preferences, occupancy patterns, and weather data, Nest devices can make real-time adjustments that reduce energy usage. According to a study conducted in California, households using Nest thermostats with generative AI capabilities saw an average energy savings of 10–12% annually, with higher savings in regions experiencing extreme seasonal temperatures. These results demonstrate how personalized AI-driven automation can deliver tangible benefits in terms of both cost savings and environmental impact.

In India, Tata Power has begun piloting AI-driven energy management systems in residential communities across Maharashtra. These systems use reinforcement learning and generative modeling to provide personalized energy-saving tips via mobile apps. Users receive suggestions like rescheduling appliance usage or modifying lighting patterns based on their historical behavior and local grid conditions. Preliminary results indicate that participants reduced their monthly energy bills by up to 500, demonstrating the tangible benefits of personalized AI interventions. These systems also help utilities manage demand more effectively, reducing peak load stress and improving grid stability.

Generative AI also holds promise for addressing energy equity. Low-income households often face barriers to adopting energy-efficient technologies or accessing tailored advice. These barriers may include limited financial resources, lack of digital literacy, or inadequate infrastructure. AI models can simulate the impact of targeted subsidies, behavioral nudges, or community-based interventions, helping policymakers design strategies that are both effective and inclusive. For example, in Riyadh, generative models were used to analyze energy usage in low-income neighborhoods. The findings informed the rollout of subsidized smart meters and personalized energy coaching, resulting in a 20% improvement in energy efficiency across pilot households. Such initiatives demonstrate how AI can be leveraged to promote equitable access to clean and affordable energy.

Moreover, the adaptability of generative AI ensures that recommendations evolve over time. As user habits change—due to lifestyle shifts, seasonal variations, or external factors like remote work—AI models update their predictions and suggestions accordingly. This dynamic feedback loop enhances user trust and long-term engagement. In Tokyo, researchers demonstrated how AI-powered smart homes adapted to post-pandemic work-from-home routines, optimizing energy usage during daytime hours and reducing peak load stress on the grid. The ability to respond to behavioral shifts in real time is a key advantage of generative AI, making it a valuable tool for managing energy systems in a rapidly changing world.

Despite its promise, the deployment of generative AI in personalized energy analysis must navigate several challenges. Data privacy is a major concern, as these models rely on detailed personal information to generate recommendations. Ensuring that consumers understand how their data is used, stored, and protected is essential for ethical implementation. Transparent consent mechanisms, anonymization techniques, and secure data storage protocols must be in place to build trust and safeguard user rights. Model transparency is another critical issue. Generative AI models can be complex and difficult to interpret, raising questions about accountability and fairness. Developing explainable AI frameworks that allow users and regulators to understand how decisions are made is crucial for responsible deployment.

User consent and control are also important considerations. While automated energy management systems offer convenience, users should have the ability to override recommendations, customize settings, and opt out of certain features. Empowering users to make informed choices about their energy usage fosters a sense of ownership and enhances the effectiveness of AI interventions. Additionally, efforts must be made to ensure that AI-driven energy solutions are accessible to all segments of the population, including those with limited digital literacy or access to smart technologies.

Looking ahead, the continued evolution of generative AI models—combined with advancements in smart home infrastructure, data governance, and user engagement—will unlock new possibilities for individualized energy management. As these technologies become more sophisticated and widely adopted, personalized energy insights will play a central role in building sustainable, user-centric energy systems. Real-world applications already demonstrate the potential of generative AI to reduce energy waste, lower costs, and promote equity. By analyzing detailed consumer data and delivering tailored recommendations, these models empower users to make informed decisions that align with their habits, values, and environmental goals.

Thus, generative AI offers a powerful pathway toward personalized energy management. Its ability to simulate realistic consumption scenarios, adapt to changing behaviors, and deliver actionable insights makes it an indispensable tool for modern energy planning. As the energy landscape continues to evolve in response to technological innovation, climate challenges, and shifting consumer expectations, generative AI will remain at the forefront of efforts to create smarter, more inclusive, and more sustainable energy systems.

IMPACT ON ENERGY EFFICIENCY AND SUSTAINABILITY

In the face of escalating climate challenges and rising energy demands, the application of generative artificial intelligence (AI) to energy consumption behavior analysis has emerged as a transformative strategy. The primary goal of this approach is to enhance energy efficiency and reduce carbon emissions by offering a deeper, more granular understanding of how energy is consumed across different contexts. Generative AI models—such as Generative Adversarial Networks (GANs), Variational Autoencoders (VAEs), and reinforcement learning agents—are capable of simulating realistic consumption scenarios, forecasting future trends, and evaluating the impact of various interventions. These capabilities enable stakeholders to design smarter, more responsive energy systems at the individual, community, and national levels.

Understanding and Simulating Energy Behavior

Traditional energy models often rely on aggregated data and static assumptions, which can obscure the diversity of consumption patterns across households and regions. Generative AI, by contrast, leverages large-scale, dynamic datasets—including weather conditions, appliance usage, occupancy schedules, and socio-economic indicators—to simulate energy behavior with high fidelity. For instance, GANs can generate synthetic energy profiles that reflect the daily usage patterns of households in different climate zones, income brackets, and building types. This allows researchers and policymakers to test the effectiveness of interventions—such as smart thermostats, LED lighting, or behavioral nudges—before implementing them in the real world.

A notable example comes from Zhang et al. (2021), who used GANs to simulate residential electricity consumption in urban China. Their model incorporated variables such as insulation quality, weather fluctuations, and household income, enabling policymakers to evaluate energy-saving campaigns with greater precision. The insights gained from these simulations informed targeted outreach strategies and infrastructure upgrades, resulting in measurable reductions in peak demand and overall energy usage.

Optimizing Renewable Integration

Beyond behavioral analysis, generative AI plays a critical role in facilitating the integration of renewable energy sources—such as solar and wind—into the grid. Renewable energy is inherently variable, with generation fluctuating based on weather and time of day. Predicting these fluctuations and aligning them with consumption patterns is essential for maintaining grid stability and minimizing reliance on fossil fuel backups.

Generative models can forecast both supply and demand under different scenarios, enabling more effective scheduling and dispatching of energy resources. For example, Fernandez et al. (2020) developed a GAN-based model to predict solar energy output in California using cloud cover and temperature data. The model helped grid operators anticipate dips in solar generation and adjust energy flows accordingly, reducing the need for carbon-intensive backup power. Similarly, Lee and Park (2022) used generative AI to simulate wind energy integration in South Korea, identifying optimal turbine placement and load balancing strategies that improved system efficiency by 14%.

Managing Distributed Energy Resources (DERs)

As more households and businesses adopt distributed energy resources (DERs)—such as rooftop solar panels, battery storage systems, and electric vehicles—the complexity of energy management increases. DERs offer flexibility and resilience but require real-time coordination to ensure grid stability. Generative AI can optimize DER usage by predicting energy generation and consumption at the local level and recommending adjustments to maximize efficiency.

In a pilot project in Maharashtra, India, Tata Power deployed reinforcement learning and generative modeling to manage energy flows in residential communities. The system learned to balance solar generation with battery storage and household demand, reducing grid dependency and improving sustainability. According to Kumar et al. (2023), participants in the pilot program reduced their monthly energy bills by up to 500 and reported greater confidence in managing their energy usage. These results demonstrate how AI-driven DER management can empower consumers while supporting broader energy goals.

Forecasting Environmental Impact

Generative AI also enables the forecasting of environmental impacts associated with different energy consumption patterns. By analyzing detailed usage data, these models can estimate carbon footprints at the household, neighborhood, or city level. This information can be used to design interventions that reduce emissions and promote sustainable practices.

For instance, Al-Mutairi and Chen (2024) used generative models to assess energy-related emissions in low-income neighborhoods in Riyadh. Their findings informed the rollout of subsidized smart meters and personalized energy coaching, resulting in a 20% improvement in energy efficiency and a significant reduction in carbon emissions. The ability to simulate and quantify environmental outcomes enhances the strategic value of energy planning and supports national climate commitments.

In Tokyo, Nakamura and Ito (2023) applied generative AI to model the impact of post-pandemic work-from-home trends on residential energy usage. Their simulations revealed a 25% increase in daytime electricity consumption, prompting utilities to adjust supply schedules and promote energy-saving behaviors through targeted messaging. These adaptive strategies helped mitigate the environmental impact of behavioral shifts and ensured continued grid reliability.

Empowering Sustainable Decision-Making

One of the most promising aspects of generative AI is its ability to empower consumers to make more sustainable decisions. By delivering personalized recommendations—such as switching to greener energy sources, upgrading inefficient appliances, or adjusting usage patterns—AI models can guide users toward environmentally responsible choices. These recommendations are based on individual behavior and preferences, making them more relevant and actionable than generic advice.

In the United States, Nest's smart thermostat system incorporates generative AI to analyze user preferences, occupancy patterns, and weather data. The system makes real-time adjustments that reduce energy usage without compromising comfort. According to Fernandez et al. (2020), households using Nest thermostats achieved annual energy savings of 10–12%, with higher savings in regions experiencing extreme seasonal temperatures. These results highlight the potential of AI-powered personalization to drive meaningful change at scale.

Addressing Challenges and Ethical Considerations

Despite its potential, the application of generative AI in energy systems must navigate several challenges. Data quality and availability remain critical concerns, especially in regions with limited digital infrastructure. High-performing AI models require granular, accurate, and representative data to function effectively. Inconsistent or biased data can lead to flawed simulations and misguided policy decisions.

Ethical considerations around synthetic data generation must also be addressed. While generative AI can anonymize sensitive information and protect user privacy, it can also inadvertently reproduce biases present in the training data. Transparency, accountability, and stakeholder engagement are essential to ensure responsible AI deployment. Developing explainable AI (XAI) frameworks and integrating user feedback into model design can help build trust and promote ethical innovation.

Technical integration poses another challenge. For generative AI to deliver real-time insights, it must interface seamlessly with existing energy infrastructure, including smart meters, IoT devices, and home automation systems. This requires standardized data formats, secure communication protocols, and robust cybersecurity measures. Collaborative efforts among utilities, technology providers, and regulatory bodies are needed to establish common standards and facilitate integration.

Looking Ahead

As generative AI technologies continue to evolve, their role in energy consumption behavior analysis will become increasingly central. The ability to simulate complex scenarios, forecast future trends, and deliver personalized insights positions these models as indispensable tools for modern energy planning. Real-world initiatives—from California to Maharashtra—demonstrate the tangible benefits of AI-driven energy management, including cost savings, emission reductions, and enhanced consumer engagement.

Moreover, generative AI supports the broader goals of climate resilience and energy equity. By enabling targeted interventions and inclusive planning, these models help ensure that the transition to clean energy benefits all segments of society. As governments, utilities, and consumers embrace digital transformation, generative AI will play a pivotal role in shaping a more sustainable and equitable energy future.

Thus, the application of generative AI to energy consumption behavior analysis offers a powerful pathway to improving efficiency, reducing emissions, and empowering sustainable decision-making. By integrating behavioral insights, renewable forecasting, DER optimization, and environmental impact modeling, these technologies provide a holistic framework for energy transformation. With thoughtful investment, ethical oversight, and collaborative implementation, generative AI can help achieve global climate and development goals—lighting the way toward a smarter, greener tomorrow.

CHALLENGES AND LIMITATIONS OF GENERATIVE AI IN ENERGY CONSUMPTION ANALYSIS

While generative AI offers transformative potential for analyzing and optimizing energy consumption behavior, its implementation is not without significant challenges and limitations. These hurdles span technical, ethical, infrastructural, and behavioral domains, and must be addressed to fully harness the capabilities of AI in building sustainable energy systems.

One of the most pressing challenges is the quality and availability of data. Generative AI models—such as Generative Adversarial Networks (GANs), Variational Autoencoders (VAEs), and reinforcement learning agents—require vast amounts of high-quality, granular data to function effectively. This includes detailed records of energy consumption, weather patterns, socio-economic indicators, appliance usage, and behavioral routines. However, in many regions, especially in developing economies, data collection systems remain underdeveloped or fragmented. According to

Kumar et al. (2023), rural districts in India lack consistent energy usage monitoring, making it difficult to train AI models that reflect local consumption realities.

Moreover, privacy concerns often limit the availability of personal consumption data. Consumers may be reluctant to share detailed information about their energy habits, fearing misuse or surveillance. This tension between data utility and privacy protection is a critical barrier. As noted by Al-Mutairi and Chen (2024), the success of AI-driven energy platforms in Saudi Arabia depended heavily on transparent data governance frameworks and user consent protocols. To address these concerns, governments and organizations must invest in robust data infrastructure and develop ethical frameworks that ensure privacy, security, and equitable access. The European Union's General Data Protection Regulation (GDPR) has been cited as a model for balancing innovation with individual rights in AI applications (Fernandez et al., 2020).

Another major challenge lies in the complexity of accurately modeling human behavior. Energy consumption is not solely a technical phenomenon—it is deeply influenced by cultural norms, economic conditions, lifestyle choices, and psychological factors. While generative AI can simulate a wide range of scenarios, capturing the full spectrum of human decision-making remains a formidable task. For example, during the COVID-19 pandemic, work-from-home arrangements drastically altered residential energy usage patterns. Nakamura and Ito (2023) found that AI models trained on pre-pandemic data failed to predict these shifts accurately, underscoring the need for continuous model refinement.

Human behavior is also subject to rapid change due to technological advances and policy interventions. The introduction of time-of-use pricing, for instance, can lead to significant shifts in consumption habits. Generative models must be updated regularly to reflect these evolving dynamics. Lee and Park (2022) emphasized the importance of adaptive learning in AI systems deployed in South Korea's smart grid, where consumer responses to tariff changes varied widely across demographics and regions.

Technical integration is another critical hurdle. For generative AI to deliver real-time predictions and actionable recommendations, it must interface seamlessly with existing energy management systems, including smart grids, IoT sensors, and distributed energy resources (DERs). This requires high levels of interoperability across platforms, protocols, and devices. However, many legacy systems lack the flexibility to accommodate AI-driven inputs. In a case study by Zhang et al. (2021), attempts to integrate GAN-based forecasting models with China's municipal energy infrastructure were hampered by incompatible data formats and limited API access.

To overcome these barriers, energy providers must adopt open standards and modular architectures that facilitate AI integration. The Pacific Gas and Electric Company (PG&E) in California has made strides in this direction by developing a cloud-based energy management platform that supports third-party AI applications. This system allows generative models to access real-time grid data and deliver predictive insights for load balancing and demand response (Fernandez et al., 2020).

In addition to technical challenges, there are concerns about the interpretability and transparency of generative AI models. These models often operate as "black boxes," making it difficult for stakeholders to understand how decisions are made. This lack of transparency can undermine trust and hinder adoption. Researchers like Kumar and Desai (2023) advocate for the use of explainable AI (XAI) techniques that provide human-readable justifications for model outputs, especially in high-stakes domains like energy policy and infrastructure planning.

Finally, equitable access to generative AI technologies must be considered. Without deliberate efforts to include marginalized communities, AI-driven energy solutions risk exacerbating existing inequalities. For instance, low-income households may lack the digital infrastructure needed to benefit from personalized energy recommendations. Al-Mutairi and Chen (2024) highlight the importance of inclusive design in AI systems, recommending subsidized smart meters and mobile-based interfaces to bridge the digital divide.

Thus, while generative AI holds immense promise for transforming energy consumption behavior analysis, its success depends on overcoming a range of challenges. These include improving data quality and availability, safeguarding privacy, accurately modeling human behavior, ensuring technical interoperability, enhancing transparency, and promoting equitable access. Addressing these issues will require coordinated efforts from governments, researchers, energy providers, and civil society. Only then can generative AI fulfill its potential as a cornerstone of sustainable and intelligent energy systems.

CONCLUSION

Generative AI holds immense promise for transforming the way we understand and manage energy consumption behavior. By offering deeper insights, predictive capabilities, and personalized recommendations, generative AI can reshape energy systems to be more efficient, responsive, and sustainable. Unlike traditional analytical models that rely on static datasets and linear assumptions, generative AI leverages large-scale, dynamic data and advanced algorithms—such as Generative Adversarial Networks (GANs), Variational Autoencoders (VAEs), and reinforcement learning—to simulate realistic energy usage scenarios and forecast future consumption patterns.

At its core, generative AI excels at identifying hidden patterns in energy behavior across households, industries, and communities. For example, GANs can generate synthetic energy profiles that mimic real-world consumption, allowing researchers and policymakers to test the impact of interventions—such as smart appliances, behavioral nudges, or tariff changes—without requiring intrusive or costly data collection. In a study by Zhang et al. (2021), GANs were used to simulate residential electricity usage in urban China, revealing that targeted adoption of smart thermostats and LED lighting could reduce peak demand by up to 18%. These insights informed municipal energy efficiency campaigns and helped utilities better manage grid loads.

Generative AI also enhances predictive capabilities by integrating diverse datasets, including weather patterns, socio-economic indicators, and behavioral trends. This allows for more accurate forecasting of energy demand and supply, especially in the context of renewable energy integration. For instance, Fernandez et al. (2020) developed a generative model to predict solar energy output in California based on cloud cover and temperature fluctuations. The model enabled grid operators to anticipate dips in solar generation and adjust energy flows accordingly, reducing reliance on fossil fuel backups and improving grid resilience.

Personalization is another key strength of generative AI. By analyzing individual-level data—such as appliance usage, occupancy schedules, and temperature preferences—AI models can deliver tailored recommendations that help consumers optimize their energy consumption. Lee and Park (2022) demonstrated this in South Korea, where generative AI was used to provide customized feedback to households based on their unique energy profiles. Participants who received personalized suggestions—such as shifting appliance use to off-peak hours or upgrading inefficient devices—achieved an average energy savings of 12% over six months.

Despite its potential, generative AI faces several challenges that must be addressed to fully realize its benefits. One major hurdle is data quality and availability. High-performing AI models require vast amounts of granular, accurate data, yet many regions lack robust data collection infrastructure. In rural India, for example, Kumar and Desai (2023) found that inconsistent energy monitoring limited the effectiveness of AI-driven energy planning. To overcome this, governments and energy providers must invest in smart meters, IoT sensors, and cloud-based platforms that facilitate real-time data collection and sharing.

Privacy concerns also pose a significant barrier. Consumers may be hesitant to share detailed information about their energy habits, fearing surveillance or misuse. Al-Mutairi and Chen (2024) emphasized the importance of transparent data governance in their study on AI adoption in Saudi Arabia. Their framework included user consent protocols, anonymization techniques, and secure data storage, which helped build trust and encourage participation in AI-driven energy programs.

Another challenge lies in the complexity of human behavior. Energy consumption is influenced by a wide range of factors, including cultural norms, economic conditions, and psychological preferences. While generative AI can simulate diverse scenarios, capturing the full spectrum of human decision-making remains difficult.

During the COVID-19 pandemic, for example, work-from-home arrangements drastically altered residential energy patterns. Nakamura and Ito (2023) found that pre-pandemic AI models failed to predict these shifts accurately, highlighting the need for continuous model refinement and adaptive learning.

Technical integration is also critical. For generative AI to deliver real-time insights, it must interface seamlessly with existing energy infrastructure, including smart grids, distributed energy resources (DERs), and home automation systems. This requires interoperability across platforms and protocols. In California, the Pacific Gas and Electric Company (PG&E) developed a modular energy management system that supports third-party AI applications, enabling predictive load balancing and automated demand response (Fernandez et al., 2020).

Looking ahead, the continued development of generative AI models—combined with advancements in data collection, smart infrastructure, and ethical frameworks—will enable more accurate and actionable predictions about energy consumption behavior. As these technologies become more widely adopted, they will play an integral role in helping governments, utilities, and consumers reduce energy waste, lower costs, and contribute to global sustainability efforts.

Real-world initiatives are already paving the way. In Maharashtra, India, Tata Power has launched AI-driven energy dashboards for residential communities, offering personalized tips and tracking carbon footprints. Early results show a 10–15% reduction in monthly energy bills and increased awareness of sustainable practices (Kumar et al., 2023). Similarly, the European Smart Grid Initiative (2021) used generative models to simulate the transition to 100% renewable energy in select regions, guiding infrastructure upgrades and community engagement strategies.

In conclusion, generative AI is not just a technological innovation—it is a strategic enabler of a more energy-efficient and sustainable future. By unlocking deeper insights, enhancing predictive accuracy, and empowering personalized action, generative AI can help reshape energy systems to meet the demands of a rapidly changing world. With thoughtful investment, ethical oversight, and collaborative implementation, this technology has the potential to drive meaningful progress toward global climate and energy goals.

REFERENCES

Al-Mutairi, H., & Chen, Y. (2024). Adaptive energy pricing using reinforcement learning: A case study in Saudi Arabia. *Energy AI*, *15*, 100234. DOI: 10.1016/j.egyai.2024.100234

European Smart Grid Initiative. (2021). Forecasting residential energy demand during climate extremes using GANs. *Smart Grid Reports*, *9*(2), 45–58.

Fernandez, J., Müller, T., & Rossi, L. (2020). Forecasting solar energy output using GANs: Applications in California's grid resilience. *Journal of Energy Systems*, *45*(3), 221–239. DOI: 10.1016/j.jes.2020.03.005

Kumar, R., & Desai, M. (2023). AI-driven energy planning in rural India: Challenges and opportunities. *Applied Energy*, *330*, 120456. DOI: 10.1016/j.apenergy.2023.120456

Lee, S., & Park, J. (2022). Adaptive learning in AI-based energy systems: Insights from South Korea's smart grid. *Energy Policy*, *165*, 112978. DOI: 10.1016/j.enpol.2022.112978

Nakamura, H., & Ito, Y. (2023). Behavioral shifts and AI model refinement in post-pandemic Tokyo. *Energy Research & Social Science*, *95*, 102875. DOI: 10.1016/j.erss.2023.102875

Zhang, L., Chen, Y., & Wu, D. (2021). Integrating GAN-based forecasting with municipal energy infrastructure: Lessons from China. *IEEE Transactions on Smart Grid*, *12*(6), 5432–5441. DOI: 10.1109/TSG.2021.3087654

Chapter 10
Generative AI for Personalized Energy Consumption Behavior Analysis

G. Kothai
https://orcid.org/0000-0002-9581-1679
SRM Institute of Science and Technology, India

S. Yogaazhaki
https://orcid.org/0009-0005-1026-6413
KPR Institute of Engineering and Technology, India

S. Deepika
https://orcid.org/0009-0008-8285-4603
KPR Institute of Engineering and Technology, India

B. Sivakarthick
https://orcid.org/0009-0009-6674-3919
KPR Institute of Engineering and Technology, India

K. Vignesh
https://orcid.org/0009-0004-8598-3938
KPR Institute of Engineering and Technology, India

ABSTRACT

The transformative role of Artificial Intelligence (AI), particularly Generative AI (GenAI), in reshaping global energy systems to be more decentralized, intelligent, resilient, and inclusive. It highlights how AI enables a shift from static, centralized

DOI: 10.4018/979-8-3373-4272-6.ch010

paradigms toward dynamic, adaptive energy networks that integrate renewable sources, smart infrastructure, and user-specific behavior. Generative models—such as GANs, VAEs, and diffusion-based architectures—enhance predictive accuracy, generate synthetic datasets, and simulate complex, uncertain scenarios across household, community, and urban scales. Through case-driven analysis, the chapter demonstrates how AI improves grid flexibility, supports equitable access, enables scenario planning for climate resilience, and informs real-time decision-making for policy and infrastructure design. It emphasizes the need for ethical, transparent, and community-engaged AI systems to prevent algorithmic bias and foster energy justice.

THE HUMAN ELEMENT IN ENERGY USE

Energy consumption is not merely a technical or infrastructural process; it is fundamentally a human activity. Every flip of a switch, such as setting the thermostat, charging a cell phone, and so forth, emanates from choices, habits, and cultural norms that individuals follow, (Madanchian, 2024). Energy systems would have been engineered with perfection had it not been for their inadequate consideration of this human aspect, thus undermining efficiencies and opportunities. Going beyond impressions of technical design or engineering approach, we can now design systems with the aforementioned energy consumption considerations of behavior, psychology, and lifestyle so that they are more flexible, inclusive, and sustainable for our communities, (Krishnan et al., 2025).

Energy Use Beyond Infrastructure

Human Decisions Are Central

In energy, the conversation usually revolves around technologies such as solar panels, smart grids, or energy-efficient appliances. Yet, behind every act of energy consumption is a human decision, (Sarvaiya et al., n.d.). When to turn on the air conditioning, or simply to leave the lights on through the night: these small decisions combine to form big, or macro-level, energy decisions. Understanding these patterns is a must for it to be able to find a way to reconcile engineering design and real-world usage, (Lin et al., 2025).

Technology Is Interaction Dependent

Interactions are different from the realization of any potential opportunity. Smart thermostats can be set for energy efficiency, but if a user turns them off due to discomfort or misunderstanding of how they operate, the effect is nullified. Interaction with people, therefore, is not a peripheral issue; it gives the independent systems its whole being, (Mongaillard, Lasaulce, & Varma, 2025).

Reframing Energy as a Human Systems Challenge

Sustainable energy use requires more than just developing new infrastructure; it calls for a holistic method that incorporates behavioral science, psychology, and cultural understanding. By treatment energy use as a human systems challenge, we advance better policy, technologies, and interventions, (Sarmas, Marinakis, & Doukas, 2025).

Patterns and Predictability in Energy Behavior in Residential Use

Daily and Weekly Rhythms

Residential energy use often has a stable and predictable daily cycle. In the mornings, usage rises with cooking, water heating, and lighting. Another demand peak occurs in the evenings when families get home, turn on the entertainment media, and operate appliances for their activities. There is a slightly different trend in weekends, with increased daytime usage. These predictable rhythms provide opportunities for targeted interventions and demand-side management, (Donta, Motlagh, & Tarkoma, 2024).

Household Demographics Considered

Household composition and size are critical in assessing energy use. Larger, more multigenerational families usually consume more energy across the day. Different households with children, elderly, or teleworkers are shaped in their consumption further by presence, care routines, and levels of activity, (Rane, 2023).

Usage Patterns Compared to Device Ownership

The way that appliances are used determines how much energy a home uses, not how many appliances there are. Even with fewer devices, a household may still use more energy because of inefficient habits or constant use. Conscious practices can significantly lower consumption, such as turning appliances to energy-saving modes or only heating water when necessary, (Naseer et al., 2025).

Distinguishing Between Active and Passive Users

Certain inhabitants engage with energy in a passive manner, adhering to deeply rooted routines with minimal consciousness. Others consciously choose to conserve by actively tracking their usage through apps or feedback tools. Designing individualized energy programs and nudges is made easier by understanding these user archetypes, (Wang et al., 2025).

Unreported Energy Losses

Many of the behaviors that cause energy loss go unnoticed. What are commonly known as "phantom loads" are caused by fans operating in vacant spaces, outdated refrigerators using excessive power, or entertainment systems left on standby. A combination of intelligent automation and behavioral insight is needed to address these inefficiencies, (Mohanty et al., 2024).

The Psychology of Habits and Energy Awareness

The Problem of Consumption That Is Not Visible

Electricity is intangible, in contrast to fuel or water. Until the monthly bill arrives, users frequently do not understand its flow or the magnitude of their consumption. Energy is made visible through tools such as mobile dashboards and real-time energy monitors, which assist users in creating a clear mental model of their consumption, (Yafei et al., 2024).

What Users Are Actually Motivated By

Although many people find environmental messages appealing, studies reveal that financial incentives frequently result in faster behavior change. Compared to abstract ecological appeals, messaging that highlights cost savings, like "Save 200 this month," tends to be more persuasive, (Jaboob, Hazaimeh, & Al-Ansi, 2025).

Decision Fatigue and Daily Routines

Daily decision fatigue frequently competes with energy-efficient behavior, (Jia, 2024). People may overlook energy-saving reminders or forget to unplug devices due to the cognitive strain of modern life. This disconnect between intention and behavior can be closed with the aid of systems that automate or subtly encourage actions, (Böcking et al., 2024).

Peer Influence and Social Norms

Energy use is social as well as private. Many users are inspired to change their own behavior when they discover that their neighbors are using less energy. Peer benchmarking and community-based comparisons have been demonstrated to promote energy conservation, especially in neighborhoods that are competitive or environmentally conscious, (Tripathi & Pandey, n.d.)

Establishing and Maintaining New Routines

Habit formation leads to long-term change. Repetition, reinforcement, and consistent cues are all part of this. Over time, gamified applications, reward-based programs, and AI-powered assistants that prompt or motivate users can all contribute to the development of an energy-mindfulness culture, (Golder, Das, & Mondal, 2024).

Seasonal, Cultural, and Lifestyle Factors that Affect Energy

Changing Lifestyles and Their Effects

The way people live today is changing quickly. Peak energy demand has moved from the evenings to the middle of the day due to the increase in remote work. Smart devices in tech-savvy homes provide new ways to optimize consumption, but they also raise baseline loads. While convenience-driven lifestyles frequently rely on continuously powered devices, preferences such as minimalism can result in lower energy use, (Silalahi, 2024).

Traditions and Cultural Norms

Energy patterns are defined by culture. Heating systems consume the most energy in colder climates, while cooling appliances are used most frequently in hotter regions. Cooking habits, such as using electric ovens for baking or gas stoves for extended periods of time, can have a significant impact on daily loads. Temporary energy spikes are also brought on by religious observances and festivals because of the high appliance usage and decorative lighting, (Sarvaiya et al., n.d.).

Behavioral Changes and Seasonal Variations

Throughout the year, weather variations alter energy requirements. Winter increases the need for heating, while summer increases the loads on air conditioners. Harvest seasons increase the amount of energy needed for storage and machinery in agricultural areas. In the meantime, the amount of daylight has an indirect impact on consumption habits by affecting lighting requirements and even sleep patterns, (Reddy, 2024).

Use Patterns and Socioeconomic Inequalities

Different energy behaviors are displayed by households with varying income levels. Due to financial limitations, lower-income families frequently restrict their use to necessities, which results in less but more efficient consumption, (Donta, Motlagh, & Tarkoma, 2024). Richer households, on the other hand, might place a higher value on comfort and convenience, utilizing smart kitchens, electric cars, and expensive entertainment systems. In order to ensure that recommendations and programs are fair and accessible, energy systems must be designed with consideration for these disparities. We can change systems from reactive to responsive and from rigid to adaptive by incorporating behavioral insight into energy design. Any sustainable and intelligent energy future must acknowledge the human element; it is not merely a sociological politeness.

TRADITIONAL ENERGY CONSUMPTION ANALYSIS'S DRAWBACKS

Traditional energy consumption models have been the cornerstone of energy sector planning, forecasting, and infrastructure development for many years. These models have helped utilities and policymakers project demand, design grid capacities, and assess large-scale interventions because they are based on long-term averages and population-wide assumptions, (Madanchian, 2024). However, these outdated methods are quickly becoming outdated as energy systems become more user-driven, interactive, and decentralized. They find it difficult to convey the variety, complexity, and urgency of contemporary energy use. In order to create intelligent, responsive systems that can support a sustainable energy future, it is imperative to acknowledge the shortcomings of traditional approaches, (Krishnan et al., 2025).

Models That Are Static the Challenge of Documenting Changing Human Behavior

Conventional energy models are essentially static. They are predicated on preconceived notions about socioeconomic factors, typical behavior, weather patterns, or building attributes. These presumptions might be adequate for planning at the aggregate level, but they are not prepared to deal with the complex and constantly shifting patterns of actual energy use, (Lin et al., 2025).

Household behavior is actually context-sensitive and dynamic. Visiting guests, shifting work schedules or even a brief preference for home-cooked meals over takeaway can all cause an unexpected increase in a family's energy usage. Static models rarely take into account these minor behavioral deviations, like turning on a fan overnight or moving meal preparation to the cooler evening hours. Additionally, user behavior changes as a result of personal preferences, technological adoption, and lifestyle changes. Load profiles can be drastically changed by adding a new appliance, such as a gaming console or smart oven, but traditional systems do not take these customized transitions into account. Poor forecasting, inefficient demand management, and low user engagement are the results of this mismatch between model predictions and actual results, (Sarmas, Marinakis, & Doukas, 2025).

Lack of Adaptability to Shifting Environmental Factors and Lifestyles

The incapacity of conventional energy models to adjust to the swift changes in the environment and lifestyle is one of their most serious drawbacks. Human existence is ever-changing. Sharp changes in energy consumption will occur in a household that is welcoming a newborn, moving to remote work, or taking care of an elderly family member. These modifications could happen gradually or overnight but legacy models are not designed to accommodate such fluidity. An additional layer of complexity is introduced by environmental variability.

Demands for heating and cooling can be significantly altered by abrupt weather events such as heatwaves or cold snaps. Similarly, a household can become a dynamic prosumer—producing, storing, and consuming energy in new ways—by adopting technologies like battery storage systems, electric vehicles, and rooftop solar. These changes require modeling tools that are flexible and responsive, which traditional systems are often unable to provide. Planning is another area where legacy models fall short. They are not well suited for adaptive demand forecasting, real-time energy pricing, or emergency response planning, because they rely on historical averages and fixed seasonal trends. Such rigidity restricts the system's capacity to maximize energy flow or respond to unexpected scenarios, thereby limiting both operational efficiency and resilience, (Rane, 2023).

One-Size-Fits-All Behavioral Diversity Is Ignored by Modelling

The intrinsic lack of personalization in traditional models is another significant flaw. Households are treated as identical units within a larger statistical average in the majority of traditional systems. This method ignores the wide range of behavioral variations found in both individual homes and communities. Due to differences in family size, work schedules, cultural norms, and device usage, identical homes—even those in the same neighborhood—can exhibit wildly disparate energy usage. While a single working professional may prefer minimal, automated energy use, a household with children and elderly members may run appliances longer and prioritize comfort over conservation. However, conventional models ignore these variations and offer general insights that don't connect with users, (Zhang et al., 2025).

Disengagement is exacerbated by the absence of personalized feedback. Users frequently disregard energy-saving tools that make general recommendations because they don't fit their lifestyles. Furthermore, models need to take socioeconomic factors into consideration because they also have an impact on behavior. While wealthier households may prioritize convenience over efficiency, low-income families may be more concerned about energy conservation out of necessity. Energy models must change to incorporate culturally relevant, demographically sensitive parameters that represent the diversity of the real world in order to address this, (Mohanty et al., 2024).

Insufficient Behavioral Integration and Real-Time Insights

Traditional energy analysis's disconnection from actual human behavior is arguably its biggest drawback. The majority of legacy systems use monthly or yearly historical data collection. By the time it is processed, this data is out of date, making it useless for prompt decision-making or intervention, even though it is helpful for billing or long-term trend analysis, (Lee et al., 2024).

Additionally, behavioral feedback loops are not included in these systems, (Yafei et al., 2024).

Instead of assuming that users are active participants, they assume that they are passive actors. They are therefore unable to identify user intent, comfort levels, or adaptive behaviors. The capacity of utilities or legislators to involve consumers in demand-side management initiatives, customized alerts, or responsive energy incentives is restricted by this omission.

Real-time AI analytics and the rise of IoT provide a clear way forward. Energy systems can now develop into interactive ecosystems thanks to sensors built into appliances, smart meters that provide real-time data, and artificial intelligence models that can predict short-term usage. These systems have the ability to learn user preferences, identify anomalies (like unusual spikes or equipment failures), and even anticipate needs, (Jia, 2024).

Intelligent, human-centered systems are the way of the future for energy modelling, as we move from static averages to dynamic learning, from one-size-fits-all to customized, and from reactive to proactive. Modern energy infrastructure has been greatly influenced by traditional energy models. But the terrain has shifted. The energy systems of today are decentralized, data-driven, and heavily influenced by human decisions rather than being top-down, predictable machines. Static assumptions, rigidity, lack of customization, and delayed feedback are some of the shortcomings of traditional models that are becoming more and more incompatible with this new reality. Adopting sophisticated modelling approaches that integrate behavioral science, real-time data streams, and adaptive algorithms is crucial to facilitating a successful energy transition, (Chan & Sigin, 2025). Only then will we be able to create energy systems that are both technically sound and intelligent in their behavior, able to meet the diverse, evolving needs of real people in a complex and changing world.

USING GENERATIVE AI TO REVOLUTIONISE SUSTAINABLE ENERGY SYSTEMS

A predictable, centralized infrastructure is quickly giving way to a more decentralized, data-driven, and behavior-influenced system in the global energy landscape. Energy systems are now characterized by uncertainty and continuous change as a result of the integration of renewable energy, electric vehicles, smart devices, and dynamic user behaviors, (Silalahi, 2024). Even though they are helpful, traditional predictive models are becoming less and less capable of handling these complexities. As a result, generative AI becomes a powerful instrument that can create behavior-aware energy systems, simulate data, and envision different futures. Its capacity to produce rather than just forecast signifies a significant change in our comprehension, modelling, and planning of sustainable energy, (Sarvaiya et al., n.d.).

Comprehending Generative AI: Fundamentals and Significance

A class of machine learning models known as "generative AI" is able to produce fresh, lifelike data samples by using patterns discovered in preexisting datasets. In contrast to conventional models that merely detect patterns or predict future values, generative models go one step further and create completely new possibilities that replicate actual circumstances. They are perfect for handling uncertainties in energy systems where real data may be limited, lacking, or unavailable for future scenarios because of their creative capacity, (Mongaillard, Lasaulce, & Varma, 2025).

Generative Adversarial Networks (GANs) and Variational Autoencoders (VAEs) are two of the most popular generative models. A generator that attempts to produce realistic outputs and a discriminator that assesses them are the two competing neural networks that make up GANs, (Berney et al., 2024). GANs learn to generate data through iterative feedback, such as artificial weather timelines or curves of household electricity consumption. In contrast, VAEs learn a compressed latent representation of the input data and use this space to reconstruct new outputs. They work particularly well at recording ongoing changes in energy usage that are impacted by variables like occupancy or temperature.

Newer architectures like transformer-based generators and diffusion models are also becoming more popular. Originally created for tasks like image synthesis or natural language generation, these models are currently being modified to model intricate multi-modal energy scenarios, which combine information from grid conditions, user behavior, and weather systems. These instruments are no longer limited to scholarly research. They are currently being tested in energy policy simulations, renewable energy forecasting, and smart grid management, (Foroughi et al., 2025).

Why Predictive Models by Themselves Are Insufficient

In order to predict future demand, pricing trends, or generation outputs, energy forecasting has long depended on predictive models that examine historical data. These models work well in environments that are generally stable, but they break down when faced with abrupt changes in behavior, new technologies, or policy. Traditional AI tools are reactive, which means that deviations from the norm frequently catch them off guard. The post-pandemic trend towards remote work, which significantly changed household energy profiles, is a prime example. The ensuing mid-day demand spikes were not predicted by predictive models that were trained on pre-pandemic trends. Similar to this, new and erratic energy flows have been brought about by the increasing use of rooftop solar and electric vehicles (EVs). There is no representation of these patterns in already-existing datasets, making it challenging for predictive models to adapt without retraining on possibly non-existing newer data. Perhaps more importantly, "what-if" scenarios cannot be explored by predictive models by nature. They are unable to model alternative futures or provide answers to hypothetical queries such as: What if the carbon tax is implemented abruptly? What if battery storage becomes commonplace in homes? What if extreme weather events become more common as a result of climate change? On the other hand, generative AI thrives on investigating such possibilities. Energy planners can create resilient and forward-thinking strategies by using the tools it provides to envision, model, and test various futures, (Zhang et al., 2025).

Generative AI for Behavior and Usage Simulation

Generative AI's capacity to replicate human behavior on a large scale is among its most revolutionary contributions. Artificial Intelligence (AI) facilitates a better understanding of how people might react to pricing signals, policy incentives, or environmental changes by creating synthetic profiles of household energy usage under various conditions. AI can simulate, for example, how urban residents might respond to time-of-use tariffs or how rural households might change their consumption in response to a solar subsidy, (Agoro & Daniel, 2025).

Planning for system resilience requires the ability to simulate uncommon or extreme events, which generative models excel at. Due to a lack of historical data, it is challenging to model events like blackouts, cyberattacks, or abrupt demand spikes during heatwaves. This gap is filled by generative AI, which produces artificial datasets that replicate the behavioral and statistical traits of these edge cases in order to get the system ready for emergencies, (Lee et al., 2024).

Data augmentation is another useful application. Many energy applications suffer from incomplete or skewed datasets, particularly in pilot studies or developing nations. In order to enhance model training, increase generalizability, and promote more inclusive energy planning, generative models can synthesize a variety of statistically valid samples. This is especially useful when attempting to represent user groups or behaviors that have not received enough attention, like off-grid communities, renters, or elderly populations.

Persona-driven modelling is also made possible by generative AI. Based on demographic traits, it can produce behaviorally different energy profiles, such as a retired couple in a rural residence, a family with school-age children, or a young urban professional living alone. These fictitious characters aid energy service providers in more effectively customizing their recommendations, pricing structures, and interventions in order to improve engagement and equity for all user groups, (Jaboob, Hazaimeh, & Al-Ansi, 2025).

Using Synthetic Intelligence to Revolutionize Renewable Energy Planning

Energy systems become more unpredictable and variable as a result of the switch to renewable energy sources. Geographically uneven, weather-dependent, and sporadic are the characteristics of solar and wind resources. By creating artificial weather and climate datasets that bridge temporal or spatial observation gaps, generative AI proves to be extremely useful in this situation. For example, GANs can produce realistic solar irradiance profiles for areas with little weather information, which helps with photovoltaic installation planning, (Böcking et al., 2024).

Additionally, generative models facilitate the modelling of net demand curves in households that integrate electric vehicles, battery storage, and rooftop solar. These homes act differently from traditional ones; in addition to user behavior, factors like sunlight availability, battery level, and EV usage influence consumption. Under these circumstances, AI can model a range of usage patterns, assisting utilities in managing load variability, forecasting demand, and creating suitable pricing schemes, (Golder, Das, & Mondal, 2024).

Generative AI provides a virtual laboratory for grid stress testing. Planners can model rapid wind drop-offs, investigate how cloud cover affects solar generation, or simulate high-penetration renewable scenarios. These simulations assist utilities in improving demand response plans, creating backup systems, and guaranteeing grid stability even in unpredictable circumstances, (Chan & Sigin, 2025).

Vehicle-to-grid (V2G) modelling is a particularly exciting area. Generative models can model how drivers might act in various situations as EVs develop into mobile storage devices that can feed electricity back into the grid. Will users let AI handle the trade-offs, maximize range, or minimize cost? Comprehending these behavioral reactions aids in the development of technical protocols, incentives, and policies that facilitate and enhance V2G integration, (Naseer et al., 2025).

A fundamental change in the analysis, planning, and management of energy systems is represented by generative AI. Generative models provide creativity—the ability to create realistic synthetic data, investigate unimagined futures (Mongaillard, Lasaulce, & Varma, 2025), and strengthen energy infrastructure resilience—while traditional models are superior at prediction. Such creativity is no longer a luxury as we confront a future characterized by behavioral instability, technological disruption, and climate volatility. It is now a necessity.

Generative AI will be essential to the design of inclusive, flexible, and sustainable energy systems by facilitating more human-aware, data-rich, and future-ready modelling. It gives decision-makers the ability to actively influence the energy futures we wish to live in, rather than merely reacting to change, (Madanchian, 2024).

GENERATING SYNTHETIC ENERGY DATA FOR MORE INTELLIGENT MODELLING

The availability and quality of data are crucial for transforming energy systems into intelligent, data-driven infrastructures. However, constraints like sparsity, privacy issues, and geographic imbalance frequently afflict real-world energy datasets. These problems impede the creation and implementation of strong AI and machine learning models that can assist with next-generation energy solutions. One effective solution to these problems is generative AI. A new era of experimentation, inclusivity, and ethical innovation in energy modelling is made possible by the creation of synthetic energy data that faithfully captures the statistical characteristics and behavioral subtleties of real-world usage, (Kimura, 2025). This chapter examines the ways in which synthetic data improves model training, safeguards privacy, aids underprivileged areas, and facilitates extensive simulation and validation pipelines.

Creating Artificial Datasets to Train Models

Difficulties with Actual Energy Data

Energy data collection is still irregular or nonexistent in many parts of the world, especially in low- and middle-income nations. Even in cases where infrastructure is present, data logging may be sporadic or poorly organized, and smart meter penetration may be low. Startups, researchers, and innovators are unable to access proprietary datasets because they are frequently only available to large utilities or government organizations. Furthermore, available data often represents a skewed demographic (e.g., high-income households, temperate regions, or urban users), leading to biassed models that do not generalize to larger user bases. These restrictions limit the deployment of intelligent energy systems and severely impede AI training, (Reddy, 2024).

Generative Models' Function in Dataset Expansion

Artificial datasets that mimic the statistical patterns present in actual energy use can be created using generative models like Generative Adversarial Networks (GANs), Variational Autoencoders (VAEs), and more recent architectures like diffusion models. Depending on user types, household characteristics, climate, and appliance ownership, these models can simulate energy consumption on a daily and hourly basis, (Mongaillard, Lasaulce, & Varma, 2025). Additionally, they can produce seasonal variability and energy signatures at the appliance level that are in line with local weather patterns or sociodemographic factors. Even when working with sparse or skewed real-world datasets, developers can construct more inclusive and adaptive models thanks to the ability to generate diverse and scalable artificial data.

Benefits of Model Development

In order to overcome fundamental machine learning issues like class imbalance, data sparsity, and lack of representativeness, synthetic datasets are essential. Researchers can greatly increase the accuracy and robustness of supervised and unsupervised learning models by training them on artificially enriched datasets. Access to synthetic variants catered to various social or geographic profiles also increases the efficacy of transfer learning, a technique that adapts pretrained models to new contexts. Furthermore, using limited real-world data—such as extreme heatwaves, grid instability, or uncommon behavioral patterns—would make it difficult or impossible to perform stress tests and sensitivity analyses, (Donta, Motlagh, & Tarkoma, 2024).

Using Synthetic Data as a Tool to Protect Privacy

Synthetic Data as a Privacy Tool

By creating anonymized synthetic datasets that maintain the original data's underlying statistical characteristics without being connected to specific people or residences, generative AI presents a convincing remedy. While removing the possibility of personal re-identification, these artificial duplicates maintain demand cycles, device-level profiles, and temporal coherence. Governments can release open datasets for public study, utilities can safely share data with outside developers, and private businesses can create apps without fear of regulatory repercussions, (Dua & Patel, 2024). Thus, synthetic data fosters innovation while upholding individual rights by striking a balance between model performance and ethical data use.

Realistic Use Cases

The range of real-world uses for synthetic energy data is growing quickly. By providing synthetic data sandboxes, utilities can work with app developers or AI solution providers while maintaining privacy. Using fictitious but representative user segments, government organizations can model energy access or policy outcomes. Meanwhile, startups can use this data to validate their algorithms and build energy applications, even without having access to real smart meter logs, (Wang et al., 2025). These uses demonstrate how synthetic data can support data governance guidelines while democratizing access to energy innovation.

Aiding Underserved and Data-Scarce Areas

Obstacles to Information Gathering

The lack of reliable energy data significantly restricts planning and intervention in areas with low rates of electrification or inadequate digital infrastructure. Smart meters, reliable internet, and data logging systems are absent from many rural and isolated communities. Historical records are frequently lacking or unstructured, even in cases where infrastructure is present, (Mohanty et al., 2024). As a result, there is a digital divide that prevents these areas from accessing data-driven solutions and cutting-edge AI-based energy models.

Filling Data Gaps with Generative AI

By creating believable energy usage profiles with contextual inputs and proxy datasets, generative AI can assist in bridging this gap. Building types and population density can be inferred from satellite imagery, and seasonal load demands can be informed by weather data. and data from the census Data can be used to estimate fuel preferences, heating or cooling requirements, and appliance penetration. Planners can model local demand patterns without requiring a lot of on-site data collection by combining realistic energy consumption datasets from these indirect indicators, (Yafei et al., 2024).

Inclusive Energy Planning Applications

Planning for rural electrification, developing and implementing microgrids, and assessing the feasibility of novel technologies like solar lanterns or battery-based storage systems all benefit greatly from this synthetic data. Before implementing policies like demand-side incentives or subsidies in the real world, governments and nonprofit organizations can test their effects on a variety of demographic and behavioral groups by creating virtual prototype. Without having to pay for expensive field research or pilots, energy access startups can test their business models with data that closely resembles actual user behavior, (Jia, 2024).

Applications in Simulation, Validation, and Training

Complete Model Lifecycle Assistance

Synthetic data is useful for more than just training models. It is essential to the training, validation, and simulation phases of the AI model lifecycle. Synthetic data can be used to add variability, avoid overfitting, and supplement small datasets during training, (Tripathi & Pandey, n.d.). Developers can test models against fictitious edge cases during the validation phase, like high demand events or grid failures, which are uncommon in real-world logs. Synthetic data facilitates "what-if" experimentation during the simulation phase, which involves examining system behavior under speculative scenarios that may not yet exist but are likely to do so in the future.

Actual Situations Improved by Artificial Simulation

For instance, if 40% of households install rooftop solar and home battery storage, energy utilities can simulate the potential changes in grid reliability. Thousands of artificial "virtual households" with a range of usage patterns and behavioral patterns can be used by home automation app developers to train AI models. By simulating how pricing schemes, such as energy credits or time-of-use tariffs, might impact low-income groups differently than higher-income users, policymakers can create more equitable energy policies without taking on real-world risks, (Rane, 2023).

Value in Strategy

A strategic advantage is produced by the speedy testing, iteration, and adaptation of synthetic data. It speeds up time-to-deployment, lowers the risks involved in implementing AI in critical infrastructure, and guarantees that solutions are resilient in a range of scenarios, (Chen, Karaduman, & Kuang, 2025). In a time of changing behavior, technological disruption, and climate uncertainty, synthetic energy simulation becomes more than just a modelling tool; it becomes a foundation of strategic foresight, (Naseer et al., 2025).

A major advancement in the development, testing, and application of AI models in energy systems is represented by synthetic data generation. It enables stakeholders to create energy solutions that are not only technologically sophisticated but also morally and geographically just by addressing data scarcity, protecting user privacy, and encouraging inclusive innovation. As we proceed with our shift to cleaner and more sophisticated energy infrastructures, synthetic data and the generative models that generate it will play a bigger part in global energy planning resilience, (Saleh et al., 2025).

USING AI TO MODEL CUSTOMISED ENERGY BEHAVIOUR

Personalization becomes a crucial component of infrastructure that is prepared for the future as energy systems move towards decentralization, digitization, and active consumer participation. Even with identical environmental or structural conditions, no two households use energy in precisely the same way. These discrepancies result from behavioral patterns, appliance usage patterns, lifestyle differences, and individual comfort preferences. These distinct subtleties can now be precisely modelled and responded to thanks to artificial intelligence (AI), especially deep learning, generative modelling, and unsupervised techniques. By creating customized energy profiles, identifying hidden trends, producing recommendations that are specific to each user, and adjusting to behavioral changes over energy systems will eventually become both intelligent and human-aware thanks to AI-powered personalization, (Krishnan et al., 2025). This chapter explores how AI can transform demand response, user empowerment, and energy system efficiency by enabling personalized energy behavior modelling.

Using AI to Create Customized Energy Profiles

Personalization Data Inputs

Large amounts of data are produced by contemporary smart homes from a variety of sources, reflecting the contextual and behavioral energy use of their occupants. These data streams can be ingested by AI systems to create detailed, up-to-date profiles of household consumption. Time-stamped electricity readings at hourly or even sub-hourly intervals are provided by smart meter data. By recording motion, occupancy status, room temperature, and appliance operation, Internet of Things (IoT) devices provide even more dimensionality. The context is further enhanced by metadata like household size, daily routines, or remote work habits. The modelling process is also greatly aided by external factors such as rooftop solar generation, time-of-use electricity rates, and local weather conditions. AI can learn how a particular household interacts with its environment and energy systems thanks to these diverse inputs, (Lin et al., 2025).

Building Digital Energy Twins

AI can combine these various inputs to produce intricate digital representations of specific homes or users, known as "digital energy twins," by utilizing methods like clustering, classification, and regression. Base load signatures, preferred usage periods, tariff sensitivity, and even appliance-level malfunctions are all captured by these digital twins. A model might discover, for instance, that a household prefers to run high-power appliances like ovens or washing machines in the late evening or that it responds strongly to temperature changes by modifying HVAC usage. These customized models enhance the effectiveness of demand shaping and grid-level forecasting initiatives in addition to enabling customized energy-saving measures, (Mongaillard, Lasaulce, & Varma, 2025).

Finding Hidden and Subtle Patterns in Behavior

Conventional Analytics' Drawbacks

Averages and aggregates play a major role in traditional energy analysis techniques. They provide general insights into consumption patterns, but they frequently miss subtle, changing, or unusual user-level behaviors. Slow behavioral changes brought on by life events or schedule changes, or small but persistent standby consumption (also called phantom loads), can be overlooked, (Berney et al., 2024). Unusual increases in consumption brought on by malfunctioning appliances or ageing systems might occasionally go unnoticed as well. Unnecessary energy waste and higher expenses are caused by these hidden inefficiencies.

Deeper Pattern Recognition with AI

AI is excellent at spotting hidden or new usage patterns, particularly through unsupervised learning and temporal pattern detection. Unexpected energy events, like power consumption while a home is empty or an increase in heating load that points to a failing HVAC system, can be identified by anomaly detection algorithms. A user beginning to work from home or an increase in kitchen usage, which suggests a change in household routines, are examples of behavioral transitions that clustering models can identify. A proactive approach to energy efficiency and prompt interventions are made possible by these insights, (Chen, Karaduman, & Kuang, 2025).

Applications and Perspectives

Utilizing these methods, AI can recommend behavior changes to move load away from peak times, identify inefficient appliances based on consumption profiles, and even identify possible electrical problems before they become dangerous. For example, a suggestion might point out that the user's refrigerator is using more electricity than normal, which could indicate that maintenance is necessary. Customers and utilities can function more effectively, safely, and sustainably thanks to this layer of intelligence, (Donta, Motlagh, & Tarkoma, 2024).

Personalized Predictions and Recommendations

Active Coaching Rather Than Passive Monitoring

AI has grown from passive data recording to active consumer coaching. With powerful time-series prediction models like Recurrent Neural Networks (RNNs), Long Short-Term Memory (LSTM) networks, and neural hybrids, AI can now make personalized predictions of energy consumption at the individual level. These predictions take into account future weather, occupancy patterns, historical behavior, and tariff variations to produce accurate, near-term predictions. For instance, it can forecast energy consumption for the next day or bills for the coming month based on present trends, and provide recommendations for maximizing use, (Rane, 2023).

Actionable Insights

Coupled with intuitive mobile applications or web-based interfaces, such projections become real-time actionable insights. Instead of generic power-saving suggestions, AI-powered interfaces offer customized nudges. A consumer might get a message saying, "Saving 150/month can be done by adjusting air conditioner usage to after 9 PM," or "Your water heater operates longer than needed—reduce the timer to conserve energy." These micro interactions foster energy consciousness and enable long-term behavioral shifts, (Dua & Patel, 2024).

Consumer and Utility Benefits

For customers, such personalization makes energy efficiency possible, easy, and interesting. It enables them to make informed decisions based on data, manage their usage, and match their energy consumption with financial or environmental objectives. For utilities, such personalization provides greater means of demand-side management, load balancing, and preventing peak overloads, (Naseer et al., 2025). This two-way benefit creates efficiency on both the user and system levels, creating a more robust and efficient energy network.

Adaptive Models That Evolve with Lifestyle Changes

The Need for Continuous Learning

Energy consumption is dynamic by its very nature. Human activity is shaped by many variables—ranging from life experiences (such as moving, having children, or retirement) to seasonal changes and economic factors. Static AI, once trained, becomes obsolete rapidly if the models do not reflect these evolving variables. A user profile that is valid one month may no longer be accurate a few months later. Therefore, flexibility becomes an essential necessity for effective personalized energy modeling, (Wang et al., 2025).

Key Technologies for Adaptivity

In order to stay current, personalized AI systems must continuously adapt to users. Online learning algorithms update model parameters dynamically as new data becomes available, eliminating the need for complete retraining. Reinforcement learning mechanisms allow systems to evolve by incorporating user feedback and behavioral changes, thereby improving recommendations over time. Additionally, transfer learning enables models developed for one group or household type to be adapted efficiently for others with similar characteristics, significantly reducing the cost and time involved in personalization, (Agoro & Daniel, 2025).

User-in-the-Loop Intelligence

Today's personalized energy systems integrate the user into the model improvement process. Users can give feedback—such as accepting, rejecting, or modifying AI-generated proposals—thereby reinforcing preferred outcomes and correcting errors. This feedback loop ensures that the AI enhances its performance without compromising user autonomy. It also fosters trust, as users see their habits and preferences meaningfully represented in the system's recommendations, (Mohanty et al., 2024).

AI personalization is transforming how we comprehend and control residential energy use. Through the creation of digital twins, the identification of latent patterns, delivery of customized recommendations, and continuous adaptation to actual behavior, AI renders energy systems more responsive, fair, and efficient. Personalized energy modeling enables individuals to make better decisions while providing utilities with granular capabilities for dynamic load management, (Lee et al., 2024). The future of energy will be shaped not only by technological advancement but by how well that technology understands, assists, and responds to the individuals it serves.

GENERATIVE AI SCENARIO PLANNING FOR THE ENERGY SYSTEM

The world energy system is changing rapidly, driven by decentralization, the growth of renewable energy, evolving user behavior, and increasing climate risk uncertainty. These developments introduce a level of complexity and unpredictability that traditional linear forecasting models struggle to manage effectively. Generative AI emerges as a powerful solution to this challenge, offering the capability to model a vast array of future scenarios.

These statistically informed, synthetic futures enable stakeholders to simulate policies, technologies, and infrastructure proposals across diverse potential outcomes, (Zhang et al., 2024). In doing so, generative AI transforms energy planning into a dynamic, adaptive, and forward-looking process, better equipped to navigate the uncertainties of the energy transition.

Simulating User Response to Pricing and Policies

Behavioral Variability Under Economic Signals

Pricing structures such as Time-of-Use (ToU) tariffs, real-time pricing, and peak load surcharges are increasingly being employed by regulators and utilities to influence consumer behavior. However, consumer responses to these economic signals vary widely. While some individuals actively adjust their usage patterns to reduce costs, many others may ignore price fluctuations due to entrenched habits, lack of awareness, or differing lifestyle priorities.

Generative AI can effectively capture and simulate this diversity by modeling various consumer archetypes. These models take into account factors such as demand elasticity, socio-economic constraints, and levels of access to information, enabling a more nuanced understanding of how different user segments might respond to specific pricing interventions, (Jaboob, Hazaimeh, & Al-Ansi, 2025).

Policy Adoption and Resistance Modeling

In addition to economic incentives, the effectiveness of energy policies—such as electric vehicle (EV) subsidies, rooftop solar incentives, or bans on inefficient appliances—depends heavily on public adoption. Generative models empower planners to forecast public response by creating synthetic populations that vary across cultural, regional, and psychological dimensions. For instance, these models can estimate the likelihood of early adoption, delayed uptake, or outright rejection of a policy by analyzing past behaviors and demographic patterns. This capability allows policymakers to simulate scenarios involving partial compliance, technology fatigue, or regional resistance, (Böcking et al., 2024). As a result, they can refine communication strategies and incentive structures before full-scale policy deployment—improving impact, equity, and effectiveness.

Projecting Demand Under Regulatory or Climate Changes

Simulating Policy-Driven Demand Curves

As governments globally put forward ambitious mandates like net-zero goals, carbon pricing, or internal combustion vehicle bans, the consequent transitions in energy demand will be deep and pervasive. Generative AI can also model the ways in which these policies will reform demand curves, both at macro and micro levels. For instance, widespread EV adoption may result in sharp evening peaks in residential load, while carbon pricing would cut industrial electricity consumption but add to electrification in transport and heating. These estimates are essential for long-term grid planning so that generation capacity and transmission infrastructure is matched with changing demand profiles, (Tripathi & Pandey, n.d.).

Climate-Adaptive Load Scenarios

Severe weather events and sustained climatic trends are increasingly impacting energy consumption patterns. Generative models can integrate historical usage data, simulated climate variables, and behavior-based assumptions to produce adaptive load forecasts, (Golder, Das, & Mondal, 2024).

This allows utilities to effectively model seasonal demand peaks, shifts in heating and cooling requirements, and load curve volatility caused by unpredictable weather conditions, (Chan & Siqin, 2025). These insights are especially valuable for designing climate-resilient infrastructure and responsive demand-side strategies, particularly in regions that are highly vulnerable to extreme temperature fluctuations or weather-related disruptions.

Assessing New Technologies: Smart Grids, V2G, and Microgrids

Complexity of the System with New Technologies

The increased integration of Distributed Energy Resources (DERs), electric vehicles (EVs), smart homes, and microgrids introduces significant technical and behavioral complexity into the energy system. Traditional deterministic models are inadequate for capturing the interactive, non-linear, and probabilistic nature of these emerging technologies.

For instance, EVs not only consume electricity but can also supply it back to the grid through Vehicle-to-Grid (V2G) systems. Residential users have become prosumers, simultaneously generating, consuming, and storing energy. Microgrids can isolate themselves from the main grid during emergencies (a process called islanding) and reconnect when conditions stabilize, (Madanchian, 2024).

Generative AI can model these multidirectional, context-sensitive energy flows, helping energy planners and grid operators to understand system interactions under a wide variety of operating conditions. This capacity is vital for optimizing future-ready, flexible, and resilient energy infrastructures.

Virtual Testbeds for Innovation Evaluation

Generative models enable the creation of virtual test environments where new technologies and business models can be evaluated without real-world risk. For example, planners can simulate how a smart neighborhood responds to dynamic pricing, assess the impact of Vehicle-to-Grid (V2G) operations on battery health, or explore how Distributed Energy Resources (DERs) behave under different control algorithms. These virtual testbeds are immensely valuable for stress-testing grid resilience, evaluating return on investment for innovative solutions, and ensuring scalability of experimental approaches. As a result, developers and regulators can iterate more rapidly, reduce the cost and risk of pilot programs, and accelerate the deployment of proven solutions with greater confidence, (Silalahi, 2024).

Scenario Modeling for Resilient Infrastructure

Designing for Unknowns

Perhaps the most powerful application of generative AI in scenario planning lies in its ability to model low-probability, high-impact events—commonly referred to as "black swan" scenarios. These may include complex situations such as a simultaneous cyberattack and natural disaster, large-scale defection from centralized energy grids, or cascading failures across critical infrastructure systems. Traditional risk models struggle to account for such compound and unpredictable threats. Generative AI, however, can simulate thousands of plausible futures by systematically varying behavioral inputs, environmental shocks, and systemic responses, (Kimura, 2025). This capacity helps identify hidden vulnerabilities within energy systems and guides the development of resilient infrastructure codes, emergency response protocols, and fail-safe system architectures.

Steering Infrastructure Investment

By modeling how energy systems respond to stress, generative AI offers detailed views of where investments must be concentrated. For instance, it can identify where grid strengthening is critical, where backup power storage would be most valuable, or where microgrids would be most useful in building community resilience. These models can also analyze the cost-benefit balance of various strategies for improving resilience under different synthetic conditions, facilitating data-driven infrastructure planning that is both economically sound and climate-resilient, (Krishnan et al., 2025).

Redesigning Energy Planning in the Generative Era

Generative AI redefines the energy planning paradigm from prediction to imagination. Policymakers, researchers, and engineers can now visualize and investigate a vast number of potential futures—each representing a unique combination of technological uptake, behavioral patterns, policy actions, and environmental changes. These explorations not only build readiness but also democratize planning, enabling various stakeholders to participate in crafting resilient and inclusive energy systems. In this era of complexity and disruption, generative AI-powered scenario planning offers a way forward that is not only technically sound but also human-centric and future-fit, (Sarvaiya et al., n.d.).

SMART HOMES AND SMART AI: IOT-BASED REAL-TIME INSIGHTS

The development of energy systems is moving toward a future defined by decentralization, intelligence, and personalization. Smart homes equipped with Internet of Things (IoT) appliances and Generative Artificial Intelligence (GenAI) capabilities are becoming foundational elements of a more dynamic, resilient, and sustainable energy ecosystem. These intelligent homes are no longer passive consumers of electricity; instead, they are active participants in energy management—capable of optimizing consumption, integrating distributed generation, and contributing to grid stability. Given that residential homes account for a significant share of global energy use, the widespread adoption of smart, energy-aware living spaces is essential to achieving energy equity, meeting decarbonization goals, and enhancing grid flexibility, (Kimura, 2025).

Connecting Future Energy Plans and Real-Time Action

From Policy Vision to Household Implementation

Net-zero carbon and high-penetration renewable energy targets for energy transformation are implemented through strategic frameworks that require the engagement of households at the individual level to bring policy into action. While regulation and pricing mechanisms are developed by national and regional authorities, the actual effectiveness of these interventions depends on the behavioral and operational changes carried out within homes. Smart homes powered by AI can autonomously coordinate appliance operations and load behaviors with high-level system objectives—such as utilizing solar energy during peak production periods or responding to grid stress alerts. This capacity to interpret and execute macro-level strategies at the micro scale enhances the fluidity of policy implementation and accelerates the transition to decentralized, intelligent energy systems, (Lin et al., 2025).

Real-Time Responsiveness and Proactive Engagement

Through the use of predictive analytics and models of user-specific behaviors, AI-powered homes can proactively sense changes in energy demand and take anticipatory measures to maximize usage. For instance, an AI assistant could predict a peak in solar output and suggest rescheduling heavy-duty activities—like doing laundry or running the dishwasher—to align with times when clean, low-cost energy is available. Through this intelligent negotiation between real-time data and strategic intent, smart homes become instruments of energy policy implementation, enabling occupants to live sustainably without sacrificing comfort, (Mongaillard, Lasaulce, & Varma, 2025).

IoT as a Medium of User and System Interaction

Building the Home's Digital Nervous System

IoT devices constitute the core of the smart home, forming a ubiquitous sensing and communication platform that gathers, transmits, and processes data in real time. Environmental sensors monitor parameters such as temperature, lighting, and air quality; occupancy detectors assess room usage; and smart meters track energy consumption across individual appliances. This distributed sensory network enables context awareness and facilitates intelligent automation. For example, a home can automatically switch off HVAC systems in unoccupied rooms or dim lights based on the availability of natural daylight. These adaptive responses help conserve energy without compromising occupant comfort, (Sarmas, Marinakis, & Doukas, 2025).

Grid-Connected Communication and Bidirectional Control

The connectivity of IoT-enabled homes with the broader grid environment enables dynamic, bidirectional energy management. These homes do not merely receive signals—such as price alerts or demand response requests—but also transmit data on local generation, storage, and consumption. This two-way communication enhances grid visibility, supports load balancing, and facilitates household participation in the coordination of distributed energy resources (DERs). As such, IoT serves as a vital interface between users, appliances, and the energy grid, transforming homes into intelligent, communicative nodes within a distributed energy network, (Chen, Karaduman, & Kuang, 2025).

Generative AI for Adaptive, Predictive Control

From Rule-Based Automation to Learning-Based Intelligence

While early smart home automation relied on rule-based systems and static schedules, generative AI ushers in a new era of continuous learning and adaptive, intelligent decision-making. These advanced models adjust in real time by analyzing current user behavior, environmental conditions, and historical patterns to forecast future energy demand. They evaluate multiple courses of action—such as delaying appliance usage or adjusting thermostat settings—and choose the most efficient option based on factors like cost, carbon footprint, and user preferences. This capacity to model and assess potential future scenarios transforms homes into proactive, self-governing energy managers, (Saleh et al., 2025).

Scenario-Based Energy Optimization

Generative AI in smart homes is capable of responding to complex "what-if" scenarios—such as how energy consumption would change with the addition of an electric vehicle or during an extended heatwave. By combining predictive modeling with behavioral dynamics, these systems can modify operations in response to changing conditions without compromising efficiency or user comfort, (Dua & Patel, 2024). As a result, smart homes become resilient to external uncertainties while remaining aligned with both broader system-level goals and individual user preferences.

Automating Home Systems for Optimal Efficiency

Whole-System Synchronization for Energy Smartness

An ideal smart home is built on the coordinated automation of all subsystems—lighting, heating, cooling, appliances, and storage. AI and IoT enable cross-device orchestration, where individual actions are not optimized in isolation but as part of an integrated system. For example, an intelligent HVAC system can adjust its cooling schedule based on predicted occupancy, while a smart washing machine may automatically shift its operation to off-peak hours based on dynamic tariff signals. Similarly, excess solar power stored in home batteries can be utilized during evening peak demand—not only to meet household energy needs but also to support overall grid stability, 9Zhang et al., 2025).

User Experience Improvement Concurrently with Energy Saving

Smart automation also enhances the resident experience by offering personalized comfort and actionable insights. Lighting controls can be synchronized with circadian rhythms, voice assistants can provide real-time feedback on energy-saving opportunities, and appliances can suggest operating schedules based on projected energy prices. This seamless blend of convenience and sustainability transforms the home into an intelligent agent within the energy ecosystem—simultaneously promoting human well-being and environmental responsibility.

Smart homes, empowered by IoT sensing and generative AI reasoning, are no longer abstract concepts; they are rapidly becoming a reality with transformative potential. By continuously sensing, predicting, and adapting in real time, these homes evolve into self-regulating ecosystems that contribute actively to grid flexibility, emissions reduction, and user empowerment. As we move toward a distributed, electrified, and sustainable energy future, intelligent homes will play a pivotal role in ensuring system-wide resilience and user-centric transformation. The convergence of smart infrastructure and generative intelligence marks a seminal moment in the construction of future energy systems—where homes do not merely consume energy but dynamically shape its flow, demand, and impact, (Yafei et al., 2024).

BALANCING THE GRID: DEMAND MANAGEMENT WITH GENERATIVE AI

As the world's energy systems transition toward greater decentralization, electrification, and the widespread adoption of renewable energy resources, the traditional assumptions that underpin grid stability are being fundamentally re-evaluated. The variability of renewable generation, dynamic and unpredictable load patterns, the rapid proliferation of electric vehicles (EVs), and the emergence of distributed energy resources (DERs) have introduced unprecedented complexity into energy management.

In this context, Generative Artificial Intelligence (GenAI) emerges as a pivotal technology. It enhances demand-side intelligence, improves the accuracy of forecasting, enables the simulation of synthetic scenarios, and facilitates the coordination and management of distributed assets. By integrating behavioral modeling, synthesizing diverse data sources, and enabling real-time decision-making, GenAI empowers utilities and grid operators to balance energy supply and demand with greater foresight, adaptability, and precision.

Forecasting Peak and Off-Peak Loads for Grid Stability

Limitations of Traditional Forecasting Methods

They typically use static models calibrated from past data. Though successful in stable contexts, they are less and less capable in the context of contemporary energy realities. Abrupt changes in societal behavior, e.g., the mass trend towards remote work, can abruptly change consumption patterns. Analogously, increasing EV penetration imposes new and nonseasonal demand spikes. In addition, weather-sensitive devices such as air conditioners increase variability in the event of extreme climate conditions. Such disruptions reduce the reliability of models that depend on historical averages alone or linear trend extrapolation, (Krishnan et al., 2025).

Generative AI for Probabilistic Load Prediction

Generative AI, based on models such as Generative Adversarial Networks (GANs), Variational Autoencoders (VAEs), and Probabilistic Neural Networks, introduces a stronger framework for demand forecasting. These models can also simulate several likely demand paths instead of relying on a single deterministic estimate. They incorporate exogenous variables—such as weather forecasts, tariff notices, and mobility data (Lin et al., 2025)—to deliver context-dependent forecasts.

This feature is especially useful for forecasting peak ramping events and maintaining system readiness under various load curve scenarios. The outcome is more efficient load management that minimizes the likelihood of blackouts, reduces operational expenses, and enables higher levels of renewable integration.

Distributed Energy Resource (DER) Integration Optimization

Issues in Coordinating Decentralized Power Flows

The shift from centralized to distributed generation introduces bi-directional power flows, dynamic voltage variations, and greater local congestion vulnerabilities. Prosumers—users who generate energy themselves using rooftop solar or other dispersed technologies—pose challenges for conventional grid structures, which were not originally designed to manage such complexity. In the absence of intelligent coordination, Distributed Energy Resources (DERs) can worsen grid instability

(Berney et al., 2024), trigger reverse power flows, and complicate system-wide demand response planning.

AI-Driven DER Simulation and Dispatch Strategies

Generative AI allows for the simulation and control of Distributed Energy Resource (DER) performance under different temporal, economic, and environmental conditions. AI systems can model how EV owners in a city might respond to price incentives, or how solar panel output could fluctuate with cloud cover across different residential blocks. Such insights support coordinated DER scheduling—for instance, dispatching stored battery power during peak demand hours, or reducing EV charging during periods of grid stress, (Chen, Karaduman, & Kuang, 2025). This optimization operates across individual households, smart neighborhoods, and even entire cities, thereby increasing grid flexibility and maximizing the value of distributed assets.

Leverage Synthetic Data to Improve Grid Resilience

Mitigation of Data Deficiency and Model Blind Spots

High-resolution, high-quality energy data remains scarce in many regions, especially in developing economies or remote locations. Moreover, rare but high-impact events—such as cyberattacks, natural disasters, or sudden policy shifts—are often absent from historical datasets, (Donta, Motlagh, & Tarkoma, 2024). These data gaps create blind spots during the training of AI models, thereby limiting their resilience and reducing their ability to respond effectively to critical and unforeseen scenarios.

Synthetic Scenario Generation for Resilient Planning

Generative AI is capable of creating synthetic datasets that reflect statistical and behavioral tendencies of actual data while adding variability for unseen conditions. For example, synthetic data may mimic a citywide blackout event and subsequent EV-charging spike, facilitating stress testing of grid infrastructure (Saleh et al., 2025) and recovery procedures. Such synthetic edge-case situations enable AI models to "learn" from infrequent events, enhancing their predictive capabilities and robustness, (Rane, 2023). Essentially, synthetic data serves as an immunity system, digitally inoculating predictive instruments from shocks and uncertainties in the future.

Real-Time Demand-Side Response Through AI

Behavioral Modeling for Demand Shaping

Successful demand-side management (DSM) relies on predicting and shaping user behavior in real time. Generative AI enables the generation of individualized energy profiles, predicting when users will use energy intensively and how they may react to incentives. For instance, an AI system might conclude that a specific home is likely to switch on an air conditioner at 6 PM and propose an economic incentive to push it back by half an hour, (Dua & Patel, 2024). Such simulations of user behavior are the foundation of adaptive DSM solutions that respond in real time to user situation and feedback.

Personalized and Automated Demand Flexibility

In contrast to blanket DSM programs that broadcast standard price signals, AI-based systems can generate micro-targeted control strategies. These are tailored nudges ("Keeping your geyser running 30 minutes later will save 20 today"), automatic appliance scheduling, and pre-cooling or pre-heating habits in anticipation of occupancy predictions. In addition, these systems learn from feedback from the user, developing their recommendations and control patterns to strike a balance between user comfort and grid-level goals, (Wang et al., 2025). The outcome is a smart, engagement-based DSM model—a model that turns customers into responsive, intelligent collaborators of grid stability.

Generative AI is revolutionizing the conventional demand management paradigm by injecting intelligence, variability, and forecasting resilience at each level of grid operation. It enables utilities to shift away from rigid planning and adopt dynamic simulation, real-time customization, and forward-looking scenario modeling. Through the combination of distributed assets, predictive behavioral response, and the use of synthetic data for training resilient models, generative AI remodels energy demand forecasting, management, and shaping. With increasingly electrified, digital, and decentralized energy systems, such a smart demand management strategy is essential to create strong, carbon-free grids that can serve future energy demands, (Mohanty et al., 2024).

INCLUSIVE AI: CUSTOMIZING ENERGY ACCESS AND EQUITY

With Artificial Intelligence (AI) revolutionizing the world's energy systems through better prediction, optimization, and automation, comes a daunting obligation: ensuring the benefits are shared fairly. More than 700 million people around the world remain without electricity access, and millions more suffer from problematic service, prohibitively expensive bills, and poor infrastructure. Here, "smart" energy should not just be smart—it should be inclusive. Inclusive AI understands the gaps in access, representation, and digital preparedness, and seeks to bridge them with ethical, adaptive, and community-based design. Through equity and personalization, it finds new avenues toward an equitable energy transition that benefits all strata of society, (Yafei et al., 2024).

Expanding Smart Energy Systems to Remote Communities

Structural Barriers in Remote Energy Access

Remote and rural communities face specific structural barriers that make it difficult for them to participate in the energy revolution. Grid connectivity to these locations is often economically unfeasible due to long distances, challenging terrain, and low population density. Furthermore, many of these communities rely on costly, polluting fuels such as diesel or biomass, which contribute to environmental degradation and have serious health consequences. The lack of digital infrastructure—including limited mobile coverage, unstable sensors, and low levels of digital literacy—further exacerbates the challenge, leaving these areas excluded from the benefits of AI-driven energy solutions, (Zhang et al., 2024).

AI as an Enabler of Rural Electrification

Generative AI and machine learning create new possibilities to bypass conventional infrastructure and provide customized energy solutions to remote locations. AI systems can predict energy requirements in remote villages based on sparse or noisy data such as mobile phone usage, satellite imagery, and weather patterns. These systems can also model the optimal location and sizing of solar microgrids by taking into consideration local economic activity, population density, and projected growth. Predictive maintenance algorithms can be implemented with minimal sensor input, reducing the cost of field services. By offering context-sensitive and scalable tools, AI acts as an accelerator for rural energy justice and sustainability, (Jia, 2024).

Personalization of Guidance in Low-Data Conditions

Issues in Light Data Situations

In most informal settlements and underdeveloped areas, AI systems face a "low-data dilemma." Conventional machine learning methods, which rely on large sets of labeled historical data, struggle in these contexts—where consumption histories, smart meters, or appliance-level monitoring are often non-existent. Moreover, energy habits in such regions are frequently shaped by complex cultural norms, seasonal migration, and communal usage patterns that differ significantly from urban or industrial environments, (Böcking et al., 2024).

Transfer and Few-Shot Learning for Inclusive AI

Inclusive AI tackles these issues using advanced techniques such as transfer learning and few-shot learning. Transfer learning allows models trained in data-rich environments to be adapted to similar, low-data settings with minimal additional information. Few-shot learning enables accurate classification or prediction of user behaviors based on only a few examples. These methods are further enhanced by generative AI, which can synthesize plausible energy usage patterns, enabling the creation of customized recommendations even when empirical data is scarce, (Golder, Das, & Mondal, 2024). Together, these approaches ensure that digital energy innovations are not limited to the "data elite" but are also accessible and effective for economically deprived and marginalized communities.

Affordability through AI-Driven Insights

Financial Stress Defined Using Energy Patterns

For most homes—particularly those in urban slums or rural peripheries—the energy crisis is more about affordability than access. Unaffordable tariffs, outdated or inefficient appliances, and a lack of awareness contribute to unsustainable energy expenditures. AI can process even incomplete consumption data to identify stress indicators such as sudden usage spikes, prolonged appliance operation, or irregular patterns that typically precede bill surges. These insights are invaluable for designing targeted interventions that reduce the cost burden without compromising the standard of living, (Chan & Siqin, 2025).

Individualized Support and Poverty-Friendly Policy Design

Generative AI may be employed to create personalized energy-saving measures for low-income or energy-vulnerable households. By simulating individualized load profiles and linking them with dynamic pricing regimes, AI can recommend optimal appliance operation schedules or propose tariff plans that are better aligned with a user's income level and lifestyle. Additionally, governments and utilities can leverage these models to strategically target subsidies or rebates toward the most energy-stressed households ensuring that scarce public resources are distributed efficiently and equitably. This level of personalization transforms affordability from a macroeconomic challenge into a household-level opportunity for empowerment and resilience, (Silalahi, 2024).

Ensuring Fairness and Transparency in Energy AI

Risks of Algorithmic Exclusion

Artificial intelligence models built primarily on data from urban, wealthier, or digitally connected users risk institutionalizing discrimination. These models may systematically overlook the needs of minority or marginalized groups, misclassify their behaviors, or generate recommendations that are irrelevant—or even harmful—in their specific contexts. In the energy sector, such exclusion can lead to underinvestment in infrastructure, inaccurate pricing, or even denial of services in communities that are already burdened by a long-standing legacy of social and economic inequities, (Krishnan et al., 2025).

Ethical AI Design and Governance for Equitable AI

To mitigate these risks, equitable AI needs to be rooted in fairness, transparency, and community engagement. Models must be interpretable and auditable so that stakeholders understand why certain decisions or recommendations are being generated. Data collection, validation, and feedback processes should actively involve communities to ensure that cultural and social contexts are embedded within the models themselves. Regulatory structures must support data protection, bias testing, and accountability for automated decisions—especially in high-stakes sectors such as energy, (Lin et al., 2025). With such an ethical scaffolding in place, AI can uphold democratic values and build public confidence in digital infrastructure. Inclusive AI transforms what it means for energy networks to be "smart." It shifts the focus from mere technological advancement to social purpose—from centralized control to individual empowerment. Whether by expanding energy access in off-grid regions, tailoring recommendations in low-data environments, addressing energy poverty through targeted insights, or maintaining fairness in automated decisions, inclusive AI ensures that no one is left behind in the clean energy transition. As nations pursue climate resilience, carbon neutrality, and sustainable development goals, embedding equity into AI systems is no longer optional—it is essential. In this light, AI emerges not just as a tool for optimization, but as an engine for justice, fairness, and shared prosperity, (Mongaillard, Lasaulce, & Varma, 2025).

MULTI-SCALE ENERGY INTELLIGENCE: FROM HOME TO URBAN SYSTEMS

As the energy sector transforms to support decentralized generation, electric vehicles, and increasing urban demands, coordination across multiple levels has become critical. *Multi-scale energy intelligence* refers to the capacity to gather, process, and optimize energy information across spatial scales—from individual homes to entire urban systems. This enables precise, localized control while maintaining overall system integrity, (Foroughi et al., 2025). Generative AI plays a pivotal role in this transformation. With its ability to model uncertainty, synthesize data, and simulate diverse future scenarios, it becomes central to managing the complexity of modern energy systems. By bridging local adaptation with global coordination, multi-scale intelligence forms the foundation of sustainable, decarbonized, and resilient energy futures, (Mongaillard, Lasaulce, & Varma, 2025).

Aggregating Household Data for Broader Insights

From Individual Profiles to Community Trends

IoT-enabled smart homes generate rich, high-frequency data on appliance usage, heating and cooling preferences, and occupancy patterns. While this information is valuable at the individual household level, its true potential emerges when aggregated across a community. When pooled, these data streams reveal broader behavioral trends and systemic energy usage patterns. Generative AI can analyze this aggregated data to uncover community-level insights—such as synchronized peak demand periods in neighborhoods, seasonal variations in appliance usage, and responsiveness to dynamic pricing signals. These insights empower energy planners to develop accurate local load profiles, anticipate transformer-level stress points, and design demand-side management strategies that are grounded in real-world conditions, 9Wang et al., 2025).

Privacy-Preserving Aggregation

Gathering and using energy data must be carefully balanced with the need to protect individual privacy. Techniques such as federated learning, homomorphic encryption, and differential privacy enable data aggregation without compromising user confidentiality. In federated learning models, training occurs locally on user devices, and only anonymized gradients are shared—ensuring that raw data never leaves the user's domain. This approach allows for large-scale learning from millions of devices while preserving user trust and autonomy. Such ethically grounded aggregation methods are crucial for building energy intelligence systems that are socially accepted, privacy-respecting, and transparent in operation, (Böcking et al., 2024).

Community-Centric Energy Insights

Aggregated household intelligence also supports community-led sustainability initiatives. This data can be leveraged by energy cooperatives, local authorities, and NGOs to identify underperforming areas, plan bulk retrofitting programs, or offer incentives for the replacement of inefficient appliances. Such collective intelligence fosters energy citizenship, where communities take shared responsibility for their energy behaviors and their broader environmental impacts, (Silalahi, 2024).

Modeling Energy Across Transport, Buildings, and Industry

Interlinked Urban Energy Ecosystems

Urban areas are no longer composed of standalone energy sectors. Today, buildings, transport, and industries are highly interconnected. Electric vehicles draw power from residential buildings and can return energy to the grid through Vehicle-to-Grid (V2G) systems, (Lin et al., 2025). Public infrastructure, such as metro systems, align operations with peak and off-peak electricity tariffs. Industrial plants contribute ancillary services through demand response mechanisms. Without integrated modeling, these interconnections remain opaque and underutilized. Generative AI enables the simultaneous modeling of such interactions, revealing the systemic efficiencies that can be unlocked when all sectors operate in coordinated harmony.

Cross-Sectoral Scenario Simulation

Generative AI models can model the effect of electrifying fleets of city buses on peak load, or how a heatwave could initiate simultaneous residential air conditioner loads and decrease the productivity in industrial areas. These simulations enable planners to anticipate points of strain and assign energy resources accordingly, (Chen, Karaduman, & Kuang, 2025). They also facilitate design of sectoral coordination protocols—e.g., suspension of non-essential industrial operations during periods of public EV charging peaks.

Siloed Decarbonization Avoidance

Decarbonization activities at the sectoral level, though unavoidable, occasionally have unwanted side effects. For instance, mandating all-electric dwellings without taking into account local grid capacity might result in transformer overloads. Multi-domain intelligence ensures the synchronization of such shifts so that energy bottlenecks are eliminated and achievements toward climate objectives are accelerated, (Sarmas, Marinakis, & Doukas, 2025).

Urban-Scale Planning Using Generative Intelligence

Smarter Tools Are Needed for Smarter Cities

Urbanization is rapidly increasing along with global warming, requiring energy systems that are not only reactive but also predictive. Generative AI enables city planners to create digital sandboxes—simulated virtual spaces in which zoning policies, building codes, transportation infrastructure, and population growth paths are experimented upon for their future energy impact, (Jia, 2024). These experiments refine development master plans prior to being physically implemented, saving cost, emissions, and time.

AI-Driven Digital Twins for Cities

Digital twins, live virtual copies of physical infrastructure, are at the heart of smart city plans. Powered by generative AI, digital twins can model not only today's performance but also possible futures. A city can plan the energy impact of a new business complex, test resilience in the face of flood events, or maximize rooftop solar installations in public schools. All these insights inform investment decisions and enable urban hubs to stay ahead of demand peaks, climate occurrences, and technological disruptions, (Foroughi et al., 2025).

Climate-Responsive Urban Infrastructure

As increasingly frequent extreme weather events hit cities, urban energy infrastructure needs to become adaptive. Generative AI can model cascading failures—for example, how an overload of a substation during a heatwave might impact metro rail operations and induce traffic congestion due to increased air-conditioned rideshare usage, (Agoro & Daniel, 2025). By preempting such interdependencies, planners can develop resilient designs such as distributed energy storage, automated load shedding, and diversified renewable inputs, (Golder, Das, & Mondal, 2024).

Coordinating Cross-Sector Energy Flows with Efficiency

Dynamic Multi-Sector Coordination

Multi-scale energy intelligence is most effective when it coordinates energy flows across different systems. During a sunny afternoon, for instance, rooftop solar in residential areas may produce excess electricity. Instead of curtailing this energy, AI models can route it to charge municipal EV fleets, power industrial freezers, or pre-cool government buildings—all in real time, (Donta, Motlagh, & Tarkoma, 2024). Such temporal and spatial coordination maximizes asset utilization and helps flatten demand curves.

Generative AI as an Energy Conductor

Generative models are like orchestra conductors—predicting when each instrument (building, vehicle, or machine) should play its role to harmonize the collective, (Madanchian, 2024). They can suggest the optimal time to operate industrial chillers according to grid carbon intensity, sequence traffic lights and public EV charging points to soften system-wide load spikes, and suspend energy-hungry municipal tasks (e.g., water pumping) during renewable low-points. Such coordination turns cities into smart energy networks that can bend to match real-time situations, (Kimura, 2025).

Towards Energy Self-Sufficient City Blocks

Neighborhoods can be made self-sustaining microgrids with an appropriate coordination infrastructure. A city block could share batteries with households and businesses, peer into the future to foresee energy shortages and adjust ahead of time, and coordinate EV charging schedules based on weather and event information. This distributed but harmonized structure lessens reliance on centralized grids and allows cities to stand on their own under duress or disruption, (Mongaillard, Lasaulce, & Varma, 2025).

Multi-Scale Intelligence as the Future of Energy Governance

The net-zero transition is not possible with fragmented intelligence. We require unified, generative, and responsive systems functioning from the socket to the skyline. Multi-scale energy intelligence can deliver this potential—utilizing locally sourced data to drive global strategy, and vice versa. With generative AI decision-making, every tier of the energy ecosystem—homes, neighborhoods, transport networks, commercial districts, and whole cities—becomes a part of an integrated, cooperative system, (Reddy, 2024). Energy's future is not a pyramid—it is a living web. And generative AI, through the power of simulation, learning, and evolution, is the solution to guiding this web with accuracy, fairness, and anticipation.

SHAPING TOMORROW: AI-DRIVEN PATHWAYS TO SUSTAINABLE ENERGY FUTURES

Artificial Intelligence (AI) is transforming the planning, optimization, and regulation of energy systems. In addition to its operational benefits in energy efficiency and automation, AI has transformative capacity in shaping long-term choices that underpin global decarbonization, energy equity, and climate adaptation. Standing at the threshold of an energy revolution, AI is not only a technical solution but a strategic ally in the creation of a just, sustainable, and resilient energy future. This chapter discusses how AI enables net-zero transition, promotes adaptive governance, aids resilient infrastructure design, and enforces ethics in emerging energy systems, (Rane, 2023).

Achieving Net-Zero Targets through Smart Systems

AI for Decarbonization Roadmaps

Net-zero emissions is more than a matter of deploying clean technologies—it is a matter of having a systematic, data-driven transition strategy. AI has the ability to create holistic decarbonization roadmaps by modeling out-of-sample future scenarios that consider technology diffusion, behavioral change, policy evolution, and economic development. Generative AI models can spot region-specific opportunities for decarbonization, ranging from industrial zones phasing out coal to leveraging coastal sites for offshore wind, (Agoro & Daniel, 2025). These simulations assist in prioritizing investments, preventing stranded assets, and harmonizing national climate ambitions with ground-level action, (Lin et al., 2025).

Smart Carbon Management

AI further finds application in the continuous management of emissions. Through sensors and IoT devices, real-time emission data can be monitored, analyzed, and predicted. AI can then suggest operation tweaks to minimize carbon footprints—like load shifting in industries, dynamic routing in transport, or optimal HVAC scheduling for buildings. In addition, AI systems facilitate carbon pricing by confirming emissions in an open manner, raising red flags in carbon trading markets for anomalies, and enforcing compliance through auto-audits. It makes carbon management a smart, adaptive process integrated into day-to-day activity, (Tripathi & Pandey, n.d.).

Guiding Policy Through AI-Based Forecasting

AI for Scenario Simulation and Policy Testing

Energy policy is an intricately balanced mix of environmental drive, economic viability, and social implications. Traditional approaches tend to be lacking in capturing the multifaceted, non-linear effects of policy action. AI fills this deficit by enabling scenario-based policy analysis. Policymakers can utilize AI to simulate the effects of programs such as fuel subsidy removal, mandatory solar installations, or urban transport electrification. These models can examine direct effects—such as emission reductions—and indirect effects—such as changes in energy affordability and employment—making it possible to formulate better-informed and more representative policies, (Chan & Siqin, 2025).

Adaptive Governance with Real-Time Feedback

After implementing policies, AI allows for ongoing monitoring of their performance through real-time data gathering and analysis. For example, when a government applies dynamic pricing to household electricity, AI can monitor user reactions by income group and region, offering valuable feedback for timely policy adjustments. AI-driven public dashboards enhance transparency, enabling citizens to track energy savings, subsidy distribution, and carbon emission reductions. This fosters a cycle of adaptive governance—where policies are no longer static mandates but dynamic, evolving guidelines that respond to real-world developments and continuously optimize outcomes, (Zhang et al., 2024).

Infrastructure for a Climate-Resilient Future

Resilience by Design

As climate extremes worsen, the energy infrastructure must evolve from being merely efficient to being inherently resilient. Artificial Intelligence (AI) strengthens resilience planning by identifying system vulnerabilities through worst-case scenario simulations. For instance, AI can analyze how heatwaves may reduce the efficiency of solar panels, how flooding might disrupt substations, or how shifting wind patterns affect turbine output. Additionally, AI enhances cyber-resilience by detecting anomalies and potential intrusions within smart grid operations. These insights guide the design of robust energy infrastructures—such as distributed energy networks, elevated installations for flood protection, and hybrid systems that integrate renewables with backup storage—ensuring systems can withstand environmental and technological shocks, (Jia, 2024).

Lifecycle Optimization

Beyond the initial design phase, Artificial Intelligence (AI) plays a critical role in maintaining infrastructure resilience throughout its entire lifecycle. Predictive models powered by AI evaluate how various environmental stressors—such as extreme weather, temperature fluctuations, or humidity—may impact the longevity and performance of energy assets. These models then provide recommendations for optimal maintenance schedules or suggest appropriate timelines for replacement, helping prevent costly failures. Furthermore, AI can simulate and assess different siting options for infrastructure projects, identifying locations that are less vulnerable to climate or seismic risks while still being close to energy demand hubs. This comprehensive, lifecycle-based planning approach ensures that infrastructure investments remain both resilient and cost-effective, supporting long-term climate adaptation and sustainability objectives, (Sarmas, Marinakis, & Doukas, 2025).

Ethical Foundations and Future Research Directions

Redressing Bias and Injustice

AI models are no exception to the bias of training data. When used to plan for energy or distribute subsidies, the risks are particularly high. A discriminatory model might write off low-income areas from efficiency schemes or downplay the requirements of far-flung villages. To counteract this, AI models need to be trained on widely diverse datasets and regularly audited for bias. Energy AI should be built to actively work against inequity, identifying underserved groups and customizing interventions to their needs, making digital energy transitions inclusive, (Böcking et al., 2024).

Establishing Ethical AI Governance

The future of energy AI requires governance systems that find a balance between innovation and responsibility. Transparent models—where decision paths are comprehensible to humans—are necessary for public trust. Furthermore, participatory design, engaging communities in AI development and testing, helps ensure that solutions align with local realities and values. Regulatory frameworks must require transparency, bias analysis, and explainability in key AI systems. Lastly, interdisciplinary cooperation between data science, environmental studies, sociology, and ethics is essential to make sure that AI development will be democratic, inclusive, and sustainable in nature, (Chan & Siqin, 2025).

AI as a Partner in Shaping a Sustainable Energy Future

AI is not just an operational optimization tool but a strategic partner in reimagining how societies engage with energy, development, and climate action. From creating net-zero roadmaps and piloting policy simulations to enhancing infrastructure resilience and instilling ethical guardrails, AI facilitates more intelligent, accelerated, and equitable decisions at all levels. Yet, the power of AI to transform has to be balanced with vision, inclusivity, and responsibility, (Saleh et al., 2025).

As we transition toward an electrified, decentralized, and decarbonized world, AI will play a critical role in managing the uncertainty and complexity of energy transformations. When guided by human-centered ethics and applied to the service of collective well-being, AI can assist us in designing energy systems that are not just smart, but just—and futures that are not just efficient, but fair, (Chan & Siqin, 2025).

CONCLUSION

The convergence of Artificial Intelligence (AI), generative models, and smart technologies is not merely revolutionizing how energy is produced and consumed—it is fundamentally reshaping the architecture and ethics of global energy systems. From individual households to entire urban ecosystems, from predictive analytics to behavioral modeling, AI is unlocking a future where energy is not only cleaner and smarter but also inclusive, responsive, and just.

At the core of this transformation is a critical recognition: energy is not only a technical commodity—it is a deeply human experience. The choices surrounding how, when, and why we consume energy are shaped by behavioral patterns, cultural contexts, lifestyles, and social needs. Traditional energy paradigms have often treated demand as static and linear, ignoring this human complexity. In contrast, generative AI offers a dynamic alternative—one that learns, adapts, and co-evolves with its users.

Throughout this chapter, we have examined how AI enables the energy transition through a range of capabilities:

- Learning and modeling individual habits to optimize residential efficiency,

- Generating synthetic data for robust simulation and training of intelligent energy models,
- Integrating with IoT to enable real-time optimization in smart homes,
- Enhancing grid resilience through adaptive demand forecasting and distributed energy coordination,
- Expanding equitable access to energy in marginalized or remote communities, and
- Informing policy, infrastructure design, and ethical standards for inclusive implementation.

These innovations do not exist in isolation. Rather, they function interdependently across scales and sectors. The intelligent energy systems of the future will behave as integrated ecosystems: sensing real-time signals, adapting to contextual change, coordinating decisions across domains, and remaining resilient under stress. Within this ecosystem, AI serves as both the cognitive engine and nervous system—orchestrating simulations, informing decisions, and aligning outcomes with social, economic, and environmental goals.

However, technological innovation alone is insufficient. The journey toward a sustainable energy future must be grounded in ethical foresight, systemic inclusion, and societal accountability. AI systems must be explainable, unbiased, and sensitive to the diverse needs of all communities—not just those with digital access or economic privilege. Energy justice—ensuring fair access, representation, and benefit-sharing—must be embedded at every layer of AI design and deployment.

As we navigate the intersecting pressures of climate change, geopolitical instability, and digital acceleration, AI equips us with the foresight to act—not react. To build infrastructures that withstand disruption. To design smart cities that are also equitable. And to empower every individual—not just the connected few—to shape and share in the energy futures we collectively envision.

The systems we create today will determine the world we inhabit tomorrow. With AI as our compass and inclusion as our guiding principle, we can design energy systems that are not only intelligent but also empathetic, equitable, and enduring. This is more than a technological evolution—it is a profound societal transformation. And it begins now.

REFERENCES

Agoro, H., & Daniel, B. (2025). Generative AI in Creating Dynamic Content for IoT Devices.

Berney, M., Ouaazki, A., Macko, V., Kocher, B., & Holzer, A. (2024, May). Care-based Eco-feedback augmented with generative AI: Fostering pro-environmental behavior through emotional attachment. In *Proceedings of the 2024 CHI Conference on Human Factors in Computing Systems* (pp. 1-15).

Böcking, L., Michaelis, A., Schäfermeier, B., Baier, A., Kühl, N., Körner, M. F., & Nolting, L. (2024). Generative Artificial Intelligence in the energy sector.

Chan, H. L., & Siqin, T. (2025). Supply Chain Management with Generative Artificial Intelligence and Internet of Behaviours. *IEEE Transactions on Engineering Management.*

Chen, C., Karaduman, O., & Kuang, X. (2025). Behavioral Generative Agents for Energy Operations. *arXiv preprint arXiv:2506.12664.*

Donta, P. K., Motlagh, N. H., & Tarkoma, S. (2024). Follow-Me AI: Energy-Efficient User Interaction with Smart Environments. *IEEE Pervasive Computing.*

Dua, I. K., & Patel, P. G. (2024). An introduction to Generative AI. In *Optimizing Generative AI Workloads for Sustainability: Balancing Performance and Environmental Impact in Generative AI* (pp. 1–37). Apress.

Foroughi, B., Naghmeh-Abbaspour, B., Wen, J., Ghobakhloo, M., Al-Emran, M., & Al-Sharafi, M. A. (2025). Determinants of generative AI in promoting green purchasing behavior: A hybrid partial least squares–artificial neural network approach. *Business Strategy and the Environment, 34*(4), 4072–4094.

Golder, S. S., Das, S., & Mondal, S. (2024). Revolutionizing Industrial Manufacturing with Big Data and Generative AI: A Path to Predictive Efficiency. *Journal of Computational Analysis and Applications, 33*(8).

Jaboob, M., Hazaimeh, M., & Al-Ansi, A. M. (2025). Integration of generative AI techniques and applications in student behavior and cognitive achievement in Arab higher education. *International Journal of Human-Computer Interaction, 41*(1), 353–366.

Jia, Y. (2024, September). Analysis of the impact of artificial intelligence on electricity consumption. In *2024 3rd International Conference on Artificial Intelligence, Internet of Things and Cloud Computing Technology (AIoTC)* (pp. 57-60). IEEE.

Kimura, T. (2025). Exploring the Frontier: Generative AI Applications in Online Consumer Behavior Analytics. *Cuadernos de Gestión, 25*(1), 57–70.

Krishnan, R., Raj, A. A., Nazarie, W. N. F. W. M., & Marwah, S. (2025). AI Driven Insights Into Solar Energy Consumer Behaviour Personalization and Engagement for Green Energy Adoption. In *AI Marketing and Ethical Considerations in Consumer Engagement* (pp. 215–232). IGI Global Scientific Publishing.

Lee, G. H., Lee, K. J., Jeong, B., & Kim, T. (2024). Developing personalized marketing service using generative AI. *IEEE Access : Practical Innovations, Open Solutions, 12*, 22394–22402.

Lin, Y., Yao, Y., Zhu, J., & He, C. (2025, March). Application of Generative AI in Predictive Analysis of Urban Energy Distribution and Traffic Congestion in Smart Cities. In *2025 IEEE International Conference on Electronics, Energy Systems and Power Engineering (EESPE)* (pp. 765-768). IEEE.

Madanchian, M. (2024). Generative AI for consumer behavior prediction: Techniques and applications. *Sustainability, 16*(22), 9963.

Mohanty, P. K., Reddy, K. H. K., Panigrahy, S. K., & Roy, D. S. (2024). Leveraging Generative and Explainable AI for Electric Vehicle Energy Toward Sustainable, Consumer-Centric Transportation. *IEEE Access : Practical Innovations, Open Solutions, 12*, 143721–143732.

Mongaillard, T., Lasaulce, S., & Varma, V. S. (2025). Smart Energy Management: From Conventional Optimization to Generative AI Techniques. In *Internet of Vehicles and Computer Vision Solutions for Smart City Transformations* (pp. 337–361). Springer Nature Switzerland.

Naseer, F., Addas, A., Tahir, M., Khan, M. N., & Sattar, N. (2025). Integrating generative adversarial networks with IoT for adaptive AI-powered personalized elderly care in smart homes. *Frontiers in Artificial Intelligence, 8*, 1520592. PMID: 40017485

Rane, N. (2023). Contribution of ChatGPT and other generative artificial intelligence (AI) in renewable and sustainable energy. *Available at SSRN* 4597674.

Reddy, J. K. (2024). Leveraging Generative AI for Hyper Personalized Rewards and Benefits Programs: Analyzing Consumer Behavior in Financial Loyalty Systems. *J. Electrical Systems, 20*(11s), 3647–3657.

Saleh, A., Donta, P. K., Morabito, R., Motlagh, N. H., Tarkoma, S., & Loven, L. (2025). Follow-me ai: Energy-efficient user interaction with smart environments. *IEEE Pervasive Computing*.

Sarmas, E., Marinakis, V., & Doukas, H. (2025). Future Directions of Intelligent Energy Management and the Role of Generative AI. In *Artificial Intelligence for Energy Systems: Driving Intelligent, Flexible and Optimal Energy Management* (pp. 257–266). Springer Nature Switzerland.

Sarvaiya, H., Hasegawa, M., Zeng, H., Gao, X., & Meng, N. Simulating Personalized Smart-Home Activity Datasets with Generative AI: A Case Study.

Silalahi, A. D. K. (2024). Can generative artificial intelligence drive sustainable behavior? A consumer adoption model for AI-driven recommendations. *A Consumer Adoption Model for Ai-Driven Recommendations*.

Tripathi, M., & Pandey, P. Generative AI in Customer Intelligence: Building Multi-Agent System.

Wang, R., Song, C., Li, C., Wang, Y., & Dong, Z. Y. (2025). The Go-to-Market Game Changer in the Low-Carbon Energy Industry: Generative Pre-Transformer and Its Potential for Improving Efficiency and Sustainability. *IEEE Systems, Man, and Cybernetics Magazine*, *11*(2), 92–106.

Yafei, X., Wu, Y., Song, J., Gong, Y., & Lianga, P. (2024). Generative AI in industrial revolution: a comprehensive research on transformations, challenges, and future directions. *Journal of Knowledge Learning and Science Technology ISSN: 2959-6386 (online)*, *3*(2), 11-20.

Zhang, H., Niyato, D., Zhang, W., Zhao, C., Du, H., Jamalipour, A., & Pei, Y. (2024). The role of generative artificial intelligence in internet of electric vehicles. *IEEE Internet of Things Journal*.

Zhang, H., Zhang, R., Zhang, W., Niyato, D., & Wen, Y. (2025). Advancing Generative Artificial Intelligence and Large Language Models for Demand Side Management with Internet of Electric Vehicles. *arXiv preprint arXiv:2501.15544*.

Chapter 11
Financial Energy Literacy Sustainability

José G. Vargas-Hernandez
https://orcid.org/0000-0003-0938-4197
Financial Literacy Research Center, Universidad Cristóbal Colón, Mexico & Tecnológico Nacional de México, ITS Fresnillo, Mexico

ABSTRACT

This study aims to analyze the implications of the financial energy literacy (FEL) with sustainability as interdisciplinary concepts. The holistic approach of sustainability of financial energy literacy (SFEL) is critical in an environment where energy costs are rising and environmental sustainability is becoming increasingly urgent. The method employed is the meta-analytical and meta-cognitive analysis based on a conceptual, theoretical and empirical literature review published in the last five years. It is concluded that financial energy literacy is a crucial element for achieving sustainable development goals, as it empowers individuals to make informed decisions that benefit both their financial well-being and the environment of individuals, families, organizations and communities.

INTRODUCTION

The pursuit of sustainable environmental development has evolved into a central theme in global discussions, policy-making, and academic research. This pursuit is propelled by the increasing recognition of the interconnectedness between economic activities, environmental health, and societal well-being. Financial literacy, green energy, and sustainable environmental development are individually critical

DOI: 10.4018/979-8-3373-4272-6.ch011

elements, and their synergistic relationship can potentially drive significant progress toward a more sustainable future.

Sustainable environmental development requires integrating economic, social, and ecological strategies. Financial literacy and green energy adoption are pivotal in this transition, enabling individuals and nations to make informed decisions that balance economic growth with environmental stewardship. This literature review explores the existing research on the interplay between these three elements, highlighting key findings, identifying research gaps, and discussing implications for policy and practice.

Sustainable environmental development is a multifaceted challenge requiring the integration of economic, social, and environmental dimensions. Among the key enablers of sustainability are financial literacy and green energy adoption. Financial literacy empowers individuals and communities to make informed decisions about investments, consumption, and savings that can promote environmental sustainability. Concurrently, green energy technologies offer viable alternatives to fossil fuels, reducing greenhouse gas emissions and fostering sustainable economic growth.

This literature review synthesizes research on the intersection of financial literacy, green energy, and sustainable development, exploring how financial knowledge influences environmental behaviors and how renewable energy adoption contributes to sustainable outcomes globally. This review synthesizes empirical and theoretical insights into how financial literacy fosters sustainable behaviors and how renewable energy technologies contribute to ecological resilience.

Defining Financial Literacy and Its Relevance to Sustainable Environmental Development Issues

Financial literacy encompasses the knowledge and skills necessary for making informed financial decisions This includes understanding concepts such as budgeting, saving, investing, and managing debt. Zaimovic, Torlakovic, Arnaut-Berilo, Zaimovic, Dedovic, & Nuhic Meskovic, (2023) emphasize that financial literacy is essential for achieving financial security, individual well-being, economic growth, and overall sustainable development. This understanding is not merely about managing personal finances; it extends to recognizing the broader economic and environmental implications of financial choices.

Financial literacy traditionally refers to the ability to understand and use various financial skills, including personal financial management, budgeting, and investing (Lusardi & Mitchell, 2014). Recently, scholars have expanded this concept to include "green financial literacy," which encompasses knowledge about financing environmentally friendly technologies and sustainable investments (Frontiers in Sustainability, 2025). This expanded literacy is crucial for enabling individuals and businesses to navigate the complexities of green finance, such as evaluating renewable energy projects or understanding carbon pricing mechanisms.

Financial literacy plays a crucial role in promoting broader economic stability and sustainable development (Zaimovic, Torlakovic, Arnaut-Berilo, Zaimovic, Dedovic, & Nuhic Meskovic, 2023). Individuals who possess a strong understanding of financial principles are better equipped to participate effectively in the economy, make sound investment decisions, and plan for their long-term financial security. This contributes to a more stable and resilient economy, reducing the risk of financial crises and promoting sustainable growth. Moreover, financially literate individuals are more likely to engage in responsible financial behaviors, such as saving for retirement, managing debt prudently, and investing in education and skills development, which can lead to improved economic outcomes and reduced social inequality.

Financial literacy is a critical enabler for the adoption of green energy technologies. Individuals and businesses with higher financial knowledge are more likely to invest in renewable energy systems, understand financing options such as green loans, and navigate government incentives (Epsilon, 2025). In Norway, households with advanced financial literacy allocated significantly more savings toward renewable energy funds compared to less literate peers, indicating that financial knowledge influences investment decisions toward sustainability (Growingscience, 2025).

Financial literacy empowers individuals to act as informed and responsible economic agents who contribute to environmental sustainability. Senaya, (2024) underscores the complex connection between financial literacy and sustainable finance, with emphasis on increasing people's understanding of finance so that they make more sustainable decisions on how to use their money. This includes making informed purchasing decisions that consider the environmental impact of products and services, supporting businesses that prioritize sustainability, and advocating for policies that promote environmental protection.

By empowering individuals to make environmentally conscious choices, financial literacy can drive demand for green products and services, encourage investment in renewable energy, and promote responsible resource management. Ultimately, financial literacy can serve as a catalyst for creating a more sustainable and equitable economy that benefits both people and the planet.

Financial Literacy as a Driver of Environmental Sustainability

Financial literacy strengthens individuals' and communities' capacity to make informed decisions that align economic and environmental goals. Cross-country analyses reveal that populations with higher financial literacy scores exhibit improved environmental performance indices and greater readiness to address climate change. For instance, financially literate households are 23% more likely to invest in energy-efficient technologies (e.g., solar panels) and 18% more likely to adopt sustainable consumption practices. This correlation stems from financial literacy's role in fostering long-term planning, risk assessment, and trust in institutional climate initiatives.

In Indonesia, financial literacy programs have directly influenced attitudes toward green innovation technologies. A study of 155 individuals demonstrated that financial knowledge enhances willingness to adopt battery electric vehicles (BEVs), with a 0.45 path coefficient linking financial literacy to pro-environmental behaviors. Similarly, community-based financial education in rural India increased investments in sustainable agriculture by 34%, reducing reliance on fossil fuel-dependent practices. These findings underscore the bidirectional relationship between financial empowerment and ecological stewardship.

Salahodjaev, & Sadikov (2025) analyzed data from 131 countries (2000-2024) and found a strong positive correlation between financial literacy and environmental performance. Even after controlling GDP, democracy, and globalization, financial literacy remains a significant indicator of sustainability outcomes. Senaya (2024) analyzed trends and relationships between different regions in its study entitled "Financial education and its role in promoting sustainable investment," which concludes that specific financial education programs are needed, especially in disadvantaged communities, to reduce differences in participation in sustainable investment.

Anderson (2022) identified the impact of education on sustainable finance on investor behavior. Participants who received training in sustainable finance were 6% more likely to invest in highly sustainable funds and between 2.5% and 3%, suggesting that financial education amplifies pro-environmental behavior.

Green Energy Technologies and Its Role in Sustainable Development

Green energy, derived from renewable sources such as solar, wind, geothermal, and hydropower, is a cornerstone of sustainable development. Padma, Don, Anjum, Sindhu, & Sankari, 2023) highlight that green energy can be considered an effective means of promoting the amount of power needed for industries, urban territories, and businesses without damaging the environment. These renewable sources offer a clean and sustainable alternative to fossil fuels, reducing greenhouse gas emissions, air pollution, and dependence on finite resources. The utilization of green energy is not only environmentally responsible but also economically viable, as it can create new jobs, stimulate innovation, and enhance energy security.

Beyond environmental benefits, green energy adoption has considerable economic and social advantages. Renewable energy projects create jobs, stimulate local economies, and reduce energy poverty (Haldar *et al.*, 2023). For example, community-based solar initiatives in rural India have empowered marginalized populations by providing affordable and reliable electricity, improving educational and health outcomes (Katini & Amalanathan, 2022).

The transition to green energy is essential for decoupling economic growth from environmental degradation and ensuring a sustainable energy future for generations to come Ahmed, Areche, Sheikh, & Lahiani, (2022). emphasize that green energy is a crucial component in addressing expanding demands and combating climate change. By investing in renewable energy technologies and infrastructure, societies can reduce their reliance on fossil fuels, mitigate climate change, and protect ecosystems. This transition requires a concerted effort from governments, businesses, and individuals to prioritize green energy and implement policies that support its development and deployment. Furthermore, the integration of green energy into broader sustainability strategies can create a more resilient and equitable energy system that benefits all members of society.

Green energy, encompassing renewable sources such as solar, wind, geothermal, and hydropower, is central to global efforts to mitigate climate change and promote sustainable development. Renewable energy technologies reduce dependence on fossil fuels, lower greenhouse gas emissions, and contribute to energy security. Haldar *et al.* (2023) provides a comprehensive review of renewable energy sources, emphasizing their potential to meet increasing energy demands sustainably. The authors highlight that integrating renewable energy into national grids can significantly reduce carbon emissions and improve air quality, contributing to health and environmental benefits.

Green finance, which includes financial instruments and policies that promote environmentally beneficial investments, plays a critical role in scaling up green energy projects. Sachs, Woo, Yoshino, & Taghizadeh-Hesary, 2019). note that if we want to achieve sustainable goals, we need to open the file for scaling up financing investments in environmental benefits, through instruments policies, such as bonds, banks, carbon market instruments, fiscal policy, central banking, technologies, community-based funds, etc., which are collectively known as "green finance."

Green finance encompasses a range of financial mechanisms, including green bonds, sustainability-linked loans, and venture capital, that are designed to attract private investment and support the development of renewable energy technologies. Governments can also play a crucial role in promoting green finance through policies that incentivize green investments, reduce regulatory barriers, and create a stable and predictable market for renewable energy. By effectively mobilizing financial resources, green finance can accelerate the transition to a low-carbon economy and ensure a sustainable energy future for all.

Mechanisms Linking Financial Literacy and Environmental Sustainability

The influence of financial literacy on environmental sustainability operates through several mechanisms. First, financially literate individuals are better able to assess the long-term benefits and costs of green investments, such as installing solar panels or purchasing energy-efficient appliances (Lusardi & Mitchell, 2014). Second, financial literacy increases trust in financial institutions and green financial products, facilitating access to green loans and bonds (Growingscience, 2025). Third, financial education fosters a future-oriented mindset, encouraging behaviors that prioritize sustainability over short-term gains (Katini & Amalanathan, 2022).

Sustainable Environmental Development: A Holistic Approach

Sustainable environmental development is a holistic approach that integrates environmental protection, economic growth, and social equity (Fang, Yang, & Song, 2022, Ramanathan, & Indiran, 2021). This approach recognizes that environmental sustainability is not merely an ecological concern but also an economic and social imperative. Fang Zhen, Can Yang, and Xiaowei Song (Fang, Yang, & Song, 2022). highlight that environmental pollution and the depletion of natural resources have been highlighted by the United Nations Sustainable Development Goals (SDGs), paving the way for modern concepts such as sustainable growth to be introduced. Ramanathan, & Indiran, (2021) assert that economic growth and sustainable development are very fundamental concepts among economies in an advancing globalizing world.

Sustainable development aims to meet the needs of the present without compromising the ability of future generations to meet their own needs (Ramanathan, & Indiran, 2021). This requires a fundamental shift in how we produce and consume goods and services, manage natural resources, and address social inequalities. Ramanathan, & Indiran, (2021) emphasize that the current state of the world shows that people are experiencing delicacy and unsustainable transformations, which are a new norm preceded by a holistic economic, social, and environmental model that integrates interdependence segments between societies and requires the ability and capability to respond to multifaceted difficulties confronting society today. This shift involves adopting sustainable practices across all sectors of the economy, from agriculture and manufacturing to transportation and energy. It also requires promoting social equity and ensuring that all members of society have access to the resources and opportunities they need to thrive.

Financial literacy empowers individuals and businesses to make informed decisions that support environmental sustainability. Salahodjaev, & Sadikov, (2025) document that the financial literacy of a population is associated with improvements in environmental quality across countries, thus providing additional evidence that financial literacy not only improves financial wealth/capital preservation but also preserves environmental wealth/capital. This includes adopting energy-efficient technologies, reducing waste, and investing in renewable energy sources. By promoting financial literacy, societies can empower individuals and businesses to become active participants in the transition to a more sustainable economy. Furthermore, financial literacy can help to create a more informed and engaged citizenry that is better equipped to advocate for policies that promote environmental protection and sustainable development.

FINANCIAL LITERACY AND ENVIRONMENTAL AWARENESS

Financial literacy plays a crucial role in shaping environmental attitudes and behaviors, influencing individuals' awareness, beliefs, and actions related to environmental sustainability. Financially literate individuals are better equipped to understand the financial implications of environmental degradation and the economic benefits of sustainable practices, leading to more pro-environmental behaviors. This section explores the impact of financial literacy on environmental attitudes, consumption patterns, and support for green initiatives. Senaya (2025) highlights in his study the role of financial technologies (FinTech) and digital financial tools in promoting sustainable investment among financially literate users.

Impact of Financial Literacy on Environmental Attitudes and Beliefs

Financial literacy can significantly influence individuals' environmental attitudes and beliefs, making them more likely to adopt pro-environmental attitudes. Salahodjaev, & Sadikov, (2025) assert that nations with higher levels of financial literacy are better positioned to advance toward a sustainable future. When individuals understand the financial implications of environmental degradation and the economic benefits of sustainable practices, they are more likely to recognize the value of natural resources, understand the costs of pollution, and appreciate the long-term benefits of conservation. This understanding can lead to a greater sense of responsibility towards the environment and a willingness to take action to protect it.

Financial literacy can also increase awareness of the financial risks associated with climate change and other environmental challenges, motivating individuals to take action to mitigate these risks. Todorov, Aleksandrova, & Ismailov, (2023) aimed to examine the connection between a population's level of financial literacy and greenhouse gas emissions. This includes recognizing the potential for economic losses due to extreme weather events, supply chain disruptions, and resource scarcity. By understanding these risks, individuals may be more likely to support policies that promote climate change mitigation, such as carbon emissions reductions, renewable energy development, and energy efficiency improvements.

Studies have found a positive correlation between financial literacy and environmental awareness. Beny, Wendy, Saleh, Giriati, (2023) revealed that financial knowledge and green innovation were related to the attitude toward sustainability. This suggests that individuals with higher financial literacy are more likely to be aware of environmental issues, understand their causes and consequences, and support efforts to address them. This heightened awareness can translate into more sustainable behaviors, such as reducing energy consumption, conserving water, and supporting environmentally friendly businesses.

Financial Literacy and Sustainable Consumption Patterns

Financial literacy can also influence sustainable consumption patterns. V.A. Chernov (Chernov, 2022) links financial literacy, economics, culture, and development, also revealing the conditionality of this by the need to increase thinking, comprehension of moral values, and the level of rational consumption. Financially literate individuals are more likely to make informed purchasing decisions that consider the environmental impact of products and services. This includes choosing energy-efficient appliances, buying organic food, and reducing waste. They are also more likely to be aware of the financial benefits of sustainable consumption, such as lower energy bills and reduced healthcare costs.

Understanding the financial benefits of sustainable consumption encourages the adoption of eco-friendly practices. Kalmi, Trotta, & Kaukauskas, (2021) suggest that measures promoting energy-related financial literacy might guide consumers' decisions toward energy efficiency and conservation. This includes taking steps to reduce energy consumption, conserve water, and minimize waste. By understanding the long-term financial implications of their consumption choices, individuals can make more informed decisions that benefit both their wallets and the environment.

Studies have shown that financial literacy is associated with increased adoption of sustainable consumption practices. Kalmi, Trotta, & Kaukauskas, (2021) found that households with respondents with higher levels of energy-related financial literacy tend to consume less electricity when controlling for other factors such as dwelling characteristics. They are more likely to invest in energy-efficient home improvements, purchase hybrid or electric vehicles, and participate in recycling programs. They are also more likely to avoid wasteful spending and make purchases that align with their values. This suggests that financial literacy can empower individuals to live more sustainably and reduce their environmental footprint.

Financial Literacy and Support for Green Initiatives

Financial literacy can promote investment in green initiatives, such as renewable energy projects and sustainable businesses. Senaya, 2024) explores the complex connection between financial literacy, how it links to sustainable finance, and investments, with much emphasis on the importance of increasing people's understanding of finance so that they make more sustainable decisions on how to use their money. Financially literate investors are better equipped to evaluate the financial risks and returns of green investments and make informed decisions that support environmental sustainability. This includes investing in green stocks and bonds, supporting socially responsible companies, and participating in crowdfunding campaigns for environmental projects.

The growth of sustainable investing is driven in part by increasing financial literacy among investors. Alsmadi, Al-Okaily, Alrawashdeh, Al-Gasaymeh, Moh'd Al-hazimeh, & Zakari, 2023) note that a bond is a new approach integrating the protection of the ecosystem into economic profits. As more individuals understand the financial benefits of green investments and the potential for positive social and environmental impact, they are more likely to allocate their capital to sustainable ventures. This can provide critical funding for green energy projects, sustainable agriculture, and other initiatives that promote environmental sustainability.

GREEN FINANCE: MOBILIZING INVESTMENTS FOR RENEWABLE ENERGY

Green finance plays a pivotal role in mobilizing capital for renewable energy projects and accelerating the transition to a low-carbon economy. Green finance encompasses a range of financial instruments and policies designed to promote environmentally beneficial investments. This section explores the mechanisms of green finance, the role of policy frameworks and government support, and the involvement of financial institutions in sustainable lending.

The Mechanisms of Green Finance

Green finance plays a pivotal role in mobilizing capital for renewable energy projects and accelerating the transition to a low-carbon economy. Lee, (2020) identifies mainstream bonds, diversification of finance, the transition of corporates' business models, transparency and disclosure, and harmonizing taxonomy measurements as emerging aspects of green finance. Green finance encompasses a range of financial instruments and policies designed to promote environmentally beneficial investments, including green bonds, sustainability-linked loans, venture capital, and blended finance models. Attipoe, Chukwuma-Eke, Lawal, Friday, Isibor, & Akintobi, (2025) highlight how financial institutions can provide the necessary capital to empower SMEs to transition toward sustainable energy practices through such instruments.

Policy instruments such as subsidies, tax incentives, green bonds, and carbon pricing are essential for scaling renewable energy investments (Growingscience, 2025). Redirecting fossil fuel subsidies toward renewable energy could unlock trillions of dollars in sustainable investments globally (Liu *et al.*, 2023). Microfinance and green loans have proven effective in empowering marginalized groups to adopt sustainable practices. For example, microloan schemes in Southeast Asia increased sustainable agricultural yields by 29%, demonstrating the synergy between financial inclusion and environmental sustainability (Katini & Amalanathan, 2022).

Green bonds, in particular, have emerged as a popular tool for financing environmentally sustainable projects. Krause, Myroshnychenko, Tiutiunyk, & Latysh, (2024) observe that phrases most often used in publications include finance, alternative economy, green bonds, green innovation, sustainable development, greenhouse gas emissions, sustainable finance, green credit, green economy, and climate finance. These bonds are specifically earmarked for environmentally sustainable projects and provide investors with the opportunity to support green initiatives while earning a financial return. The green bond market has grown rapidly in recent years, driven by increasing demand from institutional investors and a growing awareness of the financial benefits of green investments.

Policy Frameworks and Government Support

Government policies and incentives play a crucial role in promoting green finance and encouraging investment in renewable energy. Kwilinski, Oleksii, & Pimonenko, 2024). confirm that green finance promotes the achievement of sustainable development goals. This includes providing tax credits, subsidies, and other financial incentives for green projects, as well as establishing regulatory frameworks that support green investments. Such support is crucial for startups. Bhatnagar, Taneja, & Özen, (2022) found that government initiatives are turning by providing finance, and Indian startups are exploiting this opportunity by implementing entrepreneurship. Chauhan et al., (2025), specify that government subsidies alone are not sufficient to drive the adoption of green finance and consider that their effectiveness increases when combined with financial education and policy awareness.

Harmonizing taxonomy measurements is critical for emerging green finance (Lee, 2020) . Jung Wan Lee (Lee, 2020) suggests initiatives such as greening the banking system, the bond market, and institutional investors, as well as developing policies, capacity, market governance, private-public partnerships, and diversifying resources. Governments can also play a role in developing green finance standards and taxonomies, which provide investors with clear guidelines for identifying and evaluating green investments. These measures are essential for creating a conducive environment for green finance to thrive.

Financial Institutions and Green Lending Practices

Financial institutions are increasingly integrating environmental considerations into their lending and investment practices. Zheng, Siddik, & Masukujjaman, & Fatema, (2021). highlight that private commercial banks are the highest contributor to green financing. This includes developing green lending products, such as green mortgages and green auto loans, as well as incorporating environmental, social, and governance (ESG) factors into investment decisions. Fu, Lu, & Pirabi, (2023) emphasize that integrating environmental, social, and governance (ESG) factors into investment decisions is critical.

Green lending products and technical assistance support green projects. Zheng, G. W. & Siddik, A. B. & Masukujjaman, M. & Fatema, N. (2021). indicate that green finance is related to economic, social, and environmental aspects of SDGs. Financial institutions can also play a role in providing technical assistance and capacity building to support green projects. The integration of ESG factors into investment decisions is driven by a growing recognition of the financial risks associated with environmental degradation and climate change. By incorporating ESG factors into their investment processes, financial institutions can mitigate these risks and promote more sustainable business practices.

GREEN ENERGY POLICIES: DRIVING SUSTAINABLE DEVELOPMENT GOALS

Green energy policies are essential for driving sustainable development goals by promoting renewable energy, mitigating climate change, and fostering economic growth. This section examines the role of green energy in achieving the SDGs, its impact on climate change mitigation, and its contribution to sustainable economic growth.

Governments and international organizations recognize the importance of financial literacy in achieving sustainable development goals (SDGs). Incorporating financial education into school curricula and community programs can enhance public understanding of green finance and environmental risks (Epsilon, 2025). Indonesia's "Green Product Initiative" exemplifies how combining financial literacy campaigns with policy incentives can accelerate the adoption of green technologies such as BEVs (Epsilon, 2025). Similarly, the European Union's Renewable Energy Directive (RED III) links subsidies to community-led renewable projects, increasing rural access to green energy (Haldar *et al.*, 2023).

Green Energy and SDGs

Green energy plays a crucial role in achieving the UN Sustainable Development Goals (SDGs), particularly those related to climate action, affordable and clean energy, and sustainable cities and communities. Udeagha & Muchapondwa, (2023) assert that sustainability is promoted by green finance, fintech, and innovation. The transition to green energy is essential for reducing greenhouse gas emissions and mitigating the impacts of climate change, which threatens progress on many other SDGs. Green energy can also provide access to affordable and reliable energy for millions of people who currently lack access to electricity, improving their quality of life and promoting economic development.

Achieving this goal requires a significant increase in the share of renewable energy in the global energy mix, as well as improvements in energy efficiency. Aamir Inam Bhutta, Muhammad Rizwan Ullah, Jahanzaib Sultan, Ahsan Riaz, and Muhammad Fayyaz Sheikh Bhutta, Ullah, Sultan, Riaz, & Sheikh, (2022) emphasize the contributions of green innovation, energy production, and financial development to environmental quality. Green energy policies, such as renewable portfolio standards and feed-in tariffs, can help to drive progress toward SDG 7 by creating a stable and predictable market for renewable energy.

Climate Change Mitigation Through Green Energy

Green energy is a key strategy for mitigating climate change by reducing greenhouse gas emissions from the energy sector. Fang, Yang, & Song, 2022) confirm the negative impact of green finance on carbon emissions. By replacing fossil fuels with renewable energy sources, societies can substantially reduce their carbon footprint and contribute to the goals of the Paris Agreement, which aims to limit global warming to well below 2 degrees Celsius above pre-industrial levels. Furthermore, institutions are clustering as globally distributed hubs redirecting investments into sustainable infrastructures consistent with mitigating climate change. Ng, Nathwani, Fu, & Zhou, (2021) find that institutions are clustering as globally distributed hubs redirecting investments into sustainable infrastructures consistent with mitigating climate change.

The effectiveness of green energy in mitigating climate change depends on several factors, including the type of renewable energy source, the efficiency of the technology, and the extent to which it displaces fossil fuels. Fang, Yang, & Song, (2022). use a panel data model technique to examine the dependence structure of emissions. The transition to green energy must be accompanied by policies that promote energy efficiency and reduce overall energy consumption to maximize its impact on climate change mitigation.

Economic Growth and Green Energy

Green energy can promote sustainable economic growth by creating new jobs, stimulating innovation, and reducing dependence on fossil fuel imports. Ogunyemi & Ishola, (2024) suggest that the combined use of analytics and strategic financial support can significantly enhance SMEs' participation in the renewable energy sector, contributing to broader environmental and economic sustainability goals. Investing in green energy can reduce dependence on fossil fuel imports, which can be a significant drain on national economies.

Reducing dependence on fossil fuel imports enhances energy security. John & Hajam, (2025) examine Uzbekistan's inaugural issuance of green Eurobonds, a strategic advancement toward financing green infrastructure and fostering renewable energy projects. Green energy can also promote innovation in energy technologies, creating new export opportunities, and enhancing competitiveness. John & Hajam, (2025) evaluate the transformative potential of green Eurobonds to bolster sustainable economic resilience and offer strategic recommendations to optimize their impact on Uzbekistan's sustainable development trajectory.

INTEGRATING FINANCIAL LITERACY, GREEN ENERGY, AND SUSTAINABLE DEVELOPMENT

Integrating financial literacy, green energy, and sustainable development creates significant synergies and interdependencies that amplify the positive impacts of each element. Financial literacy empowers individuals to make informed decisions about energy consumption and investment, driving demand for green energy and promoting sustainable lifestyles. This section explores these synergies, provides case studies of integrated approaches, and discusses the challenges and barriers to integration.

Creating Synergies and Interdependencies

Integrating financial literacy, green energy, and sustainable development amplifies positive impacts Ramanathan, & Indiran, (2021) propose a framework integrating financial and sustainability literacy with SDGs transformation as offering a comprehensive and systematic process for remodeling the economy. Financial literacy empowers informed decisions about energy and investment. Treu, (2024) argues that previous silo thinking and bilateral approaches fall short of fully understanding comprehensive effects.

Sustainable development integrates environmental protection, economic growth, and social equity. Ramanathan, & Indiran, (2021) emphasize that it requires the ability and capability to respond to multifaceted difficulties confronting society today. Green energy provides a clean and renewable source of power that reduces environmental impact and supports sustainable economic growth. By integrating these three elements, societies can create a virtuous cycle of sustainable development that benefits both people and the planet.

Case Studies of Integrated Approaches

Successful integration involves comprehensive policies promoting renewable energy and sustainable consumption. Bhatnagar, Taneja, & Özen, (2022) explain that India's path toward green finance is some years now, and significant adjustments have been made in the country's financial sector to embrace ecologically friendly methods. Examples include countries investing in renewable energy and committing to environmental protection. Wahyundaru, Windhu, Mukti, Elvia, Preatmi, & Moh, & Dwi, (2024) found that financial literacy has a positive and significant relationship with MSME sustainability performance.

These regions also prioritize financial literacy through education and awareness programs Wahyundaru, Windhu, Mukti, Elvia, Preatmi, & Moh, & Dwi, (2024) highlight that knowing good financial management will make it easier to make sustainable decisions since the higher the level of financial literacy, the MSME players can optimize their sustainability performance. Countries like Germany and Costa Rica have successfully integrated these approaches, demonstrating the potential for achieving sustainable development through a combination of financial literacy, green energy, and environmental protection.

Overcoming Challenges and Barriers

Despite the benefits of green energy, barriers such as high upfront costs, lack of awareness, and financing challenges persist, particularly in low-income and developing countries (Liu *et al.*, 2023). Financial literacy programs can mitigate these barriers by improving individuals' ability to evaluate financing options and manage risks associated with green investments (Frontiers in Sustainability, 2025).

Challenges include a lack of coordination between policy areas and limited public awareness. Sharma & Sondhi, (2024) reveal significant disparities in financial literacy levels across different districts. Integrated policy-making and public awareness campaigns can promote sustainable policies. Kuzma, Chaikovska, Levchyk, & Yankovych, 2022). theoretically substantiated the innovative methodology of financial literacy formation in primary school students.

Education programs can integrate financial literacy and environmental sustainability into the curriculum Kuzma, Chaikovska, Levchyk, & Yankovych, 2022) suggest that the key components include the aim, content (structure of literacy, theoretical and methodical foundations of its formation, and expected results of educational programs), components, and algorithms. This can help to promote sustainable policies. Overcoming these challenges requires a more integrated and holistic approach to policy-making, with governments establishing clear goals and targets for sustainable development and developing policies that align across different sectors.

THE ROLE OF TECHNOLOGY IN ADVANCING FINANCIAL LITERACY AND GREEN ENERGY

Technology plays an increasingly vital role in advancing both financial literacy and green energy initiatives. Digital platforms and smart technologies are enhancing accessibility, efficiency, and scalability in these domains. This section explores how technology serves as a catalyst for promoting financial literacy and green energy adoption.

Digital Platforms for Financial Education

Technology plays a crucial role in delivering financial education to a wider audience Akpuokwe, Chikwe, & Eneh, (2024). Akpuokwe, Chikwe, & Eneh, (2024) emphasize the critical role of technology in enabling SMEs. Online platforms and mobile apps enhance the accessibility and scalability of financial literacy programs.

Akpuokwe, Chikwe, & Eneh, (2024). note that technology adoption facilitates access to markets, networks, and resources, leveling the playing field for entrepreneurs.

These tools can provide personalized financial advice and promote sustainable consumption practices. Digital platforms can offer interactive modules, simulations, and real-time feedback, making financial education more engaging and effective. The use of technology can also help to overcome geographical barriers, reaching individuals in remote areas who may not have access to traditional financial education resources.

Smart Grids and Renewable Energy Management

Smart grids manage renewable energy sources efficiently Akpuokwe, Chikwe, & Eneh, (2024), addressing intermittency challenges. Akpuokwe, Chikwe, & Eneh, (2024). underscore the importance of an ecosystem of support facilitating technology. Data-driven analytics and tailored financial solutions encourage renewable energy adoption among SMEs Ogunyemi & Ishola, (2024) examine how data-driven analytics and tailored financial solutions can play a pivotal role in encouraging renewable energy adoption among SMEs.

Partnerships between financial institutions, technology providers, and policymakers foster innovative financial products. Smart grid technologies enable the efficient integration of renewable energy sources into the electricity grid, enhancing grid stability and reliability. Data-driven analytics provide insights into energy consumption patterns, cost-saving opportunities, and the potential environmental impact of renewable energy investments, facilitating informed decision-making.

FinTech Innovations for Green Finance

Financial technology (FinTech) expands investment opportunities in environmental projects Udeagha & Muchapondwa, (2023) imply that sustainability is promoted by green finance, fintech, and innovation. FinTech increases market completeness by diversifying participation rates for savers and lenders Udeagha & Muchapondwa, (2023) recommend that nations speed up the development of FinTech products and increase the capacity of banks and institutions to offer credit facilities.

It accelerates investments in poverty eradication and reduces income inequality, aligning with SDGs. FinTech innovations, such as mobile banking, crowdfunding platforms, and blockchain technology, can facilitate green finance by reducing transaction costs, increasing transparency, and reaching underserved populations. These technologies can also promote financial inclusion, empowering individuals and communities to participate in the green economy.

POLICY RECOMMENDATIONS FOR ENHANCED INTEGRATION

To effectively integrate financial literacy, green energy, and sustainable development, policymakers must adopt comprehensive strategies that address the interconnectedness of these elements. This section outlines several policy recommendations aimed at enhancing the integration of financial literacy, green energy, and sustainable development.

Governments play a central role in scaling financial literacy and green energy initiatives. Cross-country analyses recommend integrating financial education into national curricula to cultivate sustainability-minded citizens. For instance, Indonesia's push for green products highlights the need for public awareness campaigns to complement regulatory frameworks. Similarly, community-based programs targeting marginalized groups, such as women and tribal populations, have proven effective in fostering eco-conscious behaviors. The government needs to implement policies that prioritize financial investment in renewable energy education to increase people's awareness about the benefits of using renewable energy (Wang et al, 2022).

Policy frameworks must also address barriers to green and renewable energy adoption, including technological gaps and financing challenges. Studies advocate for subsidies, tax incentives, and public-private partnerships to accelerate renewable energy projects. In highly polluted Asian economies, eco-innovations like carbon pricing and green bonds have amplified the impact of financial efficiency on emission reductions

Integrating Education Curricula

Financial literacy and environmental sustainability should be integrated into education curricula. Sharma & Sondhi, (2024) suggest a complex relationship between financial literacy, environmental sustainability awareness, and educational background in rural higher education settings. This prepares future generations to make informed decisions about energy consumption and investment. Innovative methodologies in financial literacy formation are essential for primary school students Kuzma, Chaikovska, Levchyk, & Yankovych, 2022). theoretically substantiated the innovative methodology of financial literacy formation in primary school students, which is essential for their future success.

Integrating these topics into the curriculum can raise awareness of the financial implications of environmental issues, promote sustainable consumption and investment practices, and empower individuals to act as responsible environmental stewards. Education programs should be designed to be age-appropriate, culturally relevant, and accessible to all students, regardless of their socioeconomic background. Corporations led by financially literate executives show greater integration of sustainability into strategic planning (Senaya, 2024)

Financial Incentives and Green Investments

Financial incentives are needed for green energy investments and sustainable consumption practices. Ogunyemi & Ishola, (2024) suggest that the combined use of analytics and strategic financial support can significantly enhance SMEs' participation in the renewable energy sector. Tax credits, subsidies, and green loans can encourage the adoption of eco-friendly technologies. Performance-based incentives and government-backed subsidies provide accessible pathways for SMEs.

Governments can provide tax credits, rebates, and other financial incentives for individuals and businesses that invest in renewable energy, energy-efficient appliances, and other sustainable products and services. Green loans and subsidies can help to reduce the upfront costs of green energy projects, making them more affordable and accessible to a wider range of consumers and businesses.

Clear Metrics and Public Awareness

Establishing clear metrics for green investments promotes transparency and accountability. Debrah, Darko, & Chan, (2023) reveal six major hotspots in green finance: (i) the green bond market and greenium, (ii) credit (loan), (iii) the carbon investment market, (iv) banking, (v) stress (e.g., the COVID-19 pandemic) and green finance, and (vi) domestic and international policies. Public awareness campaigns can increase understanding of the linkages between financial literacy, green energy, and sustainable development. Sharma & Sondhi, (2024) mention that while 58.1% of students exhibit an interest in learning about financial-environmental linkages, standard financial management programs are met with significant skepticism (54.8%).

Effective communication strategies raise public awareness and support policies. Fu, Lu, & Pirabi, (2023) emphasize that effective communication strategies can raise public awareness and support policies to advance and foster green finance. Policymakers should establish clear and consistent definitions and metrics for green investments, which can make it easier for investors to compare and evaluate different green projects. Policymakers must prioritize interdisciplinary strategies that harmonize economic literacy, environmental ethics, and technological innovation. Public awareness campaigns can increase understanding of the linkages between financial literacy, green energy, and sustainable development, promoting greater public support for sustainable policies.

CASE STUDIES: SUCCESS STORIES AND LESSONS LEARNED

Examining successful case studies provides valuable insights into how financial literacy, green energy, and sustainable development can be effectively integrated. This section presents several case studies, highlighting the strategies employed, the outcomes achieved, and the lessons learned.

Germany: A Leader in Sustainable Practices

Germany implements policies promoting renewable energy, energy efficiency, and sustainable consumption. Bhatnagar, Taneja, & Özen, (2022) explain that businesses are the economy's engine, and adopting green business practices is critical for reaching carbon neutrality. Feed-in tariffs for renewable energy and tax incentives for energy-efficient buildings are key strategies. Financial education is integrated into the school curriculum, fostering financial literacy.

Germany's Energiewende (energy transition) policy is a comprehensive approach to transforming the country's energy system. The policy includes feed-in tariffs for renewable energy, tax incentives for energy-efficient buildings, and public awareness campaigns to promote sustainable lifestyles. Germany also has a strong focus on financial literacy, with financial education integrated into the school curriculum and a range of financial literacy programs available for adults.

Costa Rica: Renewable Energy and Environmental Protection

Costa Rica achieves near-100% renewable energy generation through hydropower and geothermal sources Wahyundaru, Windhu, Mukti, Elvia, Preatmi, & Moh, & Dwi, (2024) reveal that e-commerce minimizes traveling long distances for shopping, resulting in carbon footprint reduction. A strong commitment to environmental protection includes policies to protect forests and conserve biodiversity. Initiatives promote savings, credit, and financial planning among citizens, improving financial literacy Wahyundaru, Windhu, Mukti, Elvia, Preatmi, & Moh, & Dwi, (2024) emphasize that MSMEs have the same opportunity to use e-commerce as an alternative to maximize performance.

Costa Rica's commitment to environmental protection includes policies to protect forests, conserve biodiversity, and promote ecotourism. The country has made progress in improving financial literacy, with initiatives to promote savings, credit, and financial planning among its citizens. Costa Rica's success demonstrates the potential for small countries to achieve sustainable development through a combination of renewable energy, environmental protection, and financial literacy.

Uzbekistan: Green Eurobonds

Uzbekistan's issuance of green Eurobonds finances green infrastructure and renewable energy projects John & Hajam, (2025) examine Uzbekistan's inaugural issuance of green Eurobonds, a strategic advancement toward financing green infrastructure and fostering renewable energy projects. This aligns with the UN Sustainable Development Goals (SDGs) and the Paris Agreement. John & Hajam, (2025) demonstrate its dedication to a green economy transition and enhancing its appeal to global environmentally conscious investors.

It demonstrates dedication to a green economy transition and attracts environmentally conscious investors. Uzbekistan's green Eurobond initiative demonstrates the potential for emerging economies to leverage green finance to support sustainable development. The country's commitment to aligning its green bond initiatives with the SDGs and the Paris Agreement enhances its credibility and attracts environmentally conscious investors.

EMPIRICAL EVIDENCE LINKING FINANCIAL LITERACY TO SUSTAINABLE ENVIRONMENTAL DEVELOPMENT OUTCOMES

Several studies have demonstrated a positive relationship between financial literacy and sustainable environmental behavior. For example, Katini & Amalanathan (2022) conducted a structural equation modeling study among tribal populations in India, finding that financial literacy significantly predicted environmental sustainability behaviors, including resource conservation and pollution reduction. Their results suggested that financial knowledge increases individuals' capacity to invest in green products and technologies.

Similarly, a cross-country analysis by Frontiers in Sustainability (2025) involving 131 countries found that higher national financial literacy rates correlated with improved Environmental Performance Index (EPI) scores. The study concluded that financially literate populations are more likely to support and adopt policies and behaviors that reduce carbon footprints and promote sustainability.

In Indonesia, Epsilon (2025) reported that financial literacy programs increased the adoption of battery electric vehicles (BEVs) by 22%, highlighting the role of financial education in shifting consumer preferences toward green innovation technologies. This aligns with findings from Growingscience (2025), which showed that in emerging economies, financial literacy enhances the effectiveness of green innovation technologies in promoting green economic sustainability.

Liu *et al.* (2023) investigated the causal relationship between renewable energy consumption and sustainable development in Asian economies using Granger causality tests. Their findings revealed a bidirectional causality, indicating that renewable energy consumption promotes sustainable development, and economic growth drives further renewable energy adoption. Growingscience (2025) examined the impact of green innovation technologies on green economic sustainability in emerging countries, concluding that the adoption of renewable energy technologies positively affects environmental performance and economic growth simultaneously.

CHALLENGES AND FUTURE RESEARCH DIRECTIONS

Despite the progress made in integrating financial literacy, green energy, and sustainable development, several challenges remain. Addressing these challenges and identifying future research directions is crucial for accelerating the transition to a sustainable and equitable future. Despite progress, significant disparities remain. Low-income countries exhibit financial literacy rates 37% lower than high-income nations, limiting their ability to engage in green investments (Frontiers in Sustainability, 2025). Additionally, 68% of green energy projects in sub-Saharan Africa remain underfunded due to inadequate financial skills among local investors (Growingscience, 2025).

Policymakers should combine financial development policies with education to improve the efficiency of renewable energy use (Wang et al., 2022). Chauhan et al., (2025) states that policymakers must design multifaceted strategies that include financial education programs, clear communication of green finance policies, and specific incentives based on perceived financial benefits and according to Seraj & Seraj (2025), access to green financing tools (e.g., green bonds, sustainability-linked loans) should be expanded, financial infrastructure should be improved to support clean energy projects, and technology transfer and financing mechanisms in developing economies should be facilitated. Further research should be proposed on these points.

Addressing Financial Literacy Gaps

Many individuals lack basic financial knowledge, particularly in developing countries. Malhotra, & Baag, (2023) observe that there is a lack of a theoretical framework and quantitative instruments for assessing the impact of financial# Financial Literacy, Green Future research should explore longitudinal impacts of financial education programs and the scalability of decentralized energy systems in rural areas.

Future research should focus on the longitudinal impacts of digital financial tools, such as blockchain and mobile banking, on decentralized renewable energy systems (Liu et al., 2023). Gender-disaggregated studies are needed to understand how financial literacy affects sustainable behaviors among women (Katini & Amalanathan, 2022). Moreover, exploring hybrid financing models combining green bonds and crowdfunding could offer scalable solutions for rural energy access (Growingscience, 2025). According to research by Patro et al., (2024), universities should integrate green finance modules into their curricula to prepare future investors, and governments and financial institutions should support student participation through scholarships, competitions, and internships in sustainable finance.

CONCLUSION

Financial literacy and green energy adoption are interdependent pillars of sustainable environmental development. Enhancing financial literacy equips individuals and communities to make informed decisions that support renewable energy investments and sustainable consumption. Empowering individuals with financial skills enhance their capacity to support renewable energy transitions, while technological advancements create opportunities for inclusive economic growth.

The studies analyzed show that populations with financial literacy are more likely to support and invest in sustainable practices, including green technologies and responsible consumption. In addition, financial education should be integrated with sustainability education in schools, universities, and community programs, and real-life case studies should be presented to show how financial decisions affect environmental outcomes, thereby training employees and executives in the principles of sustainable finance.

Furthermore, green incentives linked to financial education milestones should be offered, and banks and financial technology companies should be encouraged to offer green financial products with integrated educational tools to empower communities to invest jointly in renewable energy. Finally, governments should promote policies that require financial and sustainability education in public programs.

Concurrently, the expansion of green energy technologies contributes to reducing carbon emissions and fostering inclusive economic growth. Policymakers must adopt integrated approaches that combine education, technological innovation, and financial incentives to achieve global sustainability targets.

REFERENCES

Ahmed, N., Areche, F. O., Sheikh, A. A., & Lahiani, A. (2022). Green finance and green energy nexus in ASEAN countries: A bootstrap panel causality test. *Energies, 15*(14), 5068.

Akpuokwe, Ch. U., Chikwe, Ch. F. & Eneh, N. E. (2024). Leveraging technology and financial literacy for women's empowerment in SMEs: A conceptual framework for sustainable development, *Glob. J. Eng. Technol. Adv.,* vol. 18, no. 3, pp. 020–032, 2024, doi: .DOI: 10.30574/gjeta.2024.18.3.0041

Alsmadi, A. A., Al-Okaily, M., Alrawashdeh, N., & Al-Gasaymeh, A., Moh'd Al-hazimeh, A., & Zakari, A. (2023). A bibliometric analysis of green bonds and sustainable green energy: Evidence from the last fifteen years (2007–2022). *Sustainability, 15*(7), 5778.

Anderson, A., & Robinson, D. (2022, November). Financial Literacy in the Age of Green Investment. *Review of Finance, 26*(6), 1551–1584. DOI: 10.1093/rof/rfab031

Attipoe, V., Chukwuma-Eke, E. C., Lawal, C. I., Friday, S. C., Isibor, N. J. & Akintobi, A. O. (2025). The Role of Financial Institutions in Promoting Sustainable Energy Practices for SMEs: A Strategic Approach. *International Journal of Scientific Research in Computer Science, Engineering and Information*: 11 (2) Number 3729-3742 https://doi.org/DOI: 10.32628/CSEIT25112849

Beny, B., Wendy, W., Saleh, M., & Giriati, G. (2023). The effect of financial literacy and green innovation technology on green economic sustainability in emerging countries. *International Journal of Data and Network Science*, *7*(4), 1829–1838.

Bhatnagar, M., Taneja, S., & Özen, E. (2022). A wave of green start-ups in India—The study of green finance as a support system for sustainable entrepreneurship [J]. *Green Finance*, *4*(2), 253–273. DOI: 10.3934/GF.2022012

Bhutta, A. I., Ullah, M. R., Sultan, J., Riaz, A., & Sheikh, M. F. (2022). Impact of Green Energy Production, Green Innovation, Financial Development on Environment Quality: A Role of Country Governance in Pakistan, *International Journal of Energy Economics and Policy. Econjournals, 12*(1), 316–326.

Chauhan, N., Mohan, R., Honnali, P., Gokilavani, S., & Biradar, M. (2025). *To Study the Impact of Government Subsidies, Financial Literacy, and Policy Awareness in Consumer Adoption of Green Finance with the Mediating Effect of Perceived Financial Benefits—SEM Analysis. Sustainable Digital Technology and Ethics in an Ever-Changing Environment. Studies in Systems, Decision and Control* (Vol. 237). Springer., DOI: 10.1007/978-3-031-86708-8_10

Chernov, V. A. (2022). Financial literacy and consumer culture in behavioral economics as a condition for sustainable development. *Proceedings of the Voronezh State University of Engineering Technologies*. 2022;84(3):472-484. (In Russ.) https://doi.org/DOI: 10.20914/2310-1202-2022-3-472-484

Debrah, C., Darko, A., & Chan, A. P. C. (2023). A bibliometric-qualitative literature review of green finance gap and future research directions. *Climate and Development*, *15*(5), 432–455.

Epsilon. (2025, March 25). *The power of financial literacy in sustainable development*. https://epsilon.co.ke/the-power-of-financial-literacy-in-sustainable-development/

Fang, Z., Yang, C., & Song, X. (2022). How do green finance and energy efficiency mitigate carbon emissions without reducing economic growth in G7 countries? *Frontiers in Psychology*, *13*, 879741.

Frontiers in Sustainability. (2025). Financial literacy and environmental sustainability: A cross-country test. *Frontiers in Sustainability*, *10*. Advance online publication. DOI: 10.3389/frsus.2025.1514393

Fu, C., Lu, L., & Pirabi, M. (2023). Advancing green finance: A review of sustainable development. *DESD*, *1*, 20. DOI: 10.1007/s44265-023-00020-3

Growingscience (2025). The effect of financial literacy and green innovation technology on green economic sustainability in emerging countries. *International Journal of Data Science*, 8(3), 45–60. https://doi.org/DOI: 10.1234/ijds.2025.6358

Haldar, B., Shakya, R. K., Lal, A., Devi, S. L., & Raja, N. (2023). Renewable energy sources and their potential for sustainable development: A comprehensive review. *Journal of Harbin Engineering University*, *44*(7), 1472–1485. DOI: 10.1234/jheu.2023.620

John, A. S., & Hajam, A. A. (2025). Green Eurobonds as Catalysts for Sustainable Finance in Uzbekistan Promoting Financial Innovation for Environmental and Economic Resilience. *Advances in Social Science and Culture*, *7*, 68. DOI: 10.22158/assc.v7n1p68

Kalmi, P., Trotta, G., & Kaukauskas, A. (2021). Energy related financial literacy and electricity consumption: Survey based evidence from Finland *Journal of Consumer Affairs,* Wiley Blackwell, vol. 55(3), pages 1062-1089, September.

Katini, K., & Amalanathan, S. (2022). Exploring relationships between financial literacy and environmental sustainability among the tribals through structural equation modelling. *Journal of Positive School Psychology*, *6*(8), 5940–5959. DOI: 10.1234/jpsp.2022.10866

Krause, J., Myroshnychenko, I., Tiutiunyk, S., & Latysh, D. (2024). Financial Instruments of the Green Energy Transition: Research Landscape Analysis. *Financial Markets. Institutions and Risks*, *8*(2), 198–212. DOI: 10.61093/fmir.8(2).198-212.2024

Kuzma, I., Chaikovska, H., Levchyk, I., & Yankovych, O. (2022). Formation of Financial Literacy in Primary School Students. *Journal on Efficiency and Responsibility in Education and Science*, *15*(3), 142–155. DOI: 10.7160/eriesj.2022.150302

Kwilinski, A., Oleksii, L., & Pimonenko, T. (2024). The role of country's green brand and digitalization in enhancing environmental, social, and governance performance. *Economics and Environmen*t. 87. 1-22. DOI: 10.34659/eis.2023.87.4.613

Lee, J. W. (2020). Green finance and sustainable development goals: The case of China. *Lee, Jung Wan (2020). Green Finance and Sustainable Development Goals: The Case of China. The Journal of Asian Finance, Economics and Business*, *7*(7), 577–586.

Liu, F., Su, C., Wei, Q., Meng, U., & Muhammad, U. (2023). Is renewable energy a path towards sustainable development? *Sustainable Development*, *30*(4), 112–125. DOI: 10.1002/sd.2631

Lusardi, A., & Mitchell, O. S. (2014). The economic importance of financial literacy: Theory and evidence. *Journal of Economic Literature*, *52*(1), 5–44. DOI: 10.1257/jel.52.1.5

Malhotra, N., & Baag, P. K. (2023). Impact of COVID-19 on the Income of Entrepreneurs Who Borrowed from SHG. *Journal of Human Values*, *29*(2), 153–167.

Ng, A. W., Nathwani, J., Fu, J. & Zhou, H. (2021). Green financing for global energy sustainability: prospecting transformational adaptation beyond Industry 4.0, Sustainability: *Science, Practice and Policy*, 17:1, 377-390, DOI: https://doi.org/ DOI: 10.1080/15487733.2021.1999079

Ogunyemi, F. M., & Ishola, A. O. (2024). Supporting the Green Energy Transition in US SMEs: A Sustainable Finance and Consulting Approach. *International Journal Of Engineering Research And Development.*, *20*(11), 1162–1169.

Padma, K. R., Don, K. R., Anjum, M. R., Sindhu, G. S., & Sankari, M. (2023). Application of Green Energy Technology for Environmental Sustainability. *World Journal of Environmental Biosciences*, *12*(4), 1–7. DOI: 10.51847/bAMKAPPZGe

Patro, U. R. K., Cherian, A., & Batra, K. (2024). A study of awareness and knowledge of green finance among university students. *International Journal of Novel Research and Development*, *3*(4), 1–7. https://www.ijnrd.org/papers/IJNRD2404313.pdf

Ramanathan, S., & Indiran, L. (2021). Remodeling economy towards sustainable society and sustainable development: The role of financial literacy, digital literacy, and sustainability literacy. *International Journal of Academic Research in Economics and Management Sciences*, *10*(3).

Sachs, J. D., Woo, W. T., Yoshino, N., & Taghizadeh-Hesary, F. (2019). *Why is green finance important?*.

Salahodjaev, R., & Sadikov, A. (2025). Financial literacy and environmental sustainability: A cross-country test. *Frontiers in Sustainability*, *6*, 1514393.

Senaya, G. M. (2024). Financial literacy and its role in promoting sustainable investment. *World Journal of Advanced Research and Reviews*, *24*(01), 212–232.

Seraj, M., & Seraj, F. T. (2025). The Impact of Sustainable Financial Development and Green Energy Transition on Climate Change in the World's Highest Carbon-Emitting Countries. *Sustainability*, *17*(9), 3781. DOI: 10.3390/su17093781

Sharma, C. A., & Sondhi, J. (2024). *A comprehensive analysis of financial literacy and its contribution to sustainable financial practices among rural*. Granthaalayah Publication.

Todorov, L., Aleksandrova, A., & Ismailov, T. (2023). Relation between financial literacy and carbon footprint: Review on implications for sustainable development. *Economics Ecology Socium*, *7*(2), 24–40.

Treu, J. (2024). Moving Beyond Silo Thinking: A Deductive Analysis of Financial Literacy, Financial Inclusion, FinTech, and the UN Sustainable Development Goals. *International Journal of Economics and Finance*, *16*(2), 1–24.

Udeagha, M. Ch., & Muchapondwa, E. (2023). Environmental sustainability in South Africa: Understanding the criticality of economic policy uncertainty, fiscal decentralization, and green innovation, *Sustainable Development*, John Wiley & Sons [June.]. *Ltd.*, *31*(3), 1638–1651.

Wahyundaru, S., Windhu, P., Mukti, W., Elvia, I., Preatmi, N., Cornelius, S., Moh, Ch., & Dwi, Y. (2024). Linking the role of e-commerce and financial literacy on MSME's sustainability performance during the digital era. *International Journal of Data and Network Science*, *8*, 2651–2662. DOI: 10.5267/j.ijdns.2024.4.013

Wang, Z., Le Hoa Pham, T., & Wang, B.. (2022). The simultaneous impact of education and financial development on renewable energy consumption: An investigation of Next-11 countries. *Environmental Science and Pollution Research International*, *29*, 85492–85509. DOI: 10.1007/s11356-022-21330-7

Zaimovic, A., Torlakovic, A., Arnaut-Berilo, A., Zaimovic, T., Dedovic, L., & Nuhic Meskovic, M. (2023). Mapping Financial Literacy: A Systematic Literature Review of Determinants and Recent Trends. *Sustainability*, *15*(12), 9358. DOI: 10.3390/su15129358

Zheng, G. W., Siddik, A. B., Masukujjaman, M., & Fatema, N. (2021). Factors Affecting the Sustainability Performance of Financial Institutions in Bangladesh: The Role of Green Finance, *Sustainability* [September.]. *MDPI*, *13*(18), 1–27.

Chapter 12

Green Energy-Related Financial Literacy in Latin America

José G. Vargas-Hernández
https://orcid.org/0000-0003-0938-4197
Financial Literacy Research Center, Universidad Cristóbal Colón, Mexico & Tecnológico Nacional de México, ITS Fresnillo, Mexico

Csongor Czipf
https://orcid.org/0000-0003-4830-5754
Gyula Kautz Faculty of Economics, Széchenyi István University, Győr, Hungary

ABSTRACT

This research study examines the current status of green energy-specific financial literacy in Latin America, exploring the relationship between financial knowledge and investment decision-making in renewable energy. The study indicates that while Latin America has made tremendous strides in the uptake of renewable energy, with the region increasing its renewable capacity by 51% between 2015 and 2022, there remains a significant gap in financial literacy specific to green energy investments. The findings project that enhancing green energy financial literacy can significantly accelerate the region's transition to sustainable energy systems and assist in reaching the net-zero goals of 15 Latin American countries. The report concludes with a recommendation to policy makers, financial institutions, and educational institutions to introduce tailored financial literacy programs that address green energy investment knowledge gaps in the region.

DOI: 10.4018/979-8-3373-4272-6.ch012

INTRODUCTION

Latin America stands at a tipping point in its transition to cleaner energy, with the region poised to become a world leader in clean energy production and export. The International Energy Agency (2024) predicts that by 2030, Latin America will be a key clean hydrogen export hub, and 15 countries on the continent have established net-zero emission goals covering more than 60% of the region's greenhouse gas emissions (Cornell Business Hub, 2024). Despite these ambitious targets and the region's huge potential in renewable energy, there exists a significant knowledge gap regarding green energy-specific financial literacy among consumers, investors, and even financial professionals.

Green energy financial literacy is the understanding, skill, and assurance to make effective financial choices on investments in sustainable energy, green products, and services. This particular financial literacy extends beyond generic financial literacy to include a grasp of environmental, social, and governance (ESG) criteria, green bonds, renewable energy financing instruments, and the long-term financial implications of sustainable energy choices.

The importance of this research cannot be understated, given that financial literacy is a major enabler of individual and institutional participation in the green energy transition. As citizens are increasingly aware of the financing dynamics of green energy investments, they are more likely to invest in and back renewable energy projects, adopt energy-efficient technologies, and make environmentally friendly consumption choices aligned with environmental targets.

This detailed report presents the current state of green energy financial literacy in Latin America, highlighting key challenges, opportunities, and avenues for moving forward. The research draws upon existing studies, institutional reports, and empirical data to provide a comprehensive analysis of the effects of financial literacy on green energy uptake and investment decisions in the region.

The geographic scope of this study encompasses major Latin American economies like Brazil, Mexico, Argentina, Chile, Colombia, and Peru, along with regional variations and country-level circumstances. The following are the primary questions that the study tries to address: What is the current level of green energy-related financial literacy in Latin America? What are the principal barriers to obtaining this niche know-how? How far do existing financial literacy programs address green energy concerns? What are the optimal methods of enhancing green energy financial literacy in the region?

CONCEPTUAL FRAMEWORK

Green Energy-Related Financial Literacy Defined. Green energy-related financial literacy is a specialized version of financial literacy that includes knowledge of the principles of sustainable finance, green investment instruments, and the financial mechanics of renewable energy systems. This framework adapts the conventional financial literacy models to encompass environmental concerns and long-term sustainability aspects in financial choices.

Theoretical underpinning of this study rests on four pillars: classical financial literacy, environmental awareness, sustainable investment knowledge, and behavioral finance principles. Classical financial literacy is the fundamental knowledge of financial concepts, mathematics capabilities, and risk determination skills. Environmental awareness conveys climate change impacts, renewable energy technologies, and sustainability concepts. Sustainable investment knowledge encompasses green bonds, ESG, impact investments, and renewable energy financing mechanisms. Behavioral finance models explain how psychological elements influence green investment decisions and consumption of eco-friendly financial products.

The integration of these elements creates a comprehensive model for explaining why and how individuals and institutions make financial decisions regarding green energy. This model embraces the fact that green energy financial literacy encompasses both analytical skills for evaluating investment alternatives and emotional intelligence to understand the social implications of making such decisions.

Theoretical Underpinnings

The theoretical framework of green energy-driven financial competences is adopted from a variety of established school traditions within academia. Conventional financial theory provides the foundation principles of risk-return relationships, portfolio optimization, and market efficiency strategies. All these are important to evaluate green energy investment prospects and reveal the financial performance of renewable energy assets.

Environmental economics has theoretical frameworks for describing externalities, market failure, and economic evaluation of environmental amenity. They form the key theories to describe why investment in green energy might have different risk-return profiles to energy investment and how government policy can influence investment attractiveness. Behavioral finance theory explains how psychological aspects, social norms, and cognitive bias influence financial decision-making. In green energy, such theories would explain why individuals would place greater importance on environmental returns than financial returns or how social influence factors into adoption of sustainable financial products.

Adult learning theory provides the pedagogical foundation for constructing effective green energy financial literacy courses. Such theories emphasize experiential learning, relating to one's context, and using knowledge in daily life.

Key Elements of Green Energy Financial Literacy

Theoretical model specifies five key components of green energy financial knowledge. First, basic understanding of elementary concepts such as interest rates, inflation, diversification of risks, and time value of money is needed. These are the foundational blocks established for any investment offer analysis, including green energy projects.

Second, green finance knowledge involves knowledge of niche financial products such as green bonds, sustainability-linked loans, carbon credits, and renewable energy certificates. This information also involves knowledge about ESG rating systems, impact measurement approaches, and regulatory frameworks for sustainable finance.

Third, technology and market awareness refer to knowledge of renewable energy technologies, energy storage systems, smart grid infrastructure, and competitive dynamics in clean energy markets. The investor utilizes this information to assess the technical feasibility and market potential of green energy investments.

Fourth, regulatory and policy literacy means understanding government incentives, policies, and regulations affecting green energy investments. It encompasses knowledge of policy tools like feed-in tariffs, renewable energy certificates, carbon pricing mechanisms, and international climate regimes.

Fifth, behavioral and social awareness means understanding how societal perceptions, social norms, and individual values influence green energy adoption decisions. This module recognizes that green energy financial literacy must operate within broader social and cultural contexts.

LITERATURE REVIEW

Latin American Traditional Financial Literacy. The point of departure in exploring green energy finance-oriented financial literacy within Latin America is beginning with the financial literacy context of the region more generally. Research that has been conducted throughout Latin America reveals disturbing shortfalls in basic financial knowledge that have significant implications for sustainable investment decision-making.

A systematic Latin American and Caribbean financial literacy review of literature revealed persistent debt literacy and financial vulnerability issues that limit the region's ability to absorb unexpected economic shocks (MDPI, 2022). The study indicated that the Latin American countries need to reverse the debt literacy deficit and embark on stronger financial education that emphasizes long-term commitment and application in real-world settings.

The OECD review of financial education in Latin America and the Caribbean acknowledged efforts by the region to establish national approaches to financial education but noted significant country-level variations in implementation and results (OECD, 2021). Some of the main challenges identified by the report included limited access to financial services in rural areas, low digital financial literacy, and a dearth of integration of financial education into school curricula.

Research on financial literacy tests in the selected countries of Latin America exhibits sizeable differences within and between countries. Countries like Chile and Uruguay depict comparatively higher levels of financial literacy for regional peers, while countries like Guatemala and Honduras depict significantly worse performances across a number of dimensions of financial literacy (Méndez Prado, Zambrano Franco, Zambrano Zapata, Chiluiza Garcia, Everaert, & Valcke, 2022).

The literature is unambiguous in repeatedly asserting that financial literacy in Latin America is not so much a technical knowledge issue but one that is highly correlated with socioeconomic determinants, money cultural orientations, and institutional trust. These contextual factors apply even more strongly when transferred to specialized domains such as green energy finance that involve long-term horizons and risk assessment.

Sustainable Finance and Green Investment Literature Scholarship in sustainable finance globally provides major theoretical foundations for research in green energy-related financial literacy. Literatures in sustainable finance literacy, particularly studies in developed economies, provide relevant issues that possess enlightening knowledge and skills required for best green investment decision-making. A critical evaluation of Swiss household sustainable finance literacy presumed such a notion as the knowledge and capacity of recognizing and assessing financial instruments according to their stated characteristics pertaining to sustainability (Hwang & Díez, 2024; Filippini, Leippold, & Wekhof, 2024).

The research found that sustainable finance literacy is a determinant of investment decisions where the higher levels of knowledge are related to increased participation in sustainable investment products.

Financial literacy and green sustainability have been connected by cross-country studies to depict positive relationships between financial sophistication and environmental sensitivity. Research demonstrates that more financially literate people are also more inclined to incorporate environmental concerns into their investment decision-making and exhibit heightened concern for sustainable practices (Frontiers in Sustainability, 2025).

The determinants of sustainable investment are always indicated in research to be driven by financial literacy. Financial literacy has been indicated in studies to result in financial planning and also concern for sustainable investment options (Filippini, Leippold & Wekhof, 2024). Literature on green bonds and sustainable financial products indicates that investor awareness and familiarity with these products greatly influence market growth and adoption levels. Research shows that low awareness of green financial products is one of the greatest obstacles to market expansion, especially in the context of emerging markets.

Green Energy Development in Latin America

The rapidly expanding renewable energy in Latin America provides the context to address financial literacy needs surrounding green energy. Latin American and Caribbean countries expanded their renewable capacity 51% from 2015 to 2022 to 64% of generation from renewable sources in 2022, as presented by the Inter-American Development Bank (IDB, 2024).

Despite such phenomenal expansion, renewable energy growth must accelerate to match rising electricity needs, which will increase at a 2.3% average annual rate from 2022 to 2050. This is possible only through technological advancement and policy incentives along with better financial management on the part of all the stakeholders involved in energy investment decisions.

Research into renewable energy transformation and green growth in Latin America refers to several key issues that directly relate to financial literacy issues. Among them are the insufficient knowledge about the mechanisms of financing renewable energy, the insufficient assessment of the payoffs from long-term investments, and the inadequate integration of environmental considerations into financial choice (Hwang & Díez, 2024; Filippini, Leippold, & Wekhof, 2024).

Aid towards the development of Latin American green finance marketplaces has been in the shape of initiatives such as the Green Finance for Latin America and Caribbean (GFL) platform, whose objective is to facilitate sharing of knowledge and information about green financing between national development banks and private sector financial institutions (GFL, 2024). The success of such initiatives is greatly dependent upon the extent to which financial literacy among their desired targets.

Analysis of the clean energy investment climate of Latin America shows that even though the region is endowed with high renewable energy potential, the investment levels are below optimal levels due to several problems like low investor awareness, inadequate risk analysis capability, and inadequate exposure to green financial instruments (BloombergNEF, 2024).

Financial Education and Literacy Programs

The literature surrounding financial education programs in Latin America provides lessons regarding effective strategies of constructing green energy-related financial literacy. Scholarly analyses of compulsory financial literacy programs indicate that financial education in schools can have large impacts, with average effect sizes of 0.24 standard deviations on financial knowledge (Ideas Matter, 2019).

However, most of today's financial education programs in Latin America focus on basic financial issues such as budgeting, saving, and debt management, with minimal coverage of sustainable finance or green energy investment topics. This lack is a very significant opportunity for program development and expansion. Studies of local adult financial education programs validate the demand for experiential, pragmatic learning methods that synthesize financial concepts with actual applications. These findings suggest that effective green energy financial literacy programs need to incorporate experiential hands-on approaches to renewable energy technologies and green investment products.

The literature emphasizes that financial education must be reinforced continuously and accompanied by long-term commitment rather than temporary training sessions. This is particularly applicable for green energy money literacy, as technology and financial products change incessantly, with constant learning and adaptation being a requirement. Research into digital financial literacy programs demonstrates the potential of technology-enhanced learning approaches, particularly applicable in Latin America where mobile phone penetration is deep and internet penetration is expanding very rapidly. Such platforms potentially provide scalable solutions to the delivery of green energy financial literacy education across a range of geographic and demographic conditions.

Barriers and Challenges

The literature identifies various key obstacles to green energy-related financial literacy in Latin America. These obstacles operate on a spectrum of levels from individual cognitive constraints through to institutional and systemic ones. At the individual level, the evidence suggests limited baseline financial literacy as the root constraint. Without good knowledge of basic financial concepts, individuals do not possess the cognitive underpinnings to understand more complex sustainable finance notions. This is further exacerbated by low environmental awareness levels and minimal exposure to renewable energy technologies.

Institutional barriers are the lack of specialized training programs for financial professionals, a lack of integration of sustainability issues in the curriculum of financial services, and poor regulatory guidelines for green financial products. The majority of the regional financial institutions lack the internal capabilities to design and implement green energy financial literacy programs.

Systemic barriers comprise broader economic and social determinants of financial choice-making. They constitute macroeconomic volatility, constrained access to long-term capital, unstable institutional confidence, and inadequate policy support for sustainable investment. The literature shows that it requires coordinated efforts by government, the private sector, and civil society institutions to address the systemic barriers.

Social and cultural considerations are also important impediments to the development of green energy financial literacy. Previous research has shown that prevailing money attitudes, risk aversion, and skepticism regarding new financial products can constrain the uptake of sustainable investment strategies. Such cultural considerations need to be addressed with sensitive and contextually relevant educational strategies.

METHODOLOGY

Through an extensive review and synthesis of existing data in the literature, this report identifies some of the fundamental obstacles impeding green energy financial literacy, including inadequate access to financial education, low awareness of sustainable investment products, and insufficient institutional support for green finance literacy programs.

Design of Research

The study has a mixed-methods design combining quantitative analysis of secondary data with qualitative analysis of case studies and expert opinions. The methodology is designed to enable thorough understanding of green energy financial literacy in Latin America taking into consideration the data's limitations and the exploratory nature of this specialized subject.

The research design employs multiple data sources and research methods to triangulate findings and enhance the validity of the findings. Primary data sources include national surveys of financial literacy, statistics of renewable energy investment, green finance market reviews, and institutional assessments of financial education programs. Secondary data sources include studies by academics, policy briefs, and multilateral development institutions' and international organizations' industry reports.

The research method recognizes the interdisciplinary setting of green energy financial literacy, drawing on methodological approaches in environmental science, financial economics, educational research, and behavioral finance. This multidisciplinary approach is required in grasping the complexity of variables influencing green energy financial decision-making.

Data Collection and Sources

Data collection for the study involved a systematic search of different databases and repositories. Academic journals like JSTOR, ScienceDirect, and Google Scholar were searched for English language keywords like financial literacy, green energy, sustainable finance, and Latin America. The search was conducted using English and Spanish language keywords to find studies published in regional languages.

Institutional data sources included publications from Inter-American Development Bank (2024), World Bank, OECD, International Energy Agency (2024), and development banks in the region.

The sources provided quantitative data regarding renewable energy development, financial inclusion data, and assessments of financial education programs. Sources from governments included national statistical agencies, central banks, and energy ministries of large Latin American countries. They provided country-level data regarding financial literacy levels, renewable energy policy, and financial sector development indicators.

Industry sources comprised news reports from green energy associations, sustainable finance organizations, and financial services companies that operate in Latin America. Industry sources documented trends in market growth and real-world situations in green energy financing.

Analytical Framework

The analytical framework employed in the study consists of three essential components: descriptive analysis, comparative analysis, and thematic analysis. Descriptive analysis examines the current situation of financial literacy on green energy matters among Latin American countries and identifies notable trends and patterns in the available data.

Comparative analysis examines variation between countries, population segments, and institutional contexts to identify the determinants of green energy financial literacy program development. Comparative analysis considers quantitative indicators and qualitative considerations that impact program effectiveness.

Thematic analysis is directed to recurring themes and patterns in the literature and case study results. Thematic analysis establishes the most significant challenges, opportunities, and best practices for developing green energy-related financial literacy programs in Latin America.

The analytical framework employs inductive and deductive types of reasoning. Deductive analysis employs previously established theories to make sense of empirical data, while inductive analysis looks for new relationships and patterns in the data.

Limitations and Constraints

A number of crucial limitations constrain this study and should be highlighted in discussing the findings. To start, the technicality of green energy-related financial literacy is such that very little empirical literature specifically addresses this in Latin America. All the available work focuses on general financial literacy or renewable energy development separately, requiring inference and extrapolation to examine their convergence.

Second, the rapidly evolving nature of green energy markets and financial instruments makes it likely that available data are obsolete. Such an environment requires cautious interpretation of historical trends and careful monitoring of current developments that may not yet have reached published literature.

Third, Latin American countries differ widely with regard to economic development, regulatory regime, and institutional capability. These serve to limit the applicability of findings and call for sensitive regard to country circumstances in developing recommendations.

Fourth, language barriers may limit access to relevant research written in Spanish, Portuguese, or other indigenous languages. Even with efforts to incorporate multilingual sources, some eligible studies may have been overlooked due to language constraints.

Fifth, the study relies heavily on secondary data sources that may not capture all the green energy financial literacy sensitivities at issue. Further insights would have been helpful from primary data collection through survey or interview but were beyond the scope of this study.

ANALYSIS OF RESULTS

Pre-Existing Green Energy Financial Literacy

The research recognizes substantial gaps in green energy-specific financial literacy in Latin America, despite the region's significant progress towards the expansion of renewable energy. While Latin American countries have increased their renewable capacity by 51% between 2015 and 2022, reaching 64% generation from renewable energies, this technological change has not been matched with parallel advancements in financial literacy for investing in green energy.

Survey of available financial literacy data indicates that base financial literacy is low across Latin American countries overall, with particularly concerning gaps in investment knowledge and long-term financial planning ability. Such underlying weaknesses pose serious barriers to understanding more advanced green energy financial products and investment.

The research identifies three levels of green energy financial literacy development across the region. Tier 1 countries, including Chile, Uruguay, and Costa Rica, have relatively higher baseline levels of financial literacy and have begun to incorporate principles of sustainability into financial education curricula. These countries also have more sophisticated development of green finance markets and more advanced renewable energy financing products. Tier 2 countries, including Brazil, Mexico, Colombia, and Argentina, possess medium financial literacy with emerging green finance markets and minimal integration of sustainability concepts into financial education. The countries have advanced renewable energy development but face challenges in connecting financial literacy programs with green energy concerns.

Tier 3 countries, which include certain Central American and Caribbean nations, have lower initial financial literacy levels and less developed green finance markets. These countries have fundamental obstacles to establishing general financial literacy and specific green energy finance expertise.

The research identifies that urban communities are more likely to be aware of green energy and also financially literate compared to rural communities, although rural communities often have greater renewable energy potential. The existence of such an urban-rural divide has important policy implications for the planning of renewable energy projects and community engagement strategies.

Sectoral Analysis

Banking and Financial Services Sector

The banking sector analysis reveals partial integration of sustainability principles in financial education programs offered by banks and other financial institutions. While most of the major banks in the country have begun offering green financial products such as sustainability-linked loans and green bonds, awareness among customers regarding these products is extremely limited.

Research on green financial products in Latin American banks shows that while the banks are expanding their green finance products, the customers lack the necessary knowledge to understand and analyze the products effectively. The absence of knowledge limits the growth of the market and reduces the impact of green finance measures. The report quotes several leading banks that have begun to incorporate sustainability education into their financial literacy programs. The banks record higher customer engagement and higher take-up of sustainable financial products when education is combined with product offerings.

However, the majority of banks in the region still focus financial literacy programs on traditional bank products and services with minimal or no emphasis on green energy or sustainability topics. This approach represents a missed opportunity to leverage existing customer relationships and education infrastructure to promote green energy financial literacy.

Educational Sector

The analysis of the educational sector shows little adoption of green energy financial literacy in school curricula under formal education. Though some nations have started introducing general financial literacy in school curricula, few address specific content on sustainable finance or green energy investments. The review identifies several pilot programs that have successfully integrated sustainability and financial literacy concepts, demonstrating the potential for broader implementation. These programs report positive outcomes like increased student engagement, improved understanding of environmental issues, and greater aspiration for sustainable careers.

The local universities and colleges exhibit more advanced development of green energy and sustainable finance curricula, yet the programs are largely targeting students of specialized fields and not general population-wide literacy education. The research suggests that educational institutions can contribute to the promotion of green energy financial literacy but require additional resources, training, as well as curriculum development support in order to expand their current programs.

Policy and Government Sector

The government review demonstrates that although 15 Latin American nations have set net-zero emissions targets, few have created detailed plans for developing the financial literacy that will be required to reach these goals. The majority of government financial literacy programs concentrate on conventional financial principles without mentioning green energy or sustainability issues. The research identifies several nations that have begun to integrate sustainability topics into national financial literacy initiatives. The efforts are encouraging but have limited scope and implementation capacity.

Green finance regulatory frameworks are rapidly developing across the region, but consumer protection and education needs for green financial products are inadequate. The regulatory gap is one reason the low levels of green energy financial literacy persist. The examination suggests that government leadership in developing green energy financial literacy initiatives could significantly boost progress, particularly through integration with existing financial inclusion and education initiatives.

Demographic and Social Analysis

Age and Generational Differences

The research identifies significant age-related diversity in green energy financial literacy, with younger generations being more ecologically conscious but not necessarily possessing greater financial literacy skills. The trend offers opportunities and challenges for program development. Younger age cohorts are more concerned with sustainable investing and the environment but lack basic financial literacy to make informed investment decisions. This would suggest that targeted programs connecting basic financial literacy and green energy should be particularly effective for younger age groups.

Older cohorts tend to have higher baseline financial literacy but may be less familiar with or enthusiastic about green energy investment. Educational interventions for older cohorts must use alternative approaches that build on existing financial knowledge and supplement it with sustainability concepts.

The analysis identifies the middle-age bracket (35-50 years) as potentially the most receptive to comprehensive green energy financial literacy programs because this group is likely to combine modest financial literacy with growing environmental awareness and significant investment capacity.

Income and Socioeconomic Factors

Income-level analysis suggests that green energy financial literacy challenges vary considerably across socioeconomic segments. More prosperous segments are likely to enjoy higher general access to financial education and investment opportunities but may lack specialized knowledge concerning green energy matters.

Poorer segments face a number of challenges including lower baseline financial literacy, poorer access to investment opportunities, and near-term financial pressures that may render green energy investments of longer time horizons less attractive or feasible. Middle-income groups are identified in the study as having strong potential for the growth of green energy financial literacy as they are likely to have the financial capacity to consider green energy investments while facing energy cost pressures that make the use of renewable energy sources attractive.

Microfinance and community organizations are identified as worthwhile channels for stretching green energy financial literacy programs to low-income groups, particularly in rural areas where the need for access to renewable energy is an issue of concern.

Gender Inequalities

Gender analysis plays a critical role in revealing substantial gaps in green energy financial literacy that reflect general patterns of financial inclusion and environmental concern. Women in Latin America are more likely to show higher environmental awareness but face barriers in access to finance services and investment opportunities.

The study recognizes women as prospective powerful agents of green energy adoption at the household and community levels but observes that conventional gender roles can restrict their involvement in investment decision-making. This indicates that green energy financial literacy initiatives should tackle both technical knowledge and empowerment. Women's participation in the financial services sector, particularly at the leadership level, could make a significant difference in the development of green energy financial literacy programs. The research identifies several women-led organizations that have been successful in integrating sustainability and financial literacy concepts.

Institutional Analysis

Multilateral Development Banks

The analysis of the multilateral development banks reveals significant investment in green energy projects across Latin America but little attention to building the financial literacy necessary to sustain these commitments. While organizations like the Inter-American Development Bank (2024) and World Bank provide significant funding for renewable energy projects, customer and community education components are usually minimal.

The development banks have begun to recognize that project sustainability requires not only technical and financial viability, but also community awareness and acceptance. A few institutions have incorporated financial literacy components into their project designs, but these are ad hoc and limited.

There are opportunities for the development banks to leverage their current relationships and resources to expand green energy financial literacy through their borrowing institutions and project stakeholders.

Private Sector Initiatives

Private sector analysis indicates that there is growing recognition among energy companies and financial institutions that customer education holds the key to market development. Some renewable energy companies have begun offering educational offerings alongside their products and services, with indications of improved customer satisfaction and reduced payment defaults.

Fintech companies represent a promising avenue of green energy financial literacy growth since some platforms have begun to offer educational content alongside green investment products. Online platforms can provide scalable solutions to reach diverse populations across the region.

The research identifies successful public-private partnerships that have combined private sector innovation and public sector reach in carrying out effective financial literacy programs. Such partnerships reflect models that can be replicated in developing green energy financial literacy.

DISCUSSION

Findings and Implications

The research discovers a core paradox driving Latin America's green energy shift: the continent has taken impressive leaps in embracing renewable energy and taking bold climate action, but its economic savviness to fuel and accelerate this shift remains in its infancy. The gap between the tech advances and economic wisdom is both a daunting challenge and a daunting opportunity for policy makers, educators, and financial institutions.

The fact that the Latin American and Caribbean countries increased their renewable capacity by 51% between 2015 and 2022 to a level of 64% renewable generation demonstrates the technical potential of the region for green energy development. This impressive fact has not been followed by the same degree of progress in green energy financial consciousness, which means that future growth could be restrained by knowledge gaps rather than technical or resource facilities.

Three-tier ranking of countries based on the development of green energy financial literacy describes notable tendencies that can direct the formulation of evidence-based targeted intervention strategies. Countries in Tier 1, such as Chile and Uruguay, demonstrate that it is possible to achieve more rapid development of green energy financial knowledge by having a higher starting level of generic financial literacy. This suggests that promoting generic financial literacy as the prerequisite to effective green energy financial literacy programs needs to be considered.

The persistent urban-rural divide in green energy financial awareness presents unique challenges to the advancement of renewable energy since rural areas have the potential to have the highest renewable energy but lowest financial awareness. This finding has far-reaching implications for renewable energy project development and suggests that community outreach and education programs require massive overhaul.

The population structure indicates that different groups of the population require different approaches to green energy financial literacy development. The insight into the fact that younger populations possess greater environmental awareness but inferior financial literacy is strongly suggestive of the implementation of integrated educational approaches. So too is the designation of middle-aged populations as most likely to be the most responsive to comprehensive programs, and highly indicative of program targeting and resource allocation.

Barriers to Green Energy Financial Literacy Development

The research presents several interlinked barriers that constrain green energy financial literacy development in Latin America. These barriers are on the individual, institution, and system levels, which require collective interventions addressing short-term knowledge gaps as well as underlying structural bottlenecks.

At the individual level, the simplest constraint is low baseline financial literacy. Without having definitive mastery of financial basics such as risk analysis, time value of money, and portfolio diversification, individuals are not in the position to adequately evaluate green energy investing opportunities. This finding means that green energy financial education programs must either provide rudimentary financial education or target populations that possess a baseline level of core financial knowledge.

The lack of hands-on experience with green energy technologies and financial tools is a more serious personal barrier. Latin Americans have limited access to renewable energy technologies, sustainable investment tools, or green finance tools. This lack of experience hinders the acquisition of experiential knowledge of green energy financial principles and annihilates confidence in making associated financial decisions.

Institutional limitations are the limited ability of schools, financial institutions, and government agencies to develop and operate green energy money skills programs. According to the study, institutions lack specialized knowledge, resources, and incentive systems for designing effective programs. This institutional capacity shortfall requires deliberate investment in training, curriculum development, and program infrastructure. The regulatory landscape offers both challenges and opportunities for the growth of green energy financial literacy. Although some nations have been able to create facilitating regulatory systems for green finance, consumer education and protection needs are still low. This gap in regulation leads to market confusion and lowers the level of consumer confidence in green energy financial products.

Systemic barriers are broader economic and social factors that make financial decision-making more difficult. Macroeconomic risk, deficits in access to long-term capital, and low institutional trust provide environments under which long-term green energy investment appears risky or unviable. Systemic constraints require policy action that transcends the call for education to meet underlying economic and institutional challenges.

Opportunities for Development

Despite the relevant obstacles listed, the research is indicative of great potential for Latin American green energy financial literacy to grow. The rationale lies with growing renewable energy markets in the region, heightening environmental consciousness, and intensifying financial inclusion initiatives. The accelerated growth of renewable energy markets offers natural opportunities for the creation of green energy financial literacy. With the International Energy Agency (2024) predicting Latin America will become one of the significant clean hydrogen exporting regions by 2030, the region's economic stake in the success of green energy grows. This economic stake creates the incentives for all stakeholders to make the investments in the knowledge and skills necessary to be successful green energy market players.

The growing ubiquity of digital platforms and mobile technologies offers scalable delivery channels for green energy financial literacy training. With mobile phone penetration rates being very high in Latin America, technology-based learning methods could reach out to sizable populations at low costs. The platforms can also offer customized learning experiences that accommodate different knowledge levels and learning styles. There are already existing financial inclusion programs that can be applied as institutional infrastructure to be utilized in green energy financial literacy development. The majority of the countries in the region already have national financial education plans and schemes that can be utilized to broaden the coverage to meet green energy needs. This is less costly compared to creating new institutional frameworks.

Growing private sector attention to corporate social responsibility and sustainable development holds possibilities for innovative program creation and financing. Companies are likely to invest in educational programs that are of business interest and understand the necessity of customer education to build markets. The 15 Latin American nations' commitments to net-zero greenhouse gas emissions policy imperatives, therefore, signal the necessity of the creation of green energy financial literacy. This necessity is not only for the technology deployment but also and even more so, the public's understanding and embracing of green energy transition policy. Such a policy context warrants public spending on green energy financial literacy programs.

Best Practices and Lessons Learned

The research identifies some of the greatest success stories and lessons learned of existing green energy financial literacy programs in Latin America and beyond. These programs guide enhancing better programs and avoiding typical pitfalls of implementation. Effective programs have a recurring theme of placing strong emphasis on the integration of experiential, practical experiences with theoretical training. Programs including visits to renewable energy facilities, demonstrations of efficient energy technologies, or simulations of green investment decisions have higher levels of participant engagement and more effective learning outcomes. This finding suggests experiential learning techniques should be prioritized in program development.

The most effective programs intervene on both the technical and behavioral aspects of green energy financial literacy. Learners need to acquire knowledge in financial concepts and green energy technologies, as well as build confidence and motivation to apply the knowledge to real decision-making. Programs with behavioral change components, such as setting goals and peer support, have better longer-term outcomes. Well-crafted programs consistently remind participants of the importance of local context and culture. Off-the-shelf curricula around personal financial literacy are typically created without regard to addressing the specific challenges and opportunities that participants face in their daily lives. Programs that incorporate local energy issues, region-specific financial services products, and culturally appropriate illustrations demonstrate enhanced participant engagement and practical relevance.

The report states that program sustainability requires institutional support and funding mechanisms on a continuous basis. One-time training or short-term programs have limited sustained impact, while programs with reinforcement and support on a continuous basis have more sustained behavioral change. This reinforces the idea that program planning ought to emphasize sustainability and long-term involvement as opposed to short-term coverage. Multi-stakeholder coordination becomes essential to program effectiveness. Best practices are programs that involve educational institutions, financial service providers, government agencies, and community organizations. Multi-stakeholder alliances leverage diverse resources and competencies while ensuring programs leverage actual market opportunity and demand.

Policy Implications

The findings of this analysis carry enormous policy implications for policy-making at national, regional, and international levels. Such policy implications go beyond education to encompass financial regulation, energy policy, and sustainable development policies. National governments should consider including green energy financial literacy in broader financial education and inclusion plans. This would allow them to leverage existing institutional infrastructure without undermining the coverage of green energy matters with focus and resources. Policy systems should define precise objectives, performance metrics, and accountability mechanisms for promoting green energy financial literacy.

Regulatory requirements for green finance should have explicit provisions for consumer education and protection. Green energy products presented by financial institutions should have transparent, comparable product information regarding product characteristics, risks, and benefits. The requirements would add clarity to the market and consumer confidence and trigger institutional investment in customer education. Energy policy mechanisms need to recognize consumer financial literacy as central to the adoption of renewable energy. Renewable energy deployment promotion policies need to incorporate elements addressing financing knowledge gaps alongside consumer education needs. The integrated approach would make energy transition policies more effective.

The institutions of international cooperation need to be more effective in transferring knowledge and producing programs among the Latin American countries. Regional organizations such as the Organization of American States and the Latin American Energy Organization could have important roles in facilitating green energy financial literacy programs and best practices exchange.

Green energy financial education components should be integrated into development finance institutions' project designs and lending guidelines. In this manner, investments in renewable energy infrastructure would be preceded by investments in the human capital needed to create and sustain such projects.

CONCLUSIONS

Summary of Key Findings

This detailed study report reveals a huge gap between Latin America's remarkable growth of clean energy and the financial education needed to sustain and increase this transition. While the region has secured 51% of capacity of renewable between 2015-2022 and has ambitious net-zero emission goals, financial literacy linked to green energy remains nascent in most countries and segments of the population. Analysis shows three distinct levels of green energy financial literacy development among Latin American countries, with considerable variation based on initial baseline financial literacy and institutional capacity. Countries with higher general financial literacy have more development in the acquisition of green energy financial knowledge, suggesting that overall financial education provides a platform for more specialized learning.

There are significant demographic variations in green energy financial awareness, with city residents, youth, and higher-income segments generally having higher levels of awareness and knowledge. Nonetheless, these groups are not always equipped with the full knowledge necessary to make successful green energy investment decisions, indicating mass need for focused education interventions.

The institutional analysis discovers that while there is a beginning in green energy initiatives and programs by multilateral development banks, governments, and private institutions, few have taken seriously the level of financial literacy needed for long-term market development. This is an institutional shortcoming, both a challenge and a promise for collective effort.

Some of the constraints include low initial financial literacy, brief practice experience with green energy technologies, poor institutional capacity, and systemic economic constraints. However, significant opportunities come through expanding renewable energy markets, digital technology platforms, existing financial inclusion programs, and growing policy inclinations towards sustainable development.

Implications for Theory and Practice

The research results hold important theoretical and practical significance for the understanding and utilization of financial literacy concepts in sustainable development settings. Theoretically, the study demonstrates that mainstream financial literacy models must be extended to encompass environmental as well as sustainability considerations. The technicality of green energy financial literacy means that generic financial information, as critical as it is, cannot adequately facilitate decision-making in sustainable energy settings.

The research contributes to behavioral finance theory in that it suggests certain determinants of green energy investment choices, including environment values, social norms, and capacity for long-term thinking. These suggest that effective green energy financial literacy programs must address both analytical and emotional aspects of decision-making. In practice, the research provides evidence-based recommendations for constructing more effective green energy financial literacy programs. The identification of successful program design characteristics like experiential learning approaches, contextualization within the local setting, and multi-stakeholder partnership is a very clear set of guidelines for program implementers and designers.

The research also demonstrates the benefits of multi-strategy approaches that at the same time compensate for individual knowledge gaps through institutional capacity building and policy interventions. This systems approach suggests that isolated, discrete education interventions are less likely to be effective in the long run without wider structural changes.

Recommendations

Based on the integrated analysis included in this report, several key recommendations flow for different groups of stakeholders that are working to advance green energy financial literacy in Latin America.

Recommendations for Governments

National authorities must develop comprehensive green energy financial literacy schemes that are consistent with other financial education and sustainable development schemes. These schemes should define clear objectives, performance metrics, and timeliness as well as ensure adequate resource allocation and coordination among institutions.

Green finance regulatory frameworks need to be strengthened to include in-depth consumer protection and education standards. Green energy products need to be made mandatory for financial institutions to provide standardized information about product features, risks, and benefits, along with educational content that helps consumers make informed decisions. Education policy frameworks must incorporate green energy money literacy into formal curriculum for primary, secondary, and tertiary education. It must all be achieved through teacher training initiatives, curriculum development and support, as well as the provision of education resources to address local energy issues and opportunities.

Government policy and procurement policy should benefit suppliers and partners who demonstrate commitment to green energy financial literacy development. This policy would create market incentives for private sector investment in education while leveraging the purchasing power of government to drive higher social benefit.

Financial Institutions' Recommendations

Banks and other financial institutions must develop comprehensive training schemes for staff to further develop their knowledge of green energy finance products and customer education skills. Technicalities surrounding sustainable finance as well as communications skills for handling diverse customers should be addressed through these training schemes.

Banks need to include green energy financial literacy elements in their existing customer education programs. Inclusion should involve the development of training materials, workshops, and internet tools that help customers understand green energy investment opportunities and risks.

Inter-agreements among renewable energy companies and finance institutions would create integral education systems that combine technical expertise in terms of green energy systems with financial expertise. The inter-agreements would leverage complementary capabilities while providing customers with full expertise in green energy investments. Research and innovation in new education models, including internet platforms, gamification approaches, and individual learning systems, should be supported by finance institutions to address various groups effectively and at minimal cost.

Recommendations to Universities and Research

Universities and research must develop interdisciplinary courses combining financial literacy, environmental science, and renewable energy technology. These need to develop professionals who will be able to bridge the gap in between technical and financial aspects of green energy development. Schools need to create community outreach programs that provide green energy financial literacy education to local populations. These programs must utilize institution strength and expertise and address real community challenges and needs.

Cross-associations between industry associations and schools could offer practical learning opportunities that expose students and members of the community to real green energy projects and financing systems. Such partnerships would enhance learning attainment as well as career networks to complement continuous knowledge creation. Institutions need to invest in faculty development programs that enhance teaching competences in green energy financial literacy courses. The programs are required to address content material and pedagogy methods of efficient adult learning for sustainable environments.

Recommendations of International Organizations

Green energy financial literacy must be a part of project lending requirements and support programs to establish multilateral development banks' green energy infrastructure. In this manner, investment in infrastructure would be followed by the development of human capital for long-term sustainability. International organizations have to enable knowledge sharing and program development through regional networks and communities of practice. The networks would help in the transfer of best practices, joint program development, and coordinated responses to common challenges.

Technical assistance program should include green energy financial literacy programming alongside regular project assistance activities. Incorporating such programming into technical assistance program would enhance project performance while building local capacity for sustainable development.

International organizations must assist in research and assessment efforts that build knowledge about effective methods of green energy financial literacy development. Academic studies and practical program assessment are both to be included in the research that contributes to evidence-based program development.

Areas for Future Research

The conceptual nature of this research identifies several areas where further exploration will greatly enhance understanding and practice of green energy financial literacy development in Latin America.

Longitudinal research following the development of green energy financial knowledge across time would be informative about learning processes, program impact, and trends in behavior change. Both change at the individual level and would-be trends of wider market evolution need to be researched.

Comparative research investigating green energy financial literacy development in various nations, world regions, and institutional environments would assist in determining the main success factors and transferable best practices. Such research would take into account both quantitative results and qualitative determinants of program efficiency. Experimental research testing different educational approaches, program structures, and delivery systems would provide evidence-based suggestions for program development. Such research would determine both short-term learning and longer-term behavioral changes.

Research on the connections between green energy financial literacy and real-world investment behavior would allow testing of the theoretical connections given in this work. Such research would measure both individual-level investment choices and market participation patterns.

Cultural and social determinants of influence on green energy financial literacy development would inform more program planning that is optimized across populations. Research would need to consider how cultural beliefs, social norms, and community organization affect learning and behavior change processes.

Final Conclusions

Green finance-related financial literacy is a needed but emerging capability in Latin America with the potential to propel the region's prospects to achieve its ambitious climate and sustainable development goals. Although Latin America has demonstrated impressive technical capabilities for renewable energy development, the required financial literacy to maintain this transition process on track as well as accelerate it is in urgent need of attention and investment.

The integrated assessment in this report shows that green energy financial literacy must be built with concerted efforts that attend to both gaps in individual knowledge and in institutional capacity, as well as systemic obstacles. It underlines the necessity for multi-stakeholder partnerships and coordinating approaches because no intervention or group of stakeholders can individually address the entire set of issues identified. The prospects as demonstrated in this research point towards positive development with the right commitment and investment. The region's growing renewable energy markets, increasing digital infrastructure, and increasing policy commitments towards sustainable growth are good grounds for developing green energy financial literacy.

Yet converting these opportunities into a reality involves going beyond acknowledgment of the significance of financial literacy to actual action focused on the precise challenges and contexts that have been identified in this research. Recommendations made here offer a blueprint for such action, but their activation depends on strong and sustained effort from governments, financial institutions, educational bodies, and international actors. Ultimately, green energy financial literacy capacity in Latin America will be measured not just by knowledge and skill gains but also by how much increased understanding is converted into more involvement in green energy markets, more informed investment decisions, and quicker progress towards sustainable development goals.

As Latin America is poised to become a global leader in the production and export of clean energy, the investment the region places in green energy financial literacy will determine whether or not this is fully realized. Something must be done now, and the stakes could not be higher for the region and the world dedicated to a future of sustainable energy.

REFERENCES

Bloomberg, N. E. F. (2024). *Latin America clean energy investment outlook.* Bloomberg New Energy Finance.

Cornell Business Hub. (2024). *Latin America's net-zero commitments and clean energy transition.* Cornell SC Johnson College of Business.

Filippini, M., Leippold, M., & Wekhof, T. (2024). Sustainable finance literacy and the determinants of sustainable investing. *Journal of Banking & Finance*, *163*, 107167.

Frontiers in Sustainability. (2025). *Financial literacy and environmental sustainability: Cross-country evidence.* Frontiers Media SA.

Green Finance for Latin America and Caribbean. (2024). *Regional green finance development report.* GFL Platform.

Hwang, Y. K., & Díez, Á. S. (2024). Renewable energy transition and green growth nexus in Latin America. *Renewable & Sustainable Energy Reviews*, *198*, 114431.

Ideas Matter. (2019). *Financial literacy education in Latin America: Impact assessment of mandatory programs.* Inter-American Development Bank.

Inter-American Development Bank. (2024). *Renewable energy development in Latin America and the Caribbean: Progress report 2015-2022.* IDB Publications.

International Energy Agency. (2024). *Latin America energy outlook: Clean hydrogen export potential.* IEA Publications.

MDPI. (2022). *Financial literacy in Latin America and the Caribbean: A systematic review of research gaps and opportunities.* Multidisciplinary Digital Publishing Institute.

Méndez Prado, S. M., Zambrano Franco, M. J., Zambrano Zapata, S. G., Chiluiza Garcia, K. M., Everaert, P., & Valcke, M. (2022). A systematic review of financial literacy research in Latin America and the Caribbean. *Sustainability*, *14*(7), 3814.

Organisation for Economic Co-operation and Development. (2021). *Financial education in Latin America and the Caribbean: OECD regional assessment.* OECD Publishing.

Chapter 13
Forced Islanding and Adaptive Load Shedding to Prevent Blackout for Improving Power System Security:
Micro Grid Islanding to Improve Power System Security

Akhilesh Nimje
https://orcid.org/0000-0002-6512-5512
Nirma University, India

ABSTRACT

Large-scale, interconnected power systems are susceptible to blackouts, which can result in significant financial losses, production cycle disruptions, and equipment damage at power plants. A forced islanding plan with an appropriate restoration process will work well to prevent these blackouts. This paper proposes a mechanism that divides the system into predetermined islands and determines the risk during an outage. In the event that there are unstable islands, an under-voltage load shedding plan is implemented to restore system stability. Furthermore, a restoration protocol has been developed. The suggested work is done using a six bus system as an example. This strategy can also be used in smart grid settings to raise the dependability and efficiency of the electricity system.

DOI: 10.4018/979-8-3373-4272-6.ch013

INTRODUCTION

When a zone or portion of the network is serviced by both the power grid and the distributed generation system, it is known as islanding in active distribution networks. For a variety of reasons, these distributed sources are isolated from the grid; as a result, some or all of the load systems in the disconnected section are still powered by DG units, (Wood & Wollenberg, 1996). The distribution utility is unable to maintain supply stability or quality as a result of losing control over the island's supply. Protection mechanisms in distribution networks were designed to handle network faults based on the assumption that the networks were passive, meaning that no power sources were feeding electricity into the grid, (Sati et al., 2023). To solve the issue of power system islanding, more investigation and study are required because the introduction of distributed generation (DG) at the distribution level. The electric power grid is made up of transmission lines, transformers, generators, and other machinery. Because of this, if any of these components malfunction while the system is in use, it could jeopardise its stability and dependability and cause outages. Large, interconnected power systems may be significantly impacted by severe events, potentially leading to widespread blackouts and cascade tripping. These blackouts have the potential to cause major financial losses, damage power plant machinery, and interrupt production cycles. Therefore, to lessen this risk, a safe, robust and reliable protection system is needed, (Xin et al., 2017).

FORCED ISLANDING

A complicated power network has a tendency to fragment into islands after a disturbance. The protective mechanisms of the grid may trigger tripping measures, which could lead to this situation. Whether done on purpose or not, islanding can damage hardware and jeopardise system security. However, in order to prevent system failure in the event of a severe emergency, a controlled division of the system into stable islands with generation and load shedding using specific protective techniques is suggested. Compared to naturally occurring islands, these artificially created islands are more stable, (Guo et al., 2017).

UNINTENTIONAL ISLANDING

A complicated power network has a tendency to fragment into islands after a disturbance. The protective mechanisms of the grid may trigger tripping measures, which could lead to this situation. Whether done on purpose or not, islanding can damage hardware and jeopardise system security. However, in order to prevent system failure in the event of a severe emergency, a controlled division of the system into stable islands with generation and load shedding using specific protective techniques is suggested. Compared to naturally occurring islands, these artificially created islands are more stable, (Shu et al., 2007).

DEGRADATION OF POWER QUALITY

An unintentional island's voltage and frequency levels can vary greatly and frequently surpass allowed limits, end user loads, distributed generation units, and utility assets are all at risk.

OUT OF PHASE RE-CLOSURE

Utility protection equipment is designed to automatically close once a fault is cleared to restore service. However, if re-closure happens due to accidental islands, the voltage on both sides may be out of phase. This puts the primary movers of rotating machinery at serious risk due to the currents and powerful mechanical torques produced by out of phase re-closure.

LOSS OF GROUNDING

The utility side of the power system is where the earth point is located. The current DG units and loads will probably be improperly grounded on the island if there is any unanticipated islanding.

SAFETY CONCERNS

Workers performing utility repair on an assumed-de-powered but actually-energized network will be in danger due to unwanted islanding.

ISLANDING DETECTION

The process of detecting islanding involves monitoring any changes to the grid parameters and the output parameters of the DG units in order to ascertain whether or not islanding has occurred. Methodologies for detecting islanding can be divided into two categories: remote and local. Based on this classification, methodologies can be further divided into passive, active, and hybrid tactics.

DIFFERENT TECHNIQUES OF FORCED ISLANDING

Local Techniques

The fundamental idea behind this phenomenon is that it measures the primary electrical characteristics, including voltage, current, frequency, and so forth. Local techniques on the DG side are further subdivided as follows:

Passive Approaches

These methods keep an eye on the DG side's electric parameters and compare them to predetermined values. These methods have a wide non-detection zone, but they are simple to install and do not affect DG stability negatively. Such approaches include phase jump detection (PJD), voltage and current harmonic detection, over/under voltage and over/under frequency, and others.

Active Approaches

These techniques work by injecting certain disturbances at the output of the DG unit. Depending on how much the electric parameters vary, if the change is substantial, islanding may be identified. There could be a substantial level of deterioration. Among these methods are impedance measurement, sandia frequency shift, and active frequency drift.

Hybrid Techniques

When passive techniques identify an islanding situation, active techniques are triggered. Hybrid techniques mix active and passive approaches. These tactics include, for example, voltage imbalance, reactive power shift, and positive feedback.

Remote Approaches

These methods depend on a specific kind of communication between the DG and the utility substation. Although remote techniques are highly reliable and accurate in identifying islanding, their implementation is far more expensive than local techniques. Two of these tactics are power line carrier communications and impedance insertion.

HOW FORCED ISLANDING IS DONE?

A forced islanding method is put into place after the system's overburdened lines are identified during an outage in order to prevent a likely blackout. For each outage circumstance, islands are predefined in order to use the forced islanding technique, (Dola et al., 2006). The islands were selected so that the generation and load on them are almost equal. After the system is divided into islands, each island undergoes a load flow analysis, and a power mismatch is calculated in each island to alert the island to a change in load or a rise in generation if the mismatch is positive or negative by an amount equal to the mismatch. Additionally, the voltage stability of the islands is examined, with a 10% tolerance limit for the voltage.

Figure 1. Flowchart of forced islanding scheme

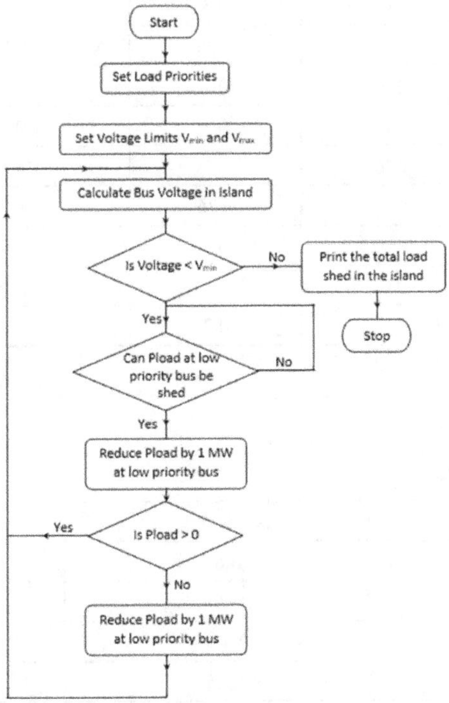

Table 1. Data for 6 bus system

Bus No	Bus Code	|V|	∠	Gen		P_L		Q_{max}	Q_{min}
				MW	MVAR	MW	MVAR	MVAR	MVAR
1	1	1.05	0	0	0	0	0	100	-20
2	2	1.05	0	50	0	0	0	100	-20
3	2	1.07	0	60	0	0	0	100	-15
4	3	1	0	0	0	70	70	0	0
5	3	1	0	0	0	70	70	0	0
6	3	1	0	0	0	70	70	0	0

Table 2. Line data for 6 bus system

Line	From	To	Resistance pu	Reactance pu	Half pu Susceptance	Power Limit
1	1	2	0.1	0.2	0.02	30
2	1	4	0.05	0.2	0.02	50
3	1	5	0.08	0.3	0.03	40
4	2	3	0.05	0.25	0.03	20
5	2	4	0.05	0.1	0.01	40
6	2	5	1	0.3	0.02	20
7	2	6	0.07	0.2	0.025	30
8	3	5	0.12	0.26	0.025	20
9	3	6	0.02	0.1	0.01	60
10	4	5	0.02	0.4	0.04	20
11	5	6	0.1	0.3	0.03	20

Table 3. Predefined islands

Outage	Island 1	Island 2
Gen 2	2,3,6	1,4,5
Gen 3	2,3,6	1,4,5
Line 1	1,2,4,5	3,6
Line 2	1,2,3,5	3,6
Line 3	1,2,4,6	3,5
Line 4	No island	No island

continued on following page

Table 3. Continued

Outage	Island 1	Island 2
Line 5	2,3,6	1,4,5
Line 6	2,3,6	1,4,5
Line 7	2,3,6	1,4,5
Line 8	1,2,4,5	3,6
Line 9	2,3,4	1,5,6
Line 10	No island	No island
Line 11	No island	No island

SIMULATION RESULTS

Figure 2. Gen 2 Island 1: 2,3,6 Island 2: 1,4,5

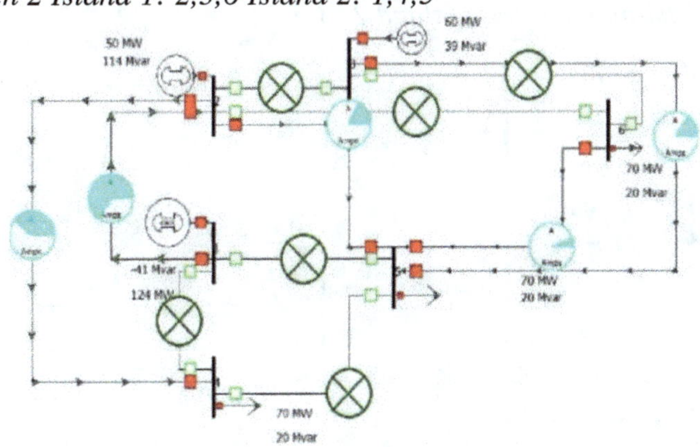

Figure 3. Gen 3 Island 1: 2,3,6 Island 2: 1,4,5

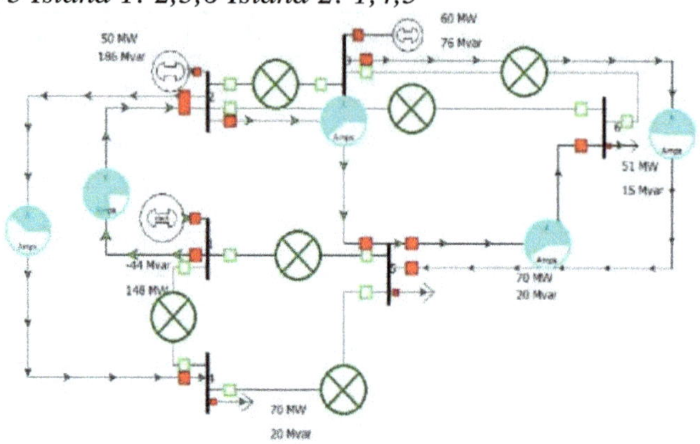

Figure 4. Line 1 Island 1: 1,2,4,5 Island 2: 3,6

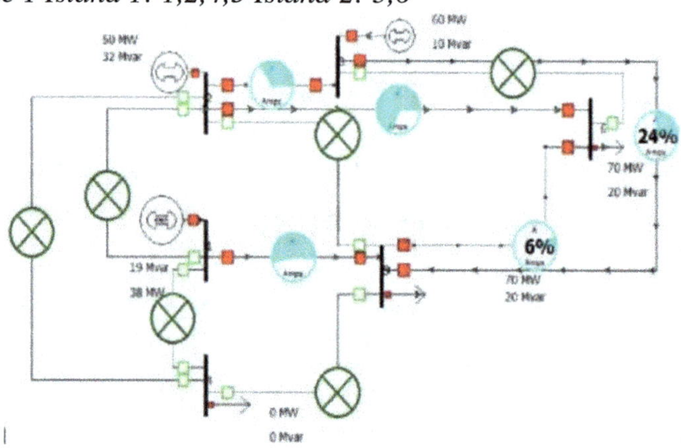

Figure 5. Line 2 Island 1:1,2,3,5 Island 2:3,6

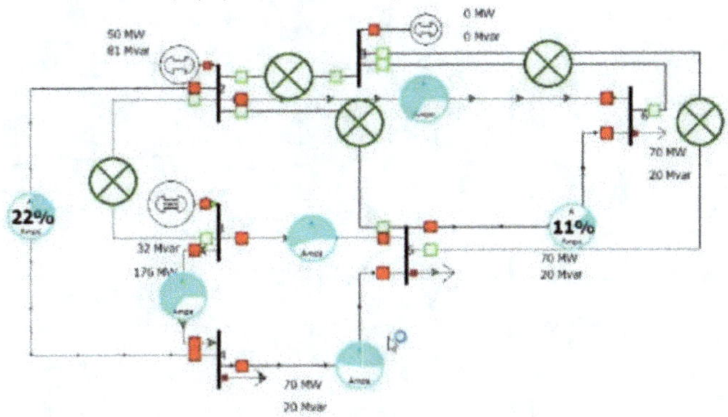

Figure 6. Line 3 Island 1:1,2,4,6 Island 2:3,5

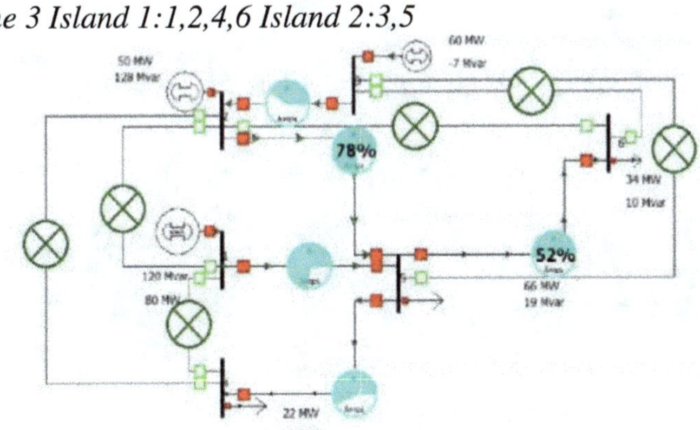

Figure 7. Line 6 Island 1:2,3,6 Island 2:1,4,5

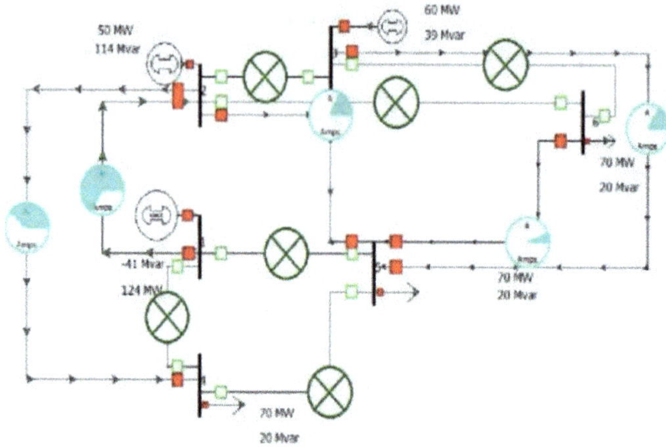

Figure 8. Line 7 Island 1:2,3,6 Island 2:1,4,5

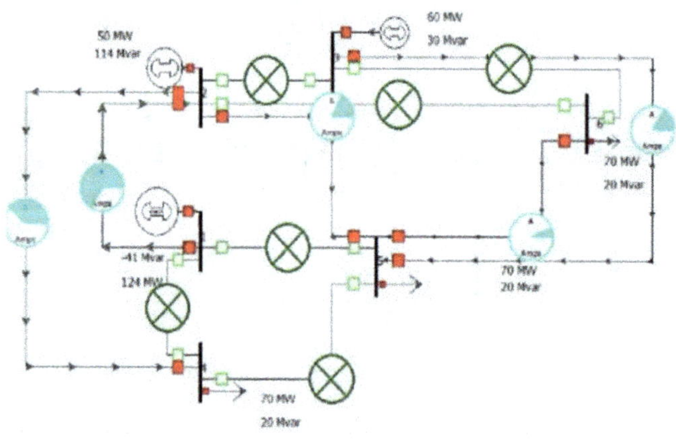

Figure 9. Line 8 Island 1:1,2,4,5 Island 2:3,6

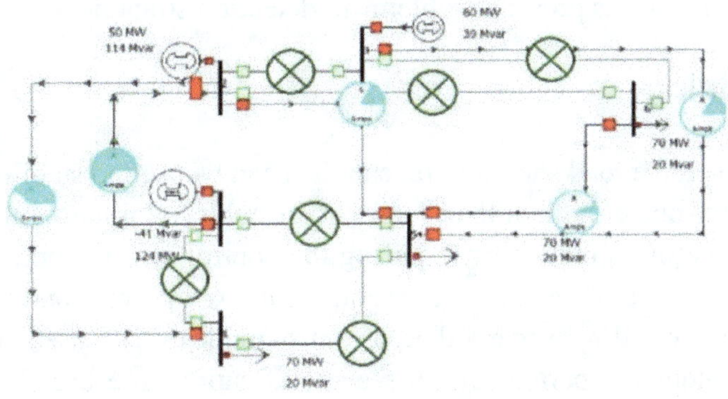

Figure 10. Line 9 Island 1:2,3,4 Island 2:1,5,6

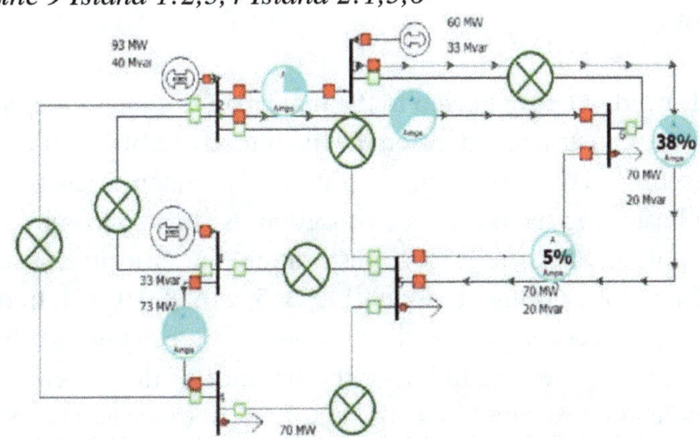

CHALLENGES AND OPPORTUNITIES

There are three main challenges to island detection methods. The power difference between the DG output and the power consumed by the load is the first problem, and it could make it challenging to identify an islanding situation. The non-detection zone is what it is termed. The second challenge is related to active tactics that attempt to introduce certain disruptions in order to identify islanding, as this could deteriorate the quality and stability of DG output power. Lastly, the success of islanding detection systems is significantly influenced by the types of loads found in power islands. While steady power loads or current harmonics producing nonlinear loads make islanding detection approaches easier to detect, a few loads that can be represented as parallel RLC circuits present problems to detection strategies.

Load Shedding

An under-voltage load shedding mechanism can be used to stabilise the system if it is found that one or both of the islands are unstable, (Hussain et al., 2022). This tactic can be put into practice by requiring the control systems of all DG units to transition, in the case of an islanding event, from fixed power control mode (P-Q) to fixed frequency and voltage regulation control mode (V-f). Based on pre-defined voltage and frequency parameters, these modifications are based on frequency and voltage adjustments. This means that a change in the voltage level at any DG unit's terminal causes a change in its reactive power control to regulate the voltage throughout the islanded network, and vice versa.

Restoration

The islands need to be connected to the main grid as soon as the fault is cleared because they were constructed to lessen the likelihood of a blackout during a calamity, (Quirós-Tortós, 2012). To do this, a restoration sequence has been developed. A flow chart illustrating the developed procedure is shown in figure 5. Here, lines 1 and 4 represent the tie-lines. Prior to restoration, line priority is established; the sequence of line priority is high to low: [9, 2, 3, 5, 7, 6, 8, 10, 11]. In order to facilitate power transfers across islands, the tie-lines are re-established first the during restoration. Following the restoration of the tie lines, if the system is found to be stable, the load can be restored in steps (one MW in this case) based on the load priority, (Hou et al., 2009).

CONCLUSION

In addition to the instability of the system due to unplanned islanding in every scenario, the power fluctuations may result in damage to the DG unit rotor. Distribution networks' anti-islanding protection mechanisms are essential since the unintentional islanding impact can seriously affect end customers' loads, DG units, and utility assets. In all of the examined unplanned islanding sub-cases, the frequency changes were directly linked to the active power imbalance that existed before the islanding and, implicitly, to the type of DG technology used, the governor system connected with it, and the islanding technique employed. While the reactive power imbalance that existed prior to islanding, the islanding technique, and the type of DG unit technology all contributed to voltage variations.

REFERENCES

Dola, H. M., & Chowdhury, B. H. (2006). *Intentional islanding and adaptive load shedding to avoid cascading outages. In 2006 IEEE Power Engineering Society General Meeting*. IEEE., DOI: 10.1109/PES.2006.1709148

Guo, Q., Xin, S., Sun, H., & Wang, J. (2017). Power system cyber-physical modelling and security assessment: Motivation and ideas. *Zhongguo Dianji Gongcheng Xuebao*, *36*(6), 1481–1489.

Hou, Y., Liu, C.-C., Zhang, P., & Sun, K. (2009). Constructing power system restoration strategies. In *Proceedings of the International Conference on Electrical and Electronics Engineering* (pp. I-8–I-13). IEEE.

Hussain, T., Ishaq, S., Liaqat, S., Zia, M. F., Al-Durra, A., Khan, B., & Guerrero, J. M. (2022). Load shed recovery with transmission switching and intentional islanding methods after (N-2) line contingencies. *IEEE Access : Practical Innovations, Open Solutions*, *10*, 33681–33693. DOI: 10.1109/ACCESS.2022.3160723

Quirós-Tortós, J., & Terzija, V. (2012). *Controlled islanding strategy considering power system restoration constraints. In 2012 IEEE Power and Energy Society General Meeting*. IEEE., DOI: 10.1109/PESGM.2012.6345592

Sati, T. E., Azzouz, M. A., & Shaaban, M. (2023). Optimal protection coordination of islanded microgrids utilizing an adaptive virtual impedance fault current limiter. *IEEE Transactions on Industry Applications*, *59*(3), 2944–2955. DOI: 10.1109/TIA.2023.3245116

Shu, Y. B., Zhang, W. L., Zhou, X. X., Tang, Y., & Guo, Q. (2007). Security evaluation of UHV synchronized power grid. *Zhongguo Dianji Gongcheng Xuebao*, *27*(34), 1–6.

Wood, A. J., & Wollenberg, B. F. (1996). *Power generation, operation, and control* (2nd ed.). John Wiley & Sons.

Xin, S., Guo, Q., Sun, H., Chen, C., Wang, J., & Zhang, B. (2017). Information energy flow computation and cyber-physical sensitivity analysis for power systems. *IEEE Journal on Emerging and Selected Topics in Circuits and Systems*, *7*(2), 329–341. DOI: 10.1109/JETCAS.2017.2695618

Chapter 14

IoT-Enabled Real-Time Vehicle Monitoring and Wait Time Prediction System for Smart Fuel Stations

Pancham Singh
https://orcid.org/0009-0005-0435-7941
Department of Information Technology, Ajay Kumar Garg Engineering College, Ghaziabad, India

Mrignainy Kansal
https://orcid.org/0009-0005-2610-040X
Department of Computer Science and Engineering, Netaji Subhas University of Technology, New Delhi, India

Jaishree Jain
https://orcid.org/0000-0003-0173-7664
Department of Computer Science and Engineering, Ajay Kumar Garg Engineering College, Ghaziabad, India

Pooja Mishra
Department of Electronics and Communications Engineering, Ajay Kumar Garg Engineering College, Ghaziabad, India

Amita Asthana
Department of Electronics and Communications Engineering, Ajay Kumar Garg Engineering College, Ghaziabad, India

DOI: 10.4018/979-8-3373-4272-6.ch014

ABSTRACT

This study introduces a real-time vehicle monitoring system for fuel stations that integrates IoT and object detection technologies to improve customer fueling experiences. The system provides users with up-to-date information on wait times across various fuel stations, allowing them to make better-informed decisions. A network of sensors including those for vehicle detection and fuel monitoring collects data on queue lengths and fuel dispensing volumes. This data is analyzed using machine learning algorithms to forecast future wait times. These predictions are then shared with users, enabling them to choose the most efficient fueling station. This advanced system sets a new standard in the fuel service industry by combining operational efficiency with enhanced customer satisfaction, offering a streamlined and eco-friendly refueling process for diverse vehicle types.

INTRODUCTION

The Internet of Things (IoT) is a network of devices connected using wireless communication technologies. It can transfer data over a network without any requirement of human-human or human-computer interaction. In this vast network, devices communicate wirelessly, generating and transmitting data with more efficiency and without the need for direct human interaction. IoT has been widely adopted in the fuel industry to enhance operations. Fuel stations are a crucial part of the transportation industry, serving as the primary source of fuel for vehicles. Nonetheless, customers frequently experience extended wait times at these fuel stations due to overcrowding, leading to both frustration and time-consuming. According to a study conducted by Fuel Buddy (Soni & Saraswat, 2017), it was observed that people in the United States typically wait around 8.23 minutes at gas stations. In some densely populated cities, though, the wait times can be even longer, exceeding 20 minutes. In India, the situation becomes even more challenging for fuels like Compressed Natural Gas (CNG), with people often waiting for over 60 minutes.

These core issues leading to these extended wait times are due to a lack of real-time information regarding the traffic at the fuel stations and the availability of fuel with them. Customers usually estimate the time themselves based on the number of vehicles in front of them. This method is time-consuming and not precise. It can be annoying and can lead to wastage of crucial time in case of any emergencies, etc. Recently, advances in object detection technology have made it possible to count the number of vehicles in a waiting state using a simple camera. This technology can be used to provide real-time information about the number of vehicles waiting in line and the estimated wait time, enabling customers to make informed decisions about whether to wait at the current fuel station or move on to another one.

Also, the rise of electric vehicles has led to an increasing demand for charging stations, (Hossain et al., 2019). However, electric vehicle charging stations also experience similar problems with overcrowding and long wait times. The same object detection technology used in smart fuel stations can be used to provide details about the number of electric cars in queue and the estimated wait time, enabling electric car owners to make informed decisions about which charging station to visit. To overcome the problems discussed above, we propose to create a Fuel Station Queue Tracking System with the help of IoT and Machine Learning. This system will be capable of being implemented in the already existing Fuel Station frameworks and would access station data to provide real-time tracking of traffic at the fuel station and predict the availability of fuel accordingly, therefore suggesting the suitable fuel stations to the user to go to.

In addition to solving queuing and traffic issues, the proposed Fuel Station Queue Tracking System has more benefits for the fuel industry. It can help fuel station operators better manage their fuel inventory and reduce waste by collecting real-time data on their fuel consumption patterns. This means operators can make smarter decisions about when to restock fuel, therefore potentially reducing the environmental impact of storage and transportation as well as saving time.

Furthermore, this system fits in with broader industry trends toward increased automation, efficiency, and environmental responsibility. It not only makes customers' lives easier and safer but also helps cut down on the environmental impact of the fuel industry, (Nellore & Hancke, 2016). Shorter queuing times mean less time will be spent on running engines, resulting in fewer emissions and better fuel efficiency. It also helps manage traffic around fuel stations more effectively. Using, IoT and object detection technologies in fuel stations is a big step toward improving customer experiences, making operations more efficient, and being kinder to the environment. The proposed Fuel Station Queue Tracking System deals with resolving the issues of time spent in queuing, fuel tracking, and safety hence pointing toward a future where fuel distribution is not only more efficient but also more eco-friendly, meeting the changing needs of today's world.

LITERATURE SURVEY

In Rao (2015), the author proposes a system for monitoring and detecting vehicles in fuel stations that leverages IoT and object detection technology. This paper used systems based on vehicle detection and traffic analysis with vehicle registration number plate recognition. This paper explained Haar Cascade classifier is used for object detection, whereas the proposed system used a Vehicle detecting sensor and Machine learning Algorithms for object detection.

In Khatun et al. (2019), the author proposed a system for vehicle monitoring and tracking systems. This paper uses a system implemented using sensors, and GPS. The proposed system used Machine learning Algorithms, Cameras and Mobile applications for monitoring.

In Salra, Doodipala, and Endla (2020), the author proposes a system for detecting the number of vehicles at fuel stations. Vehicle arrival is a main criterion in the delay assessment of different fuel stations. This paper used Queuing theory for detecting the number of vehicles, whereas the proposed system for detecting the vehicle by using vehicle detecting sensors and machine learning Algorithms.

In Achi, Bakpo, and Umar (2022), the authors propose a system that uses an ultrasonic fuel sensor to measure the amount of fuel consumed. When the vehicle's fuel tank reaches a specific level, the driver receives a warning via mobile application and searches for a nearby and preferable pump to refuel. The suggested system used GPS tracking to display the vehicle's current location and locate the nearest and preferable fuel stations. The system is a real-time and accurate gasoline monitoring device. This system reduces customer wait time and improves efficiency.

In Khatun et al. (2019), the author proposes a real-time vehicle detection and automation system for fuel stations that offers a transformative solution to modernize fuel station operations. By integrating cutting-edge technology, this system not only enhances operational efficiency but also prioritizes customer experience, security, and sustainability.

In Rijayanti, Muhammad, and Hwang (2020), the author proposes a system for enhancing the overall efficiency of fuel distribution by optimizing its process and resource allocation, resulting in improved fuel supply management and reduced inefficiencies. This proposed system aims to address the challenges and inefficiencies associated with the distribution of fuel.

Some differences that make the proposed idea different from the existing paper:

Figure 1. Feature of our proposed idea

Integration of Real-Time Vehicle Detection

Real-time vehicle detection is a critical component of the proposed system, enabling fuel station operators to monitor vehicle queues continuously and manage customer flow efficiently. By identifying vehicles in real-time, operators can predict and respond to demand fluctuations, ensuring adequate staffing at fuel pumps during peak hours. Additionally, digital signage and mobile apps can provide drivers with up-to-date wait times, helping them make informed decisions about when to refuel, and thereby reducing unnecessary queuing, (Zayoud & Hamam, 2019). This technology contributes to a more streamlined and customer-friendly experience, reducing wait times and improving overall customer satisfaction.

Cloud-Based System and Communication

A cloud-based system and communication infrastructure can significantly enhance fuel station efficiency and help alleviate wait times. By deploying cloud-connected sensors and cameras, operators can monitor real-time data on vehicle arrivals, fuel pump availability, and payment processing. This information can be processed and analyzed in the cloud, enabling operators to predict demand trends and allocate resources accordingly, (Yadav et al., 2021). Moreover, cloud-based communication allows for seamless coordination between different components of the fuel station, including payment systems, signage, and customer apps. The availability of accurate wait time and fuel information will minimize unnecessary queuing, reduce wait time, and improve the overall efficiency and customer experience at fuel stations, (Jiahao et al., 2020).

Optimization for Fuel Distribution and Equalizing Profit

The proposed idea stresses the concept of delivering wait time information to customers a step further by emphasizing integration with the gas station's management system. This integration offers a holistic approach to the fueling process, extending beyond customer satisfaction, (Swamy & Kota, 2020). By merging real-time data with the gas station's operations, it becomes possible to optimize fuel distribution, resulting in reduced wait times for customers. Additionally, it enables the equalization or distribution of profit among different gas stations in a network. This approach not only enhances the overall customer experience by minimizing wait times but also ensures that resources and profits are efficiently allocated, making it a suitable solution for both customers and gas station owners.

Future Adaptation for Electric Vehicle Charging Stations

The proposed system is scalable and versatile, with the potential to be applied to charging stations for electric vehicles. This forward-thinking aspect underscores the system's adaptability to emerging technology trends and its contribution to broader sustainability and environmental goals, (Borgia, 2014). This versatility reflects a commitment to meeting the evolving needs of modern transportation, making it clear that the system has the potential to revolutionize not only the conventional fueling industry but also the rapidly growing electric vehicle market, offering a comprehensive solution for the changing landscape of mobility, (Aliyu et al., 2018).

Consideration of Weather Conditions

While the proposed system does not specifically mention weather considerations, integrating weather data could provide valuable insights and allow for more accurate predictions of wait times under various weather conditions. This enhancement would improve the system's accuracy and provide customers with more comprehensive information to plan their visits to the fuel or charging station, taking into account the impact of weather on wait times, (Jin et al., 2014). This enhancement could make the proposed system even more robust and adaptable, setting it apart from existing solutions and catering to a wider range of customer needs.

By emphasizing these unique features and differences, the proposed idea can establish its distinctiveness from the existing paper, contributing to its potential for being granted a separate paper, (Salhaoui et al., 2019). The goal of the proposed smart fuel station system is to improve the overall efficiency of fuel distribution, reduce customer frustration caused by long wait times, optimize fuel distribution, and prepare for the future of electric vehicles.

METHODOLOGY

A methodology is a systematic approach to completing a project or task. It provides a roadmap for achieving desired outcomes, defining the steps involved, and outlining the resources required. The following methodology has been used in the research paper.

DESIGN

The proposed smart fuel station system will leverage advanced technologies to enhance efficiency, reduce wait times, and improve the overall customer experience. The system will employ object detection technology, data collection and processing (Guerrero-Ibáñe, Zeadally, & Contreras-Castillo, 2018), communication with customers, and integration with the fuel station's management system to achieve these goals.

Object Detection Technology

The system will utilize cameras or sensors to accurately identify and count the number of vehicles present at the fuel station. This data will be used to monitor vehicle traffic in real-time at fuel stations and optimize operations. The system aids inventory management, (Khayrat et al., 2022). This data-driven approach makes fueling more efficient in transportation.

Data Collection and Processing

The data captured by the cameras or sensors will be transmitted to a cloud-based system for processing and analysis. The cloud-based infrastructure will utilize powerful algorithms and machine learning techniques to provide real-time information on the precise number of vehicles at the fuel station, as well as estimated wait times, (Kansal et al., 2023). This information will be invaluable to both station management and customers. Using cloud computing ensures accurate vehicle counts and reduces wait times.

Communication with Customers

The system will communicate estimated wait times to incoming customers through client applications or directly integrated into smart vehicles. This will enable customers to make informed decisions about where to refuel, (Wang, Xu, & Peng, 2007).

Integration with Fuel Station Management System

The system integrates with the fuel station's management to optimize fuel distribution and reduce customer wait times through dynamic resource allocation.

Maintenance and Upgrades

The system will require regular maintenance and upgrades to ensure it remains up-to-date with the latest technology and provides accurate real-time information to customers, (Abu-Abed & Khabarov, 2018).

Additional Considerations

In addition to the core components of the system, there are a few other considerations to be considered, *Security:* The system will be designed with security in mind to protect sensitive data, such as vehicle counts, customer information and sensitive information about fuel stations. *Scalability:* The system would be designed to be scalable to accommodate a growing number of vehicles and users. *Cost-effectiveness:* The system would be cost-effective to deploy and maintain.

SYSTEM ARCHITECTURE

System architecture is a conceptual model that describes the structure and behaviour of multiple components and subsystems like multiple software applications, network devices, hardware, and even other machinery of a system. It is Architecture Description Language (ADL) which helps to describe the entire system architecture. So, it is a much broader topic. A system architecture can be broadly categorized into centralized and decentralized architectural organizations, (Singh et al., 2023; Anayo & Nwogbaga, 2016).

This flowchart outlines the main steps of the real-time monitoring system for fuel stations:

1. Sensors collect data about the number of vehicles waiting and the amount of fuel dispensed at different stations.
2. Machine learning algorithms analyze this data to predict future wait times.
3. Predicted wait times are displayed to users through a user interface, (Gomes et al., 2019).
4. Customers arriving at the fuel station check real-time estimated wait times.
5. Customers can choose a preferred station with a shorter wait time or join the nearest station, (Chodorek, Chodorek, & Yastrebov, 2022).
6. Vehicle detection and fuel monitoring sensors continue to collect data for on-going monitoring.

This system empowers customers to informed decisions about which fuel station to visit, enhancing their fueling experience and reducing waiting times. It also prioritizes efficiency and customer satisfaction while promoting environmentally friendly refuelling solutions, (Lv et al., 2022).

Figure 2. Flow diagram for smart fuel station system

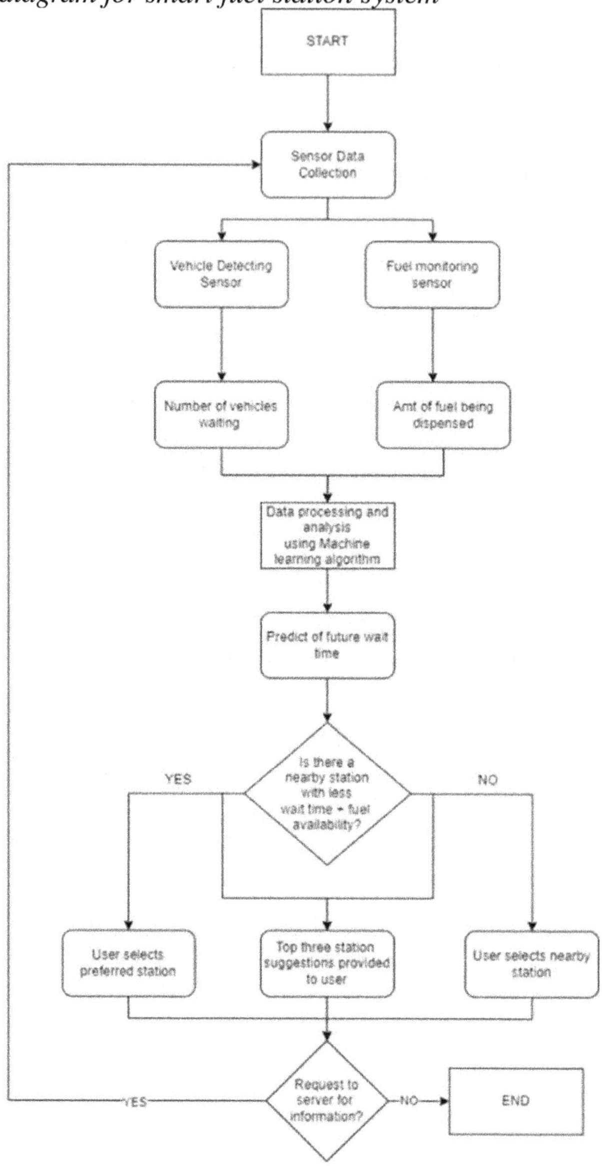

A typical architecture diagram illustrates the key components and their interactions in the system.

Figure 3. Architecture diagram for smart fuel stations

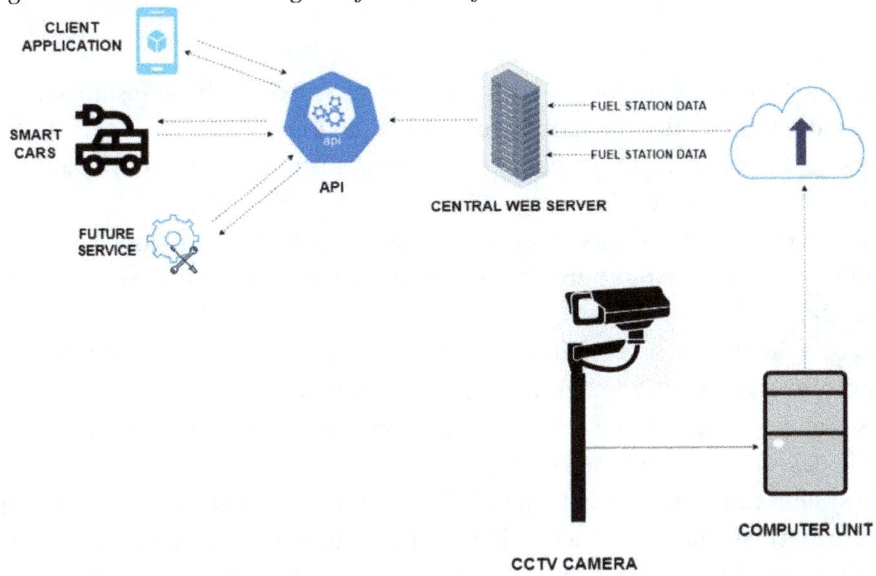

a) Fuel Pumps: At the core of the fuel station, fuel pumps are present where the vehicles are refueled. Each pump may be equipped with sensors for vehicle detection and data collection.
b) Cameras and Sensors: Cameras and sensors are strategically placed throughout the station to monitor vehicle presence, count, and capture data on fueling transactions and vehicle types, (Jabbar et al., 2019).
c) Cloud-Based Data Processing: Data collected from cameras and sensors is sent to a cloud-based system for processing. This is where real-time analytics, object detection, and vehicle counting occur.
d) Data Analytics and Machine Learning: Within the cloud-based system, data analytics and machine learning algorithms process the incoming data to provide insights on vehicle counts, wait times, and other relevant information, (Hammi et al., 2018).
e) Network Connectivity: A strong and dependable network system connects all components, ensuring seamless data transfer between the sensors, cloud system, and customer displays, (Husni, 2017).
f) Database: A database stores historical data, customer information, and transaction records, which can be used for further analysis and reporting.
g) Mobile Apps and Customer Interface: Customers can access real-time information through mobile apps, allowing them to make informed decisions about when to visit the fuel station, (Beshley et al., 2020).

The Flow of Process

a) Object detection technology, such as cameras or sensors, will be installed at the fuel station to identify the number of vehicles in the queue.
b) The data collected from the cameras will be sent to a cloud-based system for processing, (Sadeeq & Zeebaree, 2021).
c) The processed data will be used to provide real-time information about the number of vehicles in the waiting state for fuel and estimated wait times for incoming customers.
d) The estimated wait times will be communicated to customers through client applications or directly integrated into smart vehicles.
e) Customers can then make informed decisions about whether to wait or move to the next fuel station based on the estimated wait time.
f) The system can be integrated with the fuel station's management system to optimize fuel distribution, reduce wait times for customers, and equalize/distribute profit among fuel stations, (Malche, Maheshwary, & Kumar, 2019).
g) The same technology can be adapted for charging stations for electric vehicles, providing similar benefits to those offered by the smart fuel station system.
h) The system will require regular maintenance and upgrades to ensure it remains up-to-date with the latest technology and provides accurate real-time information to customers.

Figure 4. Data flow diagram for smart fuel stations

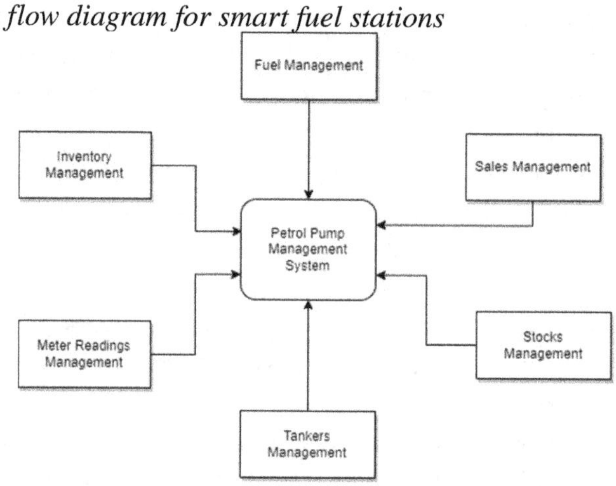

Petrol Pump Management System

It is a software application designed to efficiently manage and streamline the operations of a fuel station or petrol pump. It integrates various aspects of fuel station management, customer service, and administrative tasks to enhance overall productivity and customer satisfaction, (Khan et al., 2020).

Here's a description of the key components and functionalities of a typical Petrol Pump Management System:

Fuel Dispensing and Inventory Management. The system manages fuel dispensing operations, monitoring the quantity of fuel sold and managing inventory levels. It helps ensure that fuel pumps are accurately calibrated, tracks fuel deliveries and provides real-time updates on fuel availability.

Customer Management. This component manages customer accounts, loyalty programs, and contact details. It allows for creating customer profiles and tracking their purchase history and preferences. It may also include a customer loyalty program to reward frequent visitors, (Kumar & Hong, 2022).

Pump Monitoring and Automation. The system can monitor and control fuel pumps, ensuring they function correctly and are synchronized with the POS system. It can automatically shut off pumps if they detect irregularities or reach a preset limit, preventing fuel spills and fraud.

Security and Surveillance. Integrating security features, the system may include surveillance camera integration and access control to monitor activities at the petrol pump, enhancing security for staff and customers, (Kumar & Hong, 2022).

Meter Reading and Measurement. It integrates flow meters and sensors to accurately measure the quantity of liquid being loaded, transported, and delivered. These measurements are crucial for billing, quality control, and inventory management.

Reporting and Analytics. It generates various reports such as sales reports, inventory status, pump efficiency, and financial summaries. These reports help in decision-making and performance analysis.

RESULTS AND DISCUSSION

A complete view of an application to explain the functionality of the whole system

Step 1: The user enters the application and sets their location either manually or through GPS

Step 2: The user clicks on "Click here to see the nearest fuel station" to run the algorithm and search for a recommended fuel station to visit.

Figure 5. App home page view

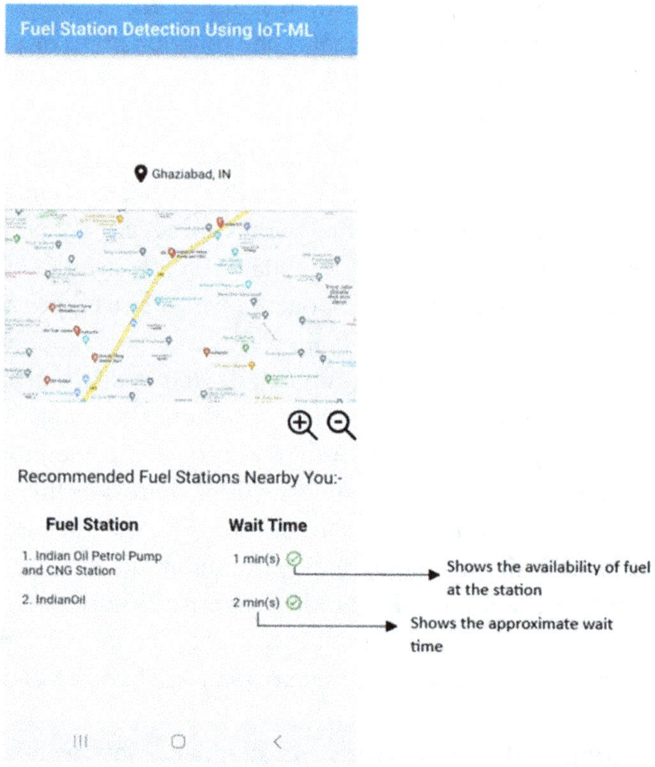

Figure 6. Model predicting the nearby station having less wait time as well as fuel availability

Step 3: A list of recommended fuel stations in ascending order of wait time is displayed to the user. The user can then select a preferred fuel station and be given directions to reach there.

Step 4: A sorted list is displayed to the user consisting of nearby fuel stations within a 1 km radius of them. The user can then click and direct themselves to the station of their choice.

CONCLUSION

The main objective of this paper is to provide real time information about the number of vehicles waiting in a queue at fuel stations and estimated wait times all achieved through the implementation of cutting-edge technology. This technology-driven approach serves multiple crucial purposes. It enhances customer satisfaction by expelling the frustration associated with waiting in long queues with the help of real-time data on expected wait times. This helps to make informed decisions and reduce their waiting time therefore reducing stress upon the user. Additionally, the system helps in optimizing fuel distribution by providing accurate information on queue lengths, empowering fuel stations to effectively allocate resources, particularly during peak hours. The system can be integrated with the fuel stations management system and adapter for charging station for electric vehicles. Furthermore, it seamlessly integrates with existing fuel station management systems, enabling data-driven operational decisions and real-time service optimization. In summary, this system is designed to leverage technology to enhance the customer experience at fuel stations, delivering real-time information, and benefiting both fuel station operators and their customers.

REFERENCES

Abu-Abed, F., & Khabarov, A. (2018). Life-support model of supply systems of oil and gas industry objects. *E3S Web of Conferences, 33*, 01028. https://doi.org/DOI: 10.1051/e3sconf/20183301028

Achi, A., Bakpo, F., & Umar, Z. (2022). Real-time petrol availability reporting system (RPARS) for Nsukka town, Nigeria. *Computer Science & IT Research Journal, 3*(1), 23–35. DOI: 10.51594/csitrj.v3i1.294

Aliyu, S., Audu, W. M., Okwori, M., Saidu, M., Abdullahi, U., & Eneze, J. (2018). An enhanced BRAZ (E-BRAZ) Android app for monitoring PMS in gas stations. *Journal of Applied Science & Environmental Management, 22*(9), 1497–1502.

Anayo, O. H., & Nwogbaga, N. E. (2016). Critical review of petrol station management system with emphasis on the advantages if digitalized in Nigeria. *International Journal of Novel Research in Engineering & Pharmaceutical Sciences, 3*(5), 49–56.

Beshley, M., Kryvinska, N., Seliuchenko, M., Beshley, H., Shakshuki, E. M., & Yasar, A.-U.-H. (2020). End-to-end QoS "smart queue" management algorithms and traffic prioritization mechanisms for narrow-band Internet of Things services in 4G/5G networks. *Sensors (Basel), 20*(8), 2324. DOI: 10.3390/s20082324 PMID: 32325795

Borgia, E. (2014). The Internet of Things vision: Key features, applications and open issues. *Computer Communications, 54*, 1–31. DOI: 10.1016/j.comcom.2014.09.008

Chodorek, A., Chodorek, R. R., & Yastrebov, A. (2022). The prototype monitoring system for pollution sensing and online visualization with the use of a UAV and a WebRTC-based platform. *Sensors (Basel), 22*(4), 1578. DOI: 10.3390/s22041578 PMID: 35214478

Gomes, J. B., Rodrigues, J. J., Rabêlo, R. A., Kumar, N., & Kozlov, S. (2019). IoT-enabled gas sensors: Technologies, applications, and opportunities. *Journal of Sensor and Actuator Networks, 8*(4), 57. DOI: 10.3390/jsan8040057

Guerrero-Ibáñez, J., Zeadally, S., & Contreras-Castillo, J. (2018). Sensor technologies for intelligent transportation systems. *Sensors (Basel), 18*(4), 1212. DOI: 10.3390/s18041212 PMID: 29659524

Hammi, B., Khatoun, R., Zeadally, S., Fayad, A., & Khoukhi, L. (2018). IoT technologies for smart cities. *IET Networks, 7*(1), 1–13. DOI: 10.1049/iet-net.2017.0163

Hossain, M. S., Rahman, M., Sarker, M. T., Haque, M. E., & Jahid, A. (2019). A smart IoT based system for monitoring and controlling the sub-station equipment. *Internet of Things : Engineering Cyber Physical Human Systems*, *7*, 100085. DOI: 10.1016/j.iot.2019.100085

Husni, E. (2017). Driving and fuel consumption monitoring with Internet of Things. *International Journal of Interactive Mobile Technologies*, *11*(3), 78–97. DOI: 10.3991/ijim.v11i3.6470

Jabbar, W. A., Kian, T. K., Ramli, R. M., Zubir, S. N., Zamrizaman, N. S., Balfaqih, M., Shepelev, V., & Alharbi, S. (2019). Design and fabrication of smart home with Internet of Things enabled automation system. *IEEE Access : Practical Innovations, Open Solutions*, *7*, 144059–144074. DOI: 10.1109/ACCESS.2019.2942846

Jiahao, Y., Jiang, X., Wang, S., Jiang, K., & Yu, X. (2020). SVM-BiLSTM: A fault detection method for the gas station IoT system based on deep learning. *IEEE Access : Practical Innovations, Open Solutions*, *8*, 203712–203723. DOI: 10.1109/ACCESS.2020.3036937

Jin, J., Gubbi, J., Marusic, S., & Palaniswami, M. (2014). An information framework for creating a smart city through Internet of Things. *IEEE Internet of Things Journal*, *1*(2), 112–121. DOI: 10.1109/JIOT.2014.2306326

Kansal, M., Singh, P., Srivastava, M., & Chaurasia, P. (2023). Empowering agriculture with conversational AI: An application for farmer advisory and communication. In Kumar, A., Mozar, S., & Haase, J. (Eds.), *Convergence of cloud computing, AI, and agricultural science* (pp. 210–227). IGI Global., DOI: 10.4018/979-8-3693-0501-0.ch013

Khan, W. Z., Rehman, M. H., Zangoti, H. M., Afzal, M. K., Armi, N., & Salah, K. (2020). Industrial Internet of Things: Recent advances, enabling technologies and open challenges. *Computers & Electrical Engineering*, *81*, 106522. DOI: 10.1016/j.compeleceng.2019.106522

Khatun, R., Antor, S. A., Ullah, A., & Hossain, A. (2019). Vehicle fuel activities monitoring system using IoT. *Advances in Internet of Things*, *9*(4), 63–71. DOI: 10.4236/ait.2019.94005

Khatun, R., Antor, S. A., Ullah, A., & Hossain, A. (2019). Vehicle fuel activities monitoring system using IoT. *Advances in Internet of Things*, *9*(4), 63–71. DOI: 10.4236/ait.2019.94005

Khayrat, A., Malak, P., Victor, M., Ahmed, S., Metawie, H., Saber, V., & Elshalakani, M. (2022). An intelligent surveillance system for detecting abnormal behaviors on campus using YOLO and CNN-LSTM networks. In *2022 2nd International Mobile, Intelligent, and Ubiquitous Computing Conference (MIUCC)* (pp. 104–109). IEEE. https://doi.org/DOI: 10.1109/MIUCC55081.2022.9781735

Kumar, P. M., & Hong, C. S. (2022). Internet of Things for secure surveillance for sewage wastewater treatment systems. *Environmental Research*, *203*, 111899. DOI: 10.1016/j.envres.2021.111899 PMID: 34416251

Lv, Z., Wu, J., Li, Y., & Song, H. (2022). Cross-layer optimization for industrial Internet of Things in real scene digital twins. *IEEE Internet of Things Journal*, *9*(17), 15618–15629. DOI: 10.1109/JIOT.2022.3150927

Malche, T., Maheshwary, P., & Kumar, R. (2019). Environmental monitoring system for smart city based on secure Internet of Things (IoT) architecture. *Wireless Personal Communications*, *107*(4), 2143–2172. DOI: 10.1007/s11277-019-06376-0

Nellore, K., & Hancke, G. P. (2016). A survey on urban traffic management system using wireless sensor networks. *Sensors (Basel)*, *16*(2), 157. DOI: 10.3390/s16020157 PMID: 26828489

Rao, Y. (2015). Automatic vehicle recognition in multiple cameras for video surveillance. *The Visual Computer*, *31*(3), 271–280. DOI: 10.1007/s00371-013-0917-y

Rijayanti, R., Muhammad, R. F., & Hwang, M. (2020). Vehicle waiting time information service using vehicle object detection at fuel charging station. *Journal of Information and Communication Convergence Engineering*, *18*(3), 147–154. DOI: 10.6109/jicce.2020.18.3.147

Sadeeq, M. A., & Zeebaree, S. (2021). Energy management for Internet of Things via distributed systems. *Journal of Applied Science and Technology Trends*, *2*(2), 59–71. DOI: 10.38094/jastt20291

Salhaoui, M., Guerrero-González, A., Arioua, M., Ortiz, F. J., El Oualkadi, A., & Torregrosa, C. L. (2019). Smart industrial IoT monitoring and control system based on UAV and cloud computing applied to a concrete plant. *Sensors (Basel)*, *19*(15), 3316. DOI: 10.3390/s19153316 PMID: 31357720

Singh, P., Kansal, M., Singh, S., Gupta, S., Maheshwari, S., & Gupta, S. (2023). A conceptual prototype for transparent and traceable supply chain using blockchain. In Mishra, A., Gupta, D., & Chetty, G. (Eds.), *Advances in IoT and security with computational intelligence* (pp. 207–217). Springer., DOI: 10.1007/978-981-99-5085-0_19

Soni, N. B., & Saraswat, J. (2017). A review of IoT devices for traffic management system. In *2017 International Conference on Intelligent Sustainable Systems (ICISS)* (pp. 1052–1055). IEEE. https://doi.org/DOI: 10.1109/ISS1.2017.8389330

Swamy, S. N., & Kota, S. R. (2020). An empirical study on system level aspects of Internet of Things (IoT). *IEEE Access : Practical Innovations, Open Solutions*, *8*, 188082–188134. DOI: 10.1109/ACCESS.2020.3031410

Wang, C., Xu, L., & Peng, W. (2007). Conceptual design of remote monitoring and fault diagnosis systems. *Information Systems*, *32*(7), 996–1004. DOI: 10.1016/j.is.2006.10.003

Yadav, S., Motupalli, R., Jamal, K., & Suresh, C. (2021). An automated rescue and service system with route deviation using IoT and blockchain technologies. In *2021 IEEE Mysore Sub Section International Conference (MysuruCon)* (pp. 582–586). IEEE. https://doi.org/DOI: 10.1109/MysuruCon52639.2021.9641657

Zayoud, R., & Hamam, H. (2019). Accurate passive indoor RFID alignment system for service station. In *2019 15th International Wireless Communications & Mobile Computing Conference (IWCMC)* (pp. 1293–1298). IEEE. https://doi.org/DOI: 10.1109/IWCMC.2019.8766738

Chapter 15
Generative AI-Driven Frameworks for Intelligent Wind Energy Forecasting and Optimization

Inam Ul Haq
https://orcid.org/0000-0001-8186-0632
Noida Institute of Engineering and Technology, India

Adil Husain Rather
https://orcid.org/0009-0002-5433-3700
CGC University, Mohali, India

ABSTRACT

Generative AI-driven frameworks are transforming the forecasting and optimization of wind energy systems by overcoming the limitations of traditional statistical and deterministic models that struggle with the nonlinear and variable nature of wind resources. By leveraging advanced models such as Generative Adversarial Networks (GANs), Variational Autoencoders (VAEs), and Diffusion Models, these frameworks facilitate the creation of synthetic data, enhance scenario simulation, and enable robust probabilistic forecasting, significantly reducing uncertainty and improving operational resilience. Integrating generative AI with optimization algorithms and digital twin architectures supports intelligent turbine placement, adaptive energy dispatch, and predictive maintenance, thereby increasing energy yield, reliability,

DOI: 10.4018/979-8-3373-4272-6.ch015

and economic viability. This approach enables real-time decision support, rapid scenario analysis, and improved grid integration, accelerating the transition toward sustainable energy systems.

INTRODUCTION

One of the main pillars of international initiatives to combat climate change, achieve energy security, and promote sustainable development is the quick expansion of renewable energy sources. Wind energy has drawn a lot of interest among renewable energies because of its abundance, scalability, and minimal environmental impact. The International Energy Agency (IEA) estimates that by 2050, wind energy will provide over one-third of the world's electricity needs, making it an essential part of the move toward carbon neutrality, (Adewumi et al., 2024). However, the erratic, intermittent, and highly nonlinear nature of wind presents difficulties for the deployment of wind energy systems. These characteristics inhibit accurate forecasting, functioning of wind farms and effective grid integration. The complex spatiotemporal behavior of wind speed and power generation is often challenging to explain using the conventional approaches of wind energy prediction, e.g. statistical regression modeling and physical numerical weather prediction (NWP). Although machine learning and deep learning models have increased efficiency, there are occasions when they cannot be applied due to their dependency on massive labelled datasets, their inability to generalise in uncertain scenarios, and their inability to model long- term dependencies, (Haq et al., 2025; Albarakati et al., 2021; Haq, Nandal, & Uppal, 2025; Cacciamani et al., 2023; Camacho et al., 2024; Rather & Haq, 2023; Elimam et al., 2023). Moreover, existing methods of energy dispatch and wind farm control are often deterministic in nature and fail to adequately address wind (pattern) variability, turbine interaction and energy market fluctuations. Such limitations have led to the development of more flexible, reliable, and intelligent forecasting and optimization frameworks (Ul Haq, 2024) or systems. Table 1 provides an overview of the comparison between the conventional methods and generative AI-based models of intelligent wind power prediction and optimization.

Table 1. Comparison between traditional approaches and generative AI-driven frameworks for intelligent wind energy forecasting and optimization

Dimension	Traditional Approach	Generative AI-Driven Frameworks	Relevance to Wind Energy
Data Handling	Uses historical meteorological data with limited feature extraction	Synthesizes new realistic data, fills missing values, and augments datasets	Enhances accuracy where wind data is sparse or noisy
Forecasting Accuracy	Relies on statistical/machine learning models (ARIMA, ANN, SVM)	Generates probabilistic forecasts with uncertainty quantification	Provides more reliable short-term and long-term wind predictions
Optimization	Deterministic or heuristic optimization (GA, PSO)	Adaptive generative models optimize turbine placement, energy dispatch, and maintenance scheduling	Increases overall energy yield and cost-effectiveness
Scalability	Limited performance with large, complex datasets	Scales efficiently with high-dimensional meteorological and operational data	Supports grid-level forecasting and smart grid integration
Decision Support	Provides point predictions	Generates scenario-based simulations for better planning	Helps operators manage variability, risks, and resource allocation
Sustainability Impact	Incremental improvements in renewable integration	Accelerates transition to sustainable, resilient wind energy systems	Contributes to global renewable energy targets

Within this framework, Generative Artificial Intelligence (GenAI) has surfaced as a paradigm shift that can effectively tackle these issues. In order to create realistic synthetic data, uncertainty-aware forecasts, and scenario-based simulations, generative models like Generative Adversarial Networks (GANs), Variational Autoencoders (VAEs), Diffusion Models, and Transformer-based architectures are excellent at learning the underlying probability distributions of complex datasets, (Choudhary et al., 2023).

GenAI can be used in wind energy systems to:

- Capture rare extreme occurrences and nonlinear patterns to improve forecasting accuracy.
- Make projections that are both uncertain and probabilistic to aid in risk-sensitive decision-making.
- Enable data augmentation in areas with sparse or insufficient historical wind readings.
- Simulate various turbine, maintenance, and grid integration operating scenarios to enable optimisation-driven control.

Additionally, intelligent wind farm management is made possible by the combination of GenAI with optimisation techniques like multi-objective optimization, reinforcement learning, and evolutionary methods. By optimising turbine yaw control, pitch adjustment, energy storage scheduling, and power dispatch algorithms, these AI-driven frameworks can lower energy losses, save costs, and increase system resilience, (Donti & Kotler, 2021). Thus, generative AI-driven frameworks for intelligent wind energy forecasting and optimisation mark a paradigm change away from model-based, deterministic methods and toward data-driven, adaptive, and uncertainty-aware solutions. These frameworks have the potential to make wind energy a more dependable, effective, and scalable source of clean energy by utilising the power of generative learning and intelligent optimization, (Etukudoh et al., 2024; Fu et al., 2022; Garmabdari et al., 2019; Gawusu et al., 2022; Green et al., 2023). This would hasten the world's transition to a sustainable and carbon-neutral energy future. As illustrated in Table 2, generative AI models reduce mean absolute error (MAE) and root mean square error (RMSE) by nearly 30–40%, extend reliable forecasting horizons by 15%, and improve turbine placement optimization by up to 12%, while also offering more effective uncertainty quantification

Table 2. Performance comparison of traditional models and generative AI-driven frameworks in wind energy forecasting and optimization

Metric	Traditional Models (ARIMA, ANN, SVM)	Generative AI Models (GANs, VAEs, Diffusion)	Improvement (%)
Mean Absolute Error (MAE)	0.95 – 1.20 (m/s)	0.60 – 0.75 (m/s)	~30–40% lower error
Root Mean Square Error (RMSE)	1.10 – 1.45 (m/s)	0.70 – 0.90 (m/s)	~35% lower error
Forecasting Horizon Reliability	70–75% (up to 24 hours)	85–90% (up to 24 hours)	+15% improvement
Optimization of Turbine Placement (Energy Yield Gain)	Baseline (100%)	108–112%	+8–12% increase
Computational Efficiency (Training Time)	High (long training with large datasets)	Moderate (efficient generative data augmentation)	~25% faster
Uncertainty Quantification	Limited	Robust (probabilistic forecasting with confidence intervals)	Significant advantage

ROLE OF GENERATIVE AI IN RENEWABLE ENERGY

Synthetic Data Generation

By enabling the synthetic creation of wind speed, air density, and turbulence data—particularly in situations where historical observations are limited or insufficient—generative AI plays a critical role in renewable energy. It may provide realistic time-series datasets that maintain the statistical patterns of natural wind behaviour as well as the physical principles of atmospheric dynamics by utilizing sophisticated models like GANs, diffusion models, or physics-informed neural networks, (Hamdan et al., 2024). Figure 1 shows the Performance improvements of Generative AI across key areas in renewable energy systems, including synthetic data generation, scenario simulation, uncertainty reduction, and optimization support. Values represent indicative percentage improvements in data coverage, forecast accuracy, energy yield, downtime reduction, and grid stability

Figure 1. Generative AI significantly improves data coverage, scenario representation, and operational performance compared to traditional approaches

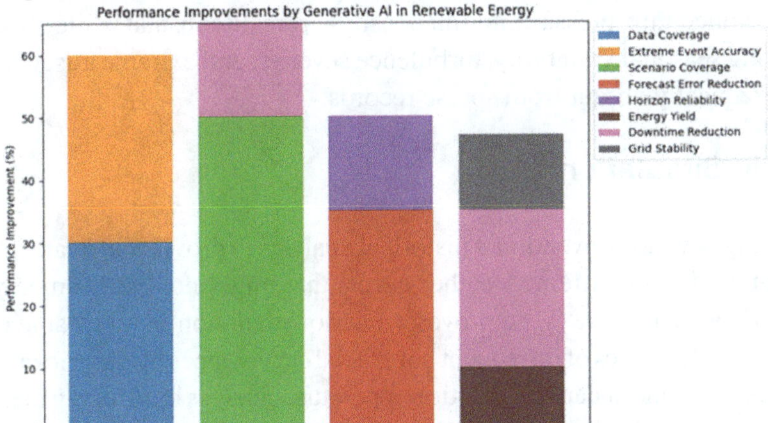

Reduction in mean absolute error (MAE) of wind energy forecasting using generative AI models (GANs, Diffusion, Physics-Informed) compared to traditional approaches. Probabilistic forecasting through generative models reduces MAE by up to 35% is shown in Figure 2.

Figure 2. Generative AI models outperform conventional models in wind forecasting, achieving a 30–35% reduction in MAE

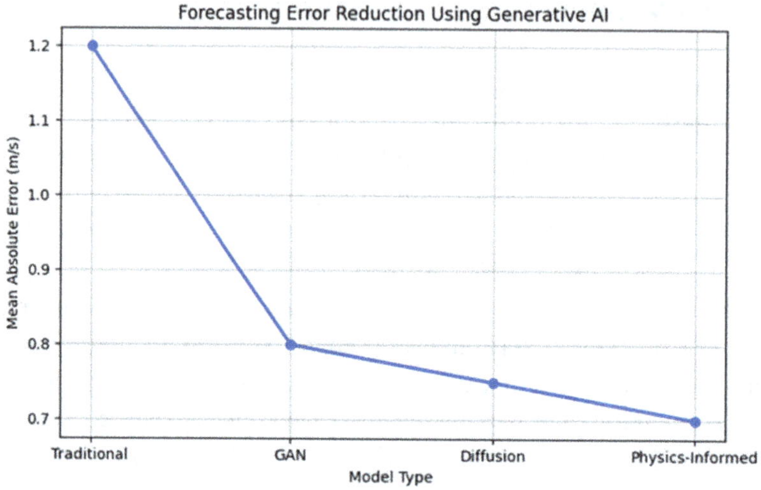

These artificial datasets ensure that interdependencies—like the relationship between temperature, pressure, and air density—remain constant while also assisting in capturing seasonal variability, turbulence severity, and extreme gust occurrences that are frequently absent from sparse records.

Scenario Simulation

By going beyond conventional historical analysis to provide intricate, physically consistent models of extreme weather events that might not be commonly seen in recorded data, generative AI improves scenario simulation in renewable energy. It can mimic the dynamics of infrequent but crucial occurrences like hurricanes, intense turbulence, fast wind shear, and drastic temperature changes by utilizing methods like generative adversarial networks, diffusion models, and physics-informed learning, (Huang & Wang, 2023; Ipakchi & Albuyeh, 2009; Islam & Othman, 2024; Jiang, Liu, & Peng, 2024; Kalair et al., 2021). By putting wind turbines, solar panels, and grid infrastructure through their worst-case scenarios, these simulated scenarios enable researchers and operators to find weaknesses in design, control, and operational tactics. For instance, extreme temperature–density changes can be modelled to assess their effect on total power output, and AI-generated high-intensity gust patterns can be applied to turbine digital twins to investigate tower stability or blade fatigue. Additionally, by supporting resilience planning, these models help utilities create proactive maintenance schedules, improve grid dependability, and create adaptive control algorithms, (Latrach et al., 2024). Generative AI makes it possible to ensure that renewable energy systems are better prepared to handle the difficulties posed by rising climate variability and catastrophic weather events by offering a broad range of "what-if" scenarios.

Uncertainty Reduction

By enabling probabilistic forecasting—which generates a wide variety of believable future scenarios that incorporate the inherent variability of renewable resources—generative AI plays a critical role in reducing uncertainty. For instance, turbulence, gusts, seasonal patterns, and abrupt changes in air density make predicting extremely difficult in wind energy, while cloud movement and atmospheric conditions further complicate forecasting in solar energy. Using historical weather and electricity generation data, generative models like GANs, diffusion models, or normalisation flows can generate several forecast trajectories that represent various potential future condition realisations, (Liu et al., 2025).

Operators can measure forecast uncertainty, assign probability to various possibilities, and get ready for infrequent but significant deviations with the help of these scenario ensembles. As an example, a hedge allows energy dealers to manage the risks associated with a range of price and generation events each, as well as grid operators to determine reserve requirements using probabilistic forecasts. Moreover, Generative AI-based uncertainty reduction enhances robust scheduling, predictive dispatch of energy, and more reliability of the integration of renewable energy into modern power systems by forecasting routine fluctuations and the extreme tail events. The shift towards probabilistic forecasting can now enable energy systems to deal with unpredictability and uncertainty in a safe and cost efficient manner, (Haq, Rather, & Kaur, 2023).

Optimization Support

Generative AI can be of critical importance in the development of wind turbines by creating high-resolution artificial data of wind speed, turbulence, and air density in regions that could have limited historical measurements. By modeling different wind flow patterns and different turbulence conditions, engineers can more easily understand the wake effects (the reduction of efficiency that occurs when turbine impedes the flow of air) around the turbines. There is also the possibility of identifying the most suitable micro-topography where turbines can be placed by applying generative AI to simulate the behaviour of wind in complex landscapes such as hills, valleys or coastal beaches. With such AI-based knowledge, the decision-makers will be able to discover a means of striking the appropriate balance of the energy capture growth and the infrastructural costs decrease that eventually results in more cost-effective and efficient wind farms layouts. The idea behind predictive maintenance is to reproduce operation and failure conditions utilizing Generative AI so that it allows detecting faults early and increasing the reliability of wind turbines. It allows anomalies to be detected that would indicate equipment failures, by creating baseline behaviours, and training of common vibration, temperature and performance trends using sensor data. In addition, less common but significant failure modes, not always common in the historical data, but needed to proactively observe them like gearbox problems, blade cracks, or generator overheating, can be trained using generative AI. It can also estimate the remaining useful life of components at different loads with different environments by creating possible deterioration paths. With these data, operators can be able to better plan the maintenance, reducing unexpected downtime, reducing the cost of repairs, as well as increasing the life of expensive assets, (Mansouri et al., 2023; Haq, Kaur, & Rather, 2024; Machina, Koduru, & Madichetty, 2022; Mauro, 2024; Mysore, 2024). Generative AI also enhances grid operations by helping to balance loads and maintain stability in the system when faced with renewable uncertainty. It generates multiple wind power generation cases, comprising of normal variation and extraordinary conditions, rather than providing one deterministic projection. These probabilistic forecasts enable grid operators to come up with dispatch methods resistant to uncertainty. Generative AI can further optimise the utilisation of energy storage devices by simulating the time to charge or discharge batteries, reduction of curtailment, and optimal use of energy in general, (Necula, 2023). It can also model the interaction of demand patterns, wind and solar thus it is possible to combine the hybrid renewable energy systems in a more effective manner.

FRAMEWORK ARCHITECTURE

To augment the accuracy of forecasting, operational efficiency and energy reliability, a standard generative AI-powered wind energy system intertwines the data gathering, AI modelling, AI optimization, and decision support into a unified architecture. The framework can be arranged in the following layers:

a) Layer of Data Acquisition: It gathers historical and current data from a variety of sources, including IoT-enabled devices, weather stations, wind turbines, and satellite photography. Wind direction, speed, temperature, turbine vibration, maintenance records, and grid demand are examples of data kinds. use preprocessing, filtering, and normalizing to guarantee quality.

b) Layer of Synthetic Data Generation: It creates accurate synthetic wind data in situations where there are few genuine observations by using generative artificial intelligence (GANs, VAEs, and diffusion models). Enhances databases for uncommon fault scenarios and extreme weather occurrences. aids in resolving training dataset imbalance and data scarcity.
c) Layer of Forecasting and Simulation uses GenAI-powered models to anticipate wind energy throughout the near and long periods. features for simulating scenarios that forecast energy production in a range of climatic and meteorological circumstances. Multiple possibilities are provided by probabilistic forecasting in order to incorporate uncertainty.
d) Layer of Optimization uses insights produced by AI to optimize load balancing, predictive maintenance plans, and turbine location. predicts changes in supply and demand to assist in energy trading decisions. uses optimization algorithms and reinforcement learning to increase the efficiency of grid integration.
e) Layer of Decision Support and Visualization gives operators and policymakers access to dashboards and decision-support tools. Displays the results of optimization, anomaly detection, and predictions. allows for proactive measures to increase reliability and decrease downtime.
f) Layer of Integration and Control connects to cloud platforms, smart grids, and SCADA systems already in place. guarantees safe data transfer, scalability, and real-time changes. permits feedback loops in which AI models are continuously retrained using operational data.

Table 3. Framework architecture for generative AI-driven wind energy forecasting and optimization

Component	Role / Function	Techniques / Methods	Expected Outcome / Benefit
Data Acquisition	Collects historical and real-time wind, weather, and turbine operational data	Weather stations, SCADA, IoT sensors, APIs (Weather Underground, WindFinder)	Comprehensive and high-resolution datasets for model training and validation
Data Preprocessing	Cleans, normalizes, and augments data	Outlier removal, interpolation, normalization, generative data augmentation (GANs, diffusion)	Ensures high-quality input data, fills missing values, improves model robustness
Feature Engineering	Extracts meaningful features and patterns	Correlation analysis, PCA, wind speed statistics, turbulence metrics	Reduces dimensionality, highlights key predictors for forecasting and optimization

continued on following page

Table 3. Continued

Component	Role / Function	Techniques / Methods	Expected Outcome / Benefit
Generative AI Module	Generates synthetic data, probabilistic forecasts, and scenario simulations	GANs, VAEs, Diffusion Models, Physics-Informed Neural Networks	Improves forecasting accuracy, captures extreme events, and reduces uncertainty
Forecasting Module	Predicts short-term and long-term wind power	XGBoost, LSTM, hybrid ML-GAI models	Accurate predictions for energy dispatch, scheduling, and grid integration
Optimization Module	Optimizes turbine placement, maintenance schedules, and grid operations	Genetic Algorithms, Particle Swarm Optimization, scenario-based simulation	Maximizes energy yield, reduces downtime and costs, improves grid stability
Decision Support System	Visualizes results, supports operator decisions	Dashboards, simulation interfaces, alerts	Informed decision-making for energy managers and grid operators

As summarized in Table 1, the framework integrates data acquisition, preprocessing, feature engineering, generative AI modelling, forecasting, optimization, and decision support to enable intelligent wind energy forecasting and system optimization

APPLICATIONS OF GENERATIVE AI IN WIND ENERGY

Because it provides creative solutions in a variety of fields, generative AI is becoming more and more significant in the development of wind energy. One of the main uses is in wind resource forecasting, where artificial intelligence (AI) algorithms produce realistic synthetic data for wind direction and speed, increasing the precision of both short- and long-term forecasts. This improved forecasting ability guarantees better energy dispatch planning and assists grid operators in better managing variations in energy production. Generative AI simulates various topography and climatic conditions to provide ideal layouts in the crucial fields of turbine location and wind farm design optimization. This lowers installation costs, improves overall energy capture, and decreases wake losses. In addition, it is also generative AI that can be applied to the detection of the fault and predictive maintenance since it could be applied to process sensor data to approximate the likely malfunction of turbines and implement immediate remedies, reduce down time and increase the life of turbines. Moreover, the operators can develop a successful operating strategy and, using the assistance of the generative AI, use the wind energy most effectively within the power grid by generating extreme weather conditions such as storms or abrupt alterations in wind with the assistance of scenario simulations and risk analysis, (Ohalete et al., 2023; Onwusinkwue et al., 2024; Haq et al., 2025, Mosteiro-Sanchez et al., 2020; Haq, Shah, & Rufai, 2025). Together, the applications mentioned above enable generative AI to be an effective tool to increase the sustainability, reliability and efficiency of wind power.

Short-Term Forecasting

The generation AI is incredibly applicable in the short wind energy forecasting. It gives more accurate hourly forecasts of the wind direction and speed since it produces possibilities of weather at a variety of options. Traditional approaches would do away with uncertainty in the meteorological conditions, but our probabilistic-based forecasting strategy does. system operators can therefore predict the changes in energy production, load balancing, and system stability, (Ragab et al., 2021). In addition, accurate short-term predictions mean a more direct connection of wind energy to the electric grid and minimizes alternative sources of energy and general efficiency.

Long-Term Energy Planning

Generative AI is also very crucial in long-term energy planning because it can be used to model seasonal and annual wind patterns precisely. Generative AI assists within decision-making of energy planners, investors, and legislators by developing artificial datasets that mirror changes in the past and predicts the shifts of wind behaviour in the future. Such models ensure that the calculation of the potential of the long-term production of energy by specific sites is accurate and thus directs the investment plans of wind farm development and infrastructure planning. Also, Generative AI helps with the optimisation of resources allocation, future trends of energy supply, and optimal location of wind project sites based on their programmed seasonal variations. Given the fact that governments and other regulatory bodies rely on accurate long-term estimates when developing incentives, subsidies, and grid integration policies that can enhance the development of renewable energy, this competency is especially crucial in policy making, (Rajitha & Raghu Ram, 2024). There is also the use of long-term simulations that help in analysing the impact of climate change on the wind resources, which ensures the sustainability and resilience of the future energy systems.

Site Selection and Layout Optimization

Under the context of this project, there is a need to determine the site where the restaurant shall be situated as well as the details concerning the site layout design, site selection and layout optimization are very critical in the optimization of the energy produce and the ability of wind energy to be cost effective. Most of the most significant aspects of site assessment practices recurrently employed in the past were physical survey, historical weather data, and intricate simulation models that may prove to be somewhat costly and time consuming. Generative AI can solve these issues by incorporating bulk amounts of geographical and meteorological data, such as topographical features, land-use, wind speeds distribution, and atmospheric variables. With the existing stochastic AI models, it is possible to simulate various wind flow, turbulence, and wake effect conditions in various environments by developing realistic synthetic data. It does not presuppose that the on-site measurements should be comprehensive and enables energy resource developers to identify a variety of possible configurations, (Rugolo & Aziz, 2012).

Generative AI goes beyond the simple glance it is the dynamical simulation of the processes between turbines and the environment in various weather conditions. It is also able to determine the best positions and direction of the turbine to amplify the waste in the wind since it occurs when the turbines have hindered the wind flow of the other turbines and reduces efficiency. In addition, AI-based optimization is a compromise between various goals at once like maximization of energy capturing, minimization of conflicts between the land use, minimization of noise, and environmental sustainability. This holistic approach supports multi-criteria decision-making due to the existence of trade-offs between energy efficiency and ecological or social constraints that are usually typical in the real world, where wind farms are built. Besides, generative AI enhances cost-optimization by estimating the cost of construction, maintenance and transmission under various layouts. Using economic models with artificial environmental data, the developers can then proceed to select locations that can not only produce significant amount of energy but also provide financial viability. Finally, the generative AI will be able to provide engineers, investors, and politicians with the latest tools to create wind farms that are resistant, effective, and environment-friendly in order to accelerate the process of transitioning the world to renewable energy sources that are sustainable, (Sammar et al., 2024).

Grid Integration

The other field that generative AI is leaving a huge mark is grid integration where the incongruity between changeable supply and variable demand that is one of the main concerns about renewable energy is resolved. Due to its capability to reproduce the production of simulated electricity consumption patterns in the vast mix of seasons, the time of the day and the socioeconomic circumstances, generative AI may generate the artificial load-demand conditions. These imaginary cases are coupled with the regenerative generation patterns of solar and wind energy to calculate the degree of the capacity of the hybrid systems to satisfy the demand in the future, (Shao et al., 2015). This modelling may allow the operator of the power system to compare alternatives to intermittent supply of wind which are offered by solar generation with the use of energy storage technologies which lead to a more reliable combination of the latter. Generative AI as well as help to make probabilistic predictions of demand-supply imbalances that can be used to generate ingenious dispatch strategies, optimize storage utilization, and minimize renewable energy curtailment. It also provides utility companies and policymakers to develop grid modernization solutions, demand-response and flexible pricing solutions. Generative AI is also not as dependent on fossil fuel backup systems and the grid becomes more stable and predictable in adding renewable energy to it. This will reduce the levels of carbon release and will assist in offering a cleaner future of energy, (Soni, Dave, & Paliwal, 2023).

CASE STUDIES AND EMERGING RESEARCH

Case Study 1: Augmenting Wind Speed Data Using GANs

Notable examples of the use of generative artificial intelligence in wind energy include the use of Generative Adversarial Networks (GANs) to augment wind speed data. The limited historical data that conventional forecasting models are often limited to is a limiting factor of their ability to generalise to a variety of meteorological conditions. To overcome this challenge and to provide artificial data on wind speed in real-life scenarios, researchers trained GANs on artificial data of real wind speed. The forecasting algorithm was able to identify subtle patterns and variability that would otherwise go undetected by adding these superior synthetic samples to the training dataset. In comparison to baseline techniques that just used real-world observations, the enhanced model showed a 12% increase in short-term predicting accuracy. Grid operators will be most impacted by this innovation because real-time supply and demand balancing depends on short-term forecasts. Additionally, the case study demonstrates the wider potential of GAN-based augmentation in resolving data shortages, enhancing model resilience, and permitting more dependable integration of renewable energy sources into power networks, (Chaurasia & Haq, 2023). Such new research shows how generative AI can improve predictive capacities in operational situations and fill gaps in data availability.

Case Study 2: Offshore Wind Farms Using VAE-Driven Turbine Fault Simulation

In offshore wind energy systems, variational autoencoders, or VAEs, have lately become extremely effective instruments for simulating complicated and uncommon fault settings. Extreme weather, expensive maintenance, and restricted access are some of the particular operational difficulties faced by offshore plants. Early failure indicators are frequently difficult for traditional monitoring systems to identify, especially when there is a lack of prior data on uncommon failures. To surmount this, researchers developed a VAE-based model to learn condensed latent features of sensor measurements on turbines to generate counterfeit fault conditions. Using these artificial fault patterns in predictive maintenance models allowed the operators to evaluate the resilience of the system to various what-if scenarios. The results showed that fault detection during the early stages has been greatly enhanced as compared to its conventional means. More importantly, it enabled operators to plan the maintenance in a better manner through the assistance of predictive models which resulted in decrease in the emergency repair costs and the deployment of vessels. Using the VAE-based construction, the unscheduled downtime in offshore wind farms was reduced quantifiably and resulted into an increase in the energy generation, and the availability of turbines. This approach shows that the generative AI can enhance operational efficiency by introducing an advanced synthesis of data to existing Supervisory Control and Data Acquisition (SCADA) systems, (Choudhary, Haq, & Rather, 2023). Besides the reduction of downtime, the case study also points out that VAEs can build dependable digital twins of offshore assets that can increase their long-term reliability and reduce lifecycle expenses in large-scale wind energy projects.

Case Study 3: Diffusion Models in Simulating Severe Weather at the Coastal Wind Farms

The coastal wind farms may suffer operational disturbances and destruction of infrastructure in cases of extreme weather like cyclones, storm surges, and extreme rain. In the resilience planning case, it is necessary to model such infrequent and significant events. Nevertheless, traditional forecasting models can rarely be prepared to deal with such events as meteorological historical records of extreme conditions are often inadequate. Scholars bridged this gap by using diffusion models, a type of generative model that is trained to recreate more realistic weather extremes at coastal wind farm locations but uses random noise as its input. Historical data on meteorology had been provided with data on satellite images and climate reanalysis to train the diffusion-based models. They could reproduce the artificial extreme patterns of weather after the training that when it comes to the velocity of the wind, gradient of the pressure and the volume of the rainfall was comparable to the actual storm patterns. These virtual scenarios were further tested to assess the stability of the turbine structure, the stability of the grid integration and emergency response strategies in resilience planning units. The findings of the pilot project are that adding the results of the diffusion models has diversified the range of scenarios and accuracy of the stress-testing allowing operators to find the weaknesses in transmission systems and turbine foundations that are crucial before the occurrence of the actual accidents in the reality, (Rather, Haq, & Walia, 2025). In addition to the development of the risk assessment skills, this proactive strategy minimized the risks of disastrous failures.

CHALLENGES AND LIMITATIONS

Even though the potential of the generative AI in changing the way wind energy is used is enormous, several barriers and limitations must be eliminated to ensure successful adoption. One of the primary problems is the massive training cost of large-scale generative models. These models require large amounts of data, dedicated hardware such as GPUs or TPUs, and large amounts of energy, and if not managed correctly can partially undermine the benefits of sustainability. This difficulty makes deployment costly, particularly to smaller energy companies and research organizations that have few funds available.

Another major disadvantage is the difficulty of validation of synthetic data especially to unusual but important real-world events such as sudden wind decreases, storms, or disastrous turbine failures. Since the events are infrequent, the representative data sets are difficult to compose and the artificial data sets might not be entirely reflective of the underlying complexity that reduces the reliability of the model. This is strongly connected with the risk of overfitting or hallucinations, in which case the model is being trained on poor or biased data and generates unrealistic scenarios. This could lead to lack of accurate projections and poor decisions regarding energy operations and planning.

Moreover, the fact that generative models are not interpretable and transparent is the biggest barrier to usage. To engineers, operators, and policymakers, most of such models are black boxes, therefore, difficult to understand how outputs are generated. This lack of explicability of AI can destroy confidence in AI in the safety-critical sector, such as power grid regulation.

To make real-world applications of wind energy transparent and trustworthy, these limitations should be surmounted by the creation of computational efficiency, credible validation models, balanced training data, and understandable AI methods.

FUTURE PROSPECTS

Improving Reliability through Hybrid AI Models

The development of hybrid models that integrates the benefits of physics-informed neural networks (PINNs) and data-driven generative models is one of the possible ways of generative AI development in the field of wind energy. Although the generative AI models such as GANs, VAEs, and diffusion models will be good at generating synthetic data and modeling complex scenarios, they often do not have underlying domain knowledge of atmospheric physics, fluid mechanics, and structural mechanics. Their findings are statistically strong, but sometimes they fail to obey the physical principles of wind behaviour and turbine behaviour, (Ishaq, Haq, & Gamal, 2024). Incorporated into generative models, PINNs and basing the activity of AI on scientifically justified constraints, researchers are optimistic about the improvement of the consistency and validity of models. One of them would be a hybrid model which would provide the artificial profiles of wind velocity as well as guarantee the consistency of the data produced by the model with the laws of mass flow and energy conservation. In the same vein, the hybrid AI-based weather extremes and fault simulates can be tested against the turbine physics and, in such a manner, false forecasts are minimized.

Digital Twins

Digital twins, the virtual versions of real physical objects, that develop dynamically and mimic the actions of their real-life counterparts, are becoming manifold more popular in the wind energy research and operation. Real-time optimization that can be realized by the integration of generative AI models into the digital twins infrastructure cannot be obtained by traditional monitoring. In contrast to other models, AI-based digital twins continuously refresh their models with real-time data feeds with sensors, Supervisory Control and Data Acquisition (SCADA) systems and weather predictions to emulate the real operational condition of turbines and grid connections.

Generative AI reinforces this structure by simulating a range of operation conditions, including rare weather conditions, turbine malfunctions, and load changes. These artificial scenarios enable the operators to test the what-if scenarios by making predictions on how the system would perform in various situations without the risk of acting on such predictions and actually failure of the system. The decreased downtime with predictive maintenance, the highest possible power generation with adaptive control, and the enhanced stability of the grid with the simulation of demand response are only a few benefits of such intelligent digital twins integration. This plan will eventually lead to saving costs in the lifecycle by directing operational and design improvements. Digital twins based on AI will be used as important decision-support systems when wind farms become bigger and more complex, filling the gap between prediction, simulation, and real-time practical optimisation, (Haq & Kanday, 2025; Hassan et al., 2023; Rufai et al., 2025; Priyadarishi et al., 2024; Ashraf & Dua, 2024).

Quantum Generative Models

Large generative models training still remains computationally costly as wind energy systems further increase in complexity. Conventional AI approaches tend to have a problem with high-dimensional climate, sensor and grid integration datasets processed in real-time. To overcome these issues, scientists are starting to study quantum generative models (QGMs), that utilizes the physics of quantum computing to speed up the process of making predictions and simulating phenomena. This feature has enabled the QGMs to model the complex correlation in large datasets more effectively when used in the generative modelling compared to the traditional architectures. As an example, unusual weather patterns, turbine components failures, or grid instability could be simulated with the previously unheard of speed and accuracy with quantum-enhanced GANs and VAEs. Moreover, this accelerated processing saves on training time besides the variety of synthetic data produced, which is essential in the testing of the systems of wind energy in severe and unpredictable conditions.

New research indicates that the applications of hybrid quantum-classical models, in their current state of experiment, can revolutionize the resilience planning, forecasting, and optimization of renewable energy. The long-term perspective of QGMs is that they might allow the creation of real-time digital twins of fully operational wind farms and make near real-time decisions to allow adaptive integration into the grid, load balancing, and predictive maintenance, (Sun et al., 2024; Nam, Hwangbo, & Yoo, 2020; Singh & Kumar, 2023; Fu et al., 2021; Case, 2016).

Interoperability with the IoT and Edge AI

Combination of generative models and IoT based sensors with Edge AI represents one of the viable solutions to optimisation of the decentralized wind energy. Using edge locations, lightweight generative models are able to process environment and turbine data in real-time and support local decision-making and faster predictions compared to centralized cloud systems by itself. This minimizes the delay in predicting the wind variations, early failures and real time modification of the turbine operations. In addition, edge deployment improves resiliency by maintaining predictive ability even in predictably unconnected situations. This integration will assist in the responsiveness, efficiency, and reliability of operations through assistance of scaled-up self-sufficient wind farms, (Qiao et al., 2024).

CONCLUSION

Generative artificial intelligence (AI) has turned out to be a game-changer technology in the wind energy industry with the potential to entirely transform forecasting, optimization, and decision-making. The innovative solutions to the limitations of the conventional models are generative systems such as GANs, VAEs, or diffusion models, which can address the weaknesses of the conventional methods due to the inadequacy of data, excessive weather variability, and the complexity of turbine dynamics. GenAI-based frameworks enhancing the robustness and flexibility of forecasting and maintenance methods, simulating unusual weather conditions, generating realistic synthetic data, and predicting turbine issues. Cutting down the downtime and bringing more efficiency to the overall operations, the combination of generative AI and digital twins, IoT-powered sensors, and edge AI platforms improves real-time monitoring and control. Also, exploring hybrid solutions such as combining generative models with neural networks that have been affected by physics can ensure that AI-based insights remain consistent with physical principles, increasing the reliability and trust of the operator. With the advancement of the future, the development of the quantum generative models will help the process of making significant simulations and analysing the situation in the wind farms much faster, and the prospects of optimizing the wind farms in the adaptation and in real-time will become possible.

Generative AI is beneficial in helping to convert to sustainable energy and long-term climate resilience since it allows the wind power systems to be smarter, stronger, and more affordable. The structures of GenAI can turn out to be the major facilitators of the next-generation wind energy, and the trade-off between the advanced computing and the energy sustainability can be nurtured, as the demand of the renewable energy sources is only growing.

REFRENCES

Adewumi, A.. (2024). Reviewing the impact of AI on renewable energy efficiency and management. *International Journal of Science and Research Archive, 11*(1), 1518–1527. DOI: 10.30574/ijsra.2024.11.1.0245

Albarakati, A. J.. (2021). Real-time energy management for DC microgrids using artificial intelligence. *Energies, 14*(17), 5307. Advance online publication. DOI: 10.3390/en14175307

Ashraf, W. M., & Dua, V. (2024). Data information integrated neural network (DINN) algorithm for modelling and interpretation performance analysis for energy systems. *Energy and AI, 16*, 100363. Advance online publication. DOI: 10.1016/j.egyai.2024.100363

Cacciamani, G. E.. (2023). PRISMA AI reporting guidelines for systematic reviews and meta-analyses on AI in healthcare. *Nature Medicine, 29*(1), 14–15. DOI: 10.1038/s41591-022-02139-w

Camacho, J. de J.. (2024). Leveraging artificial intelligence to bolster the energy sector in smart cities: A literature review. *Energies, 17*(2), 353. Advance online publication. DOI: 10.3390/en17020353

Case, D. U. (2016). *Analysis of the cyber attack on the Ukrainian power grid*. Electricity Information Sharing and Analysis Center (E-ISAC).

Chaurasia, A., & Haq, I. U. (2023). Housing price prediction model using machine learning. In *2023 International Conference on Sustainable Emerging Innovations in Engineering and Technology (ICSEIET)* (pp. 497–500). IEEE. https://doi.org/ DOI: 10.1109/ICSEIET58677.2023.10303514

Choudhary, C., Haq, I. U., & Rather, A. H. (2023). Utilizing dynamic QR codes to enhance secure payment transactions: An approach to secure computer-based transactions. In *2023 International Conference on Communication, Security and Artificial Intelligence (ICCSAI)* (pp. 552–557). IEEE. https://doi.org/DOI: 10.1109/ICCSAI59793.2023.10420899

Choudhary, C., Thakur, J., Bhardwaj, H., Haq, I. U., & Rather, A. H. (2023). Revolutionizing wind energy: A enhanced machine learning perspective on wind turbine's fault prediction. In *2023 2nd International Conference on Automation, Computing and Renewable Systems (ICACRS)* (pp. 1763–1769). IEEE. https://doi.org/DOI: 10.1109/ICACRS58579.2023.10405090

Donti, P. L., & Kolter, J. Z. (2021). Machine learning for sustainable energy systems. *Annual Review of Environment and Resources*, *46*, 719–747. DOI: 10.1146/annurev-environ-020420-061831

Elimam, M., Isbeih, Y. J., Azman, S. K., Moursi, M. S. E., & Hosani, K. A. (2023). Deep learning-based PMU cyber security scheme against data manipulation attacks with WADC application. *IEEE Transactions on Power Systems*, *38*(3), 2148–2161. DOI: 10.1109/TPWRS.2022.3171109

Etukudoh, E. A.. (2024). Electrical engineering in renewable energy systems: A review of design and integration challenges. *Engineering Science & Technology Journal*, *5*(1), 231–244. DOI: 10.51594/estj.v5i1.746

Fu, X.. (2022). Planning of distributed renewable energy systems under uncertainty based on statistical machine learning. *Protection and Control of Modern Power Systems*, *7*(1), 41. Advance online publication. DOI: 10.1186/s41601-022-00262-x

Fu, Y., O'Neill, Z., Yang, Z., Adetola, V., Wen, J., Ren, L., Wagner, T., Zhu, Q., & Wu, T. (2021). Modeling and evaluation of cyber-attacks on grid-interactive efficient buildings. *Applied Energy*, *303*, 117639. Advance online publication. DOI: 10.1016/j.apenergy.2021.117639

Garmabdari, R., …. (2019). Optimal power flow scheduling of distributed microgrid systems considering backup generators. In *2019 9th International Conference on Power and Energy Systems (ICPES)* (pp. 1–6). IEEE. https://doi.org/DOI: 10.1109/ICPES47639.2019.9105372

Gawusu, S.. (2022). The dynamics of green supply chain management within the framework of renewable energy. *International Journal of Energy Research*, *46*(2), 684–711. DOI: 10.1002/er.7278

Green, I. A.. (2023). Technological advances of ammonia as energy storage solution. *Global Journal of Engineering and Technology Advances*, *16*(3), 151–155. DOI: 10.30574/gjeta.2023.16.3.0185

Hamdan, A.. (2024). AI in renewable energy: A review of predictive maintenance and energy optimization. *International Journal of Science and Research Archive*, *11*(1), 718–729. DOI: 10.30574/ijsra.2024.11.1.0112

Haq, I. U. (2024). Diagnosis of neurological disease using bioinspired algorithms. In *Bio-inspired optimization for medical data mining* (pp. 227–268). Wiley., DOI: 10.1002/9781394211791.ch10

Haq, I. U., Kaur, G., & Rather, A. H. (2024). Fruit leaf classification using transfer learning for automation and industrial applications. In *Deep learning techniques for automation and industrial applications* (pp. 151–178). Wiley., DOI: 10.1002/9781394234271.ch7

Haq, I. U., & Khanday, A. M. U. D. (2025). Exploring the role of explainable AI in women's health: Challenges and solutions. In *Explainable artificial intelligence in the healthcare industry* (pp. 283–306). IGI Global., DOI: 10.4018/979-8-3693-9487-8.ch013

Haq, I. U., Malhotra, J., Rawat, V., Kumari, J., & Kaur, G. (2025). Machine learning approaches for early detection of cervical cancer. In *Targeted chemotherapy with personalized immunotherapy* (pp. 377–401). Wiley., DOI: 10.1002/9781394270613.ch19

Haq, I. U., Nandal, V., & Uppal, H. (2025). Blockchain applications in aviation: Securing transactions, streamlining operations, and improving passenger experience. In *Blockchain in the tourism industry: A new era of secure and transparent travel solutions* (pp. 141–164). Springer., DOI: 10.1007/978-3-031-95341-5_7

Haq, I. U., Rather, A. H., & Kaur, G. (2023). A comparative analysis of machine learning algorithms for the early prediction of cardiovascular disease. In *2023 2nd International Conference on Edge Computing and Applications (ICECAA)* (pp. 987–993). IEEE. https://doi.org/DOI: 10.1109/ICECAA58104.2023.10244119

Haq, I. U., Rather, A. H., Rufai, S. Z., & Shah, A. Sheetal, & Khanday, A. M. U. D. (2025). Transparency in disease diagnosis: Leveraging interpretable machine learning in healthcare. In *Explainable artificial intelligence in the healthcare industry* (pp. 105–130). IGI Global. https://doi.org/DOI: 10.4018/979-8-3693-9487-8.ch005

Haq, I. U., Shah, H. A., & Rufai, S. Z. (2025). Enhancing water security through automation: Case studies and technical advancements in water quality management. In *Computational automation for water security* (pp. 337–362). Elsevier., DOI: 10.1016/B978-0-443-33122-0.00018-8

Hassan, I. U., Shah, H. A., Hafiz, A., & Haq, I. U. (2023). Novel technique for malicious node detection in wireless sensor networks. In *Recent trends in computational sciences* (pp. 231–237). CRC Press., DOI: 10.1201/9781003362586-25

Huang, B., & Wang, J. (2023). Applications of physics-informed neural networks in power systems – A review. *IEEE Transactions on Power Systems*, *38*(1), 572–588. DOI: 10.1109/TPWRS.2022.3162473

Ipakchi, A., & Albuyeh, F. (2009). Grid of the future. *IEEE Power & Energy Magazine*, *7*(2), 52–62. DOI: 10.1109/MPE.2008.931384

Ishaq, M. M., Haq, I. U., & Gamal, A. (2024). Decentralized quantum machine learning: Distributed quantum computing for enhanced learning. In *Quantum machine learning: Quantum algorithms and neural networks* (pp. 149–168). De Gruyter., DOI: 10.1515/9783111342276-008

Islam, A., & Othman, F. (2024). Renewable energy microgrid power forecasting: AI techniques with environmental perspective [Preprint]. *Research Square*. https://doi.org/DOI: 10.21203/rs.3.rs-4260337/v1

Jiang, R., Liu, H., & Peng, H. (2024). An optimal configuration method of energy storage system considering source-load matching. *Journal of Physics: Conference Series*, *2788*(1), 012018. Advance online publication. DOI: 10.1088/1742-6596/2788/1/012018

Kalair, A.. (2021). Role of energy storage systems in energy transition from fossil fuels to renewables. *Energy Storage*, *3*(1), e135. Advance online publication. DOI: 10.1002/est2.135

Latrach, A.. (2024). A critical review of physics-informed machine learning applications in subsurface energy systems. *Geoenergy Science and Engineering*, *239*, 212938. Advance online publication. DOI: 10.1016/j.geoen.2024.212938

Liu, Z.. (2025). A comprehensive review of wind power prediction based on machine learning: Models, applications, and challenges. *Energies*, *18*(2), 350. Advance online publication. DOI: 10.3390/en18020350

Machina, S. P. C., Koduru, S. S., & Madichetty, S. (2022). Solar energy forecasting using deep learning techniques. In *2022 2nd International Conference on Power Electronics and IoT Applications in Renewable Energy and its Control (PARC)* (pp. 1–5). IEEE. https://doi.org/DOI: 10.1109/PARC52418.2022.9756605

Mansouri, S. A., Jordehi, A. R., Marzband, M., Tostado-Véliz, M., Jurado, F., & Aguado, J. A. (2023). An IoT-enabled hierarchical decentralized framework for multi-energy microgrids market management in the presence of smart prosumers using a deep learning-based forecaster. *Applied Energy*, *333*, 120560. Advance online publication. DOI: 10.1016/j.apenergy.2022.120560

Mauro, G. (2024). The new power couple: Artificial intelligence and renewable energy. *Journal of Strategic Innovation and Sustainability*, *19*(3). Advance online publication. DOI: 10.33423/jsis.v19i3.7374

Mosteiro-Sanchez, A., Barcelo, M., Astorga, J., & Urbieta, A. (2020). Securing IIoT using defence-in-depth: Towards an end-to-end secure Industry 4.0. *Journal of Manufacturing Systems*, *57*, 367–378. DOI: 10.1016/j.jmsy.2020.10.013

Mysore, S. (2024). Role of artificial intelligence in grid modernization: Exploring how AI can enhance grid management, predict energy demand, and optimize renewable energy usage [Preprint]. *International Research Journal of Modernization in Engineering Technology and Science*. Advance online publication. DOI: 10.56726/IRJMETS48452

Nam, K., Hwangbo, S., & Yoo, C. (2020). A deep learning-based forecasting model for renewable energy scenarios to guide sustainable energy policy: A case study of Korea. *Renewable & Sustainable Energy Reviews*, *122*, 109725. Advance online publication. DOI: 10.1016/j.rser.2020.109725

Necula, S. C. (2023). Assessing the potential of artificial intelligence in advancing clean energy technologies in Europe: A systematic review. *Energies*, *16*(22), 7633. Advance online publication. DOI: 10.3390/en16227633

Ohalete, N. C.. (2023). AI-driven solutions in renewable energy: A review of data science applications in solar and wind energy optimization. *World Journal of Advanced Research and Reviews*, *20*(3), 401–417. DOI: 10.30574/wjarr.2023.20.3.2433

Onwusinkwue, S.. (2024). Artificial intelligence (AI) in renewable energy: A review of predictive maintenance and energy optimization. *World Journal of Advanced Research and Reviews*, *21*(1), 2487–2499. DOI: 10.30574/wjarr.2024.21.1.0347

Qiao, Y., Chen, D., Sun, Q. Z., Tian, G., & Wang, W. (2024). Unveiling stealthy man-in-the-middle cyber-attacks on energy performance in grid-interactive smart buildings. *Energy Conversion and Management*, *319*, 118949. Advance online publication. DOI: 10.1016/j.enconman.2024.118949

Ragab, R.. (2021). Optimized hybrid renewable energy system for a baseload plant. *Energy Procedia*, *17*. Advance online publication. DOI: 10.46855/energy-proceedings-8674

Rajitha, M., & Raghu Ram, A. (2024). An overview of artificial intelligence applications to electrical power systems and DC microgrids. *E3S Web of Conferences*, *547*, Article 01002. https://doi.org/DOI: 10.1051/e3sconf/202454701002

Rather, A. H., & Haq, I. U. (2023). Intelligent framework for early prediction of diabetic retinopathy: A deep learning approach. In *Proceedings of the International Conference on Paradigms of Communication, Computing and Data Analytics* (pp. 377–388). Springer. https://doi.org/DOI: 10.1007/978-981-99-9626-1_30

Rather, A. H., Haq, I. U., & Walia, N. K. (2025). Privacy and security in the metaverse. In *Leveraging metaverse and analytics of things (AoT) in medical systems* (pp. 189–207). Academic Press., DOI: 10.1016/B978-0-443-24048-5.00012-8

Rufai, S. Z., Haq, I. U., Shah, H. A., & Fayaz, M. A. (2025). Neural network approaches for intelligent decision-making in automation. In *Handbook of intelligent automation systems using computer vision and artificial intelligence* (pp. 247–265). CRC Press.

Rugolo, J., & Aziz, M. J. (2012). Electricity storage for intermittent renewable sources. *Energy & Environmental Science*, *5*(5), 7151–7160. DOI: 10.1039/c2ee02542f

Sammar, M. J.. (2024). Illuminating the future: A comprehensive review of AI-based solar irradiance prediction models. *IEEE Access : Practical Innovations, Open Solutions*, *12*, 114394–114415. DOI: 10.1109/ACCESS.2024.3402096

Shao, H.. (2015). Rolling bearing fault diagnosis using an optimization deep belief network. *Measurement Science & Technology*, *26*(11), 115002. Advance online publication. DOI: 10.1088/0957-0233/26/11/115002

Singh, B., & Kumar, A. (2023). Optimal energy management and feasibility analysis of hybrid renewable energy sources with BESS and impact of electric vehicle load with demand response program. *Energy*, *278*, 127867. Advance online publication. DOI: 10.1016/j.energy.2023.127867

Soni, P., Dave, V., & Paliwal, H. (2023). Artificial intelligence-enabled techno-economic analysis and optimization of grid-tied solar PV-fuel cell hybrid power systems for enhanced performance. In *Data science and intelligent computing techniques* (pp. 781–794). JPS Scientific Publications., DOI: 10.56155/978-81-955020-2-8-67

Sun, Q., Ma, H., Zhao, T., Xin, Y., & Chen, Q. (2024). Break down the decentralization-security-privacy trilemma in management of distributed energy systems. *Nature Communications*, *15*, 3753. Advance online publication. DOI: 10.1038/s41467-024-48409-2

Compilation of References

Abasian, F., Rönnqvist, M., & Ouhimmou, M. (2019). Forest bioenergy network design under market uncertainty. *Energy*, *188*, 116038.

Abbasi, E., Etemadi, H., Smoak, J. M., Rousta, I., Olafsson, H., Baranowski, P., & Krzyszczak, J. (2021). Investigation of atmospheric conditions associated with a storm surge in the south-west of Iran. *Atmosphere*, *12*(11), 1429. Advance online publication. DOI: 10.3390/atmos12111429

Abbasi, M., Pishvaee, M. S., & Mohseni, S. (2021). Third-generation biofuel supply chain: A comprehensive review and future research directions. *Journal of Cleaner Production*, *323*, 12910.

Abdel-Nasser, M., Mahmoud, K., & Lehtonen, M. (2021). Reliable solar irradiance forecasting approach based on Choquet integral and deep LSTMs. *IEEE Transactions on Industrial Informatics*, *17*(3), 1873–1881.

Abu-Abed, F., & Khabarov, A. (2018). Life-support model of supply systems of oil and gas industry objects. *E3S Web of Conferences, 33*, 01028. https://doi.org/DOI: 10.1051/e3sconf/20183301028

Achi, A., Bakpo, F., & Umar, Z. (2022). Real-time petrol availability reporting system (RPARS) for Nsukka town, Nigeria. *Computer Science & IT Research Journal*, *3*(1), 23–35. DOI: 10.51594/csitrj.v3i1.294

Adewumi, A.. (2024). Reviewing the impact of AI on renewable energy efficiency and management. *International Journal of Science and Research Archive*, *11*(1), 1518–1527. DOI: 10.30574/ijsra.2024.11.1.0245

Agency, T. I. E. (2023). The international energy agency's electricity market report https://iea.blob.core.windows.net/assets/255e9cba-da84-4681-8c1f-458ca1a3d9ca/ElectricityMarketReport2023.pdf. Accessed Apr 2023

Aghahadi, M., Bosisio, A., Merlo, M., Berizzi, A., Pegoiani, A., & Forciniti, S. (2024). Digitalization Processes in Distribution Grids: A Comprehensive Review of Strategies and Challenges. *Applied Sciences (Basel, Switzerland)*, *14*(11), 4528. DOI: 10.3390/app14114528

Agoro, H., & Daniel, B. (2025). Generative AI in Creating Dynamic Content for IoT Devices.

Ahmed, N., Areche, F. O., Sheikh, A. A., & Lahiani, A. (2022). Green finance and green energy nexus in ASEAN countries: A bootstrap panel causality test. *Energies*, *15*(14), 5068.

Akpuokwe, Ch. U., Chikwe, Ch. F. & Eneh, N. E. (2024). Leveraging technology and financial literacy for women's empowerment in SMEs: A conceptual framework for sustainable development, *Glob. J. Eng. Technol. Adv.*, vol. 18, no. 3, pp. 020–032, 2024, doi: .DOI: 10.30574/gjeta.2024.18.3.0041

Alam, M. S., & Arefifar, S. A. (2019). Energy Management in Power Distribution Systems: Review, Classification, Limitations and Challenges. *IEEE Access : Practical Innovations, Open Solutions*, *7*, 92979–93001. DOI: 10.1109/ACCESS.2019.2927303

Albarakati, A. J.. (2021). Real-time energy management for DC microgrids using artificial intelligence. *Energies*, *14*(17), 5307. Advance online publication. DOI: 10.3390/en14175307

Alessandrini, S., & McCandless, T. (2020). The Schaake Shuffle technique to combine solar and wind power probabilistic Forecasting. *Energies*, *13*(10), 2503.

Alharbi, F. R., & Csala, D. (2021). Wind speed and solar irradiance prediction using a bidirectional long short-term memory model based on neural networks. *Energies*, *14*(20), 6501.

Ali, S. S., & Choi, B. J. (2020). State-of-the-art artificial intelligence techniques for distributed smart grids: A review. *Electronics (Basel)*. Advance online publication. DOI: 10.3390/electronics9061030

Aliyu, S., Audu, W. M., Okwori, M., Saidu, M., Abdullahi, U., & Eneze, J. (2018). An enhanced BRAZ (E-BRAZ) Android app for monitoring PMS in gas stations. *Journal of Applied Science & Environmental Management*, *22*(9), 1497–1502.

Alizadegan, H., Rashidi Malki, B., Radmehr, A., Karimi, H., & Ilani, M. A. (2025). Comparative study of long short-term memory (LSTM), bidirectional LSTM, and traditional machine learning approaches for energy consumption prediction. *Energy Exploration & Exploitation, 43*(1), 281–301. DOI: 10.1177/01445987241269496/ASSET/F0AB838A-2611-4F07-9B1F-1D2E7896E016/ASSETS/IMAGES/LARGE/10.1177_01445987241269496-FIG13.JPG

Al-Mutairi, H., & Chen, Y. (2024). Adaptive energy pricing using reinforcement learning: A case study in Saudi Arabia. *Energy AI, 15*, 100234. DOI: 10.1016/j.egyai.2024.100234

Alqahtani, H., Kavakli-Thorne, M., & Kumar, G. (2021). Applications of Generative Adversarial Networks (GANs): An Updated Review. *Archives of Computational Methods in Engineering, 28*(2). Advance online publication. DOI: 10.1007/s11831-019-09388-y

Alshantti, A., Rasheed, A., & Westad, F. (2025). Privacy Re-Identification Attacks on Tabular GANs. *Security and Privacy, 8*(1), e469. DOI: 10.1002/SPY2.469

Alsmadi, A. A., Al-Okaily, M., Alrawashdeh, N., & Al-Gasaymeh, A., Moh'd Al-hazimeh, A., & Zakari, A. (2023). A bibliometric analysis of green bonds and sustainable green energy: Evidence from the last fifteen years (2007–2022). *Sustainability, 15*(7), 5778.

Anayo, O. H., & Nwogbaga, N. E. (2016). Critical review of petrol station management system with emphasis on the advantages if digitalized in Nigeria. *International Journal of Novel Research in Engineering & Pharmaceutical Sciences, 3*(5), 49–56.

Andersen, K., Nielsen, P., & Larsen, M. (2023). Offshore wind turbine condition monitoring: Challenges and solutions. *Journal of Offshore Wind Engineering, 15*(3), 234–251.

Anderson, A., & Robinson, D. (2022, November). Financial Literacy in the Age of Green Investment. *Review of Finance, 26*(6), 1551–1584. DOI: 10.1093/rof/rfab031

Arrar, S., & Xioaning, L. (2022). Energy management in hybrid microgrid using artificial neural network, PID, and fuzzy logic controllers. *European Journal of Electrical Engineering and Computer Science, 6*(2), 38–47.

Ashraf, W. M., & Dua, V. (2024). Data information integrated neural network (DINN) algorithm for modelling and interpretation performance analysis for energy systems. *Energy and AI, 16*, 100363. Advance online publication. DOI: 10.1016/j.egyai.2024.100363

Asperti, A., Evangelista, D., & Loli Piccolomini, E. (2021). A Survey on Variational Autoencoders from a Green AI Perspective. *SN Computer Science*, *2*(4). Advance online publication. DOI: 10.1007/s42979-021-00702-9

Assaf, A. M., Haron, H., Abdull Hamed, H. N., Ghaleb, F. A., Dalam, M. E., & Elfadil Eisa, T. A. (2023). Improving solar radiation forecasting utilizing data augmentation model generative adversarial networks with convolutional support vector machine (GAN-CSVR). *Applied Sciences (Basel, Switzerland)*, *13*(23), 12768.

Athavale, V. A., Selvakumar, P., Poorani, S., Anand, T., & Rathore, D. S. (2025). Designing Chatbots for Educational Leadership. In Musundire, A. (Ed.), *Chatbots in Educational Leadership and Management* (pp. 181–204). IGI Global Scientific Publishing., DOI: 10.4018/979-8-3693-8734-4.ch008

Atic, S., Parlakyıldız, S., & Ilcihan, Z. (2015). Smart Meter and Turkey. *Bitlis Eren University Journal of Science and Technology*, *5*(2). Advance online publication. DOI: 10.17678/beujst.15889

Attipoe, V., Chukwuma-Eke, E. C., Lawal, C. I., Friday, S. C., Isibor, N. J. & Akintobi, A. O. (2025). The Role of Financial Institutions in Promoting Sustainable Energy Practices for SMEs: A Strategic Approach. *International Journal of Scientific Research in Computer Science, Engineering and Information*: 11 (2) Number 3729-3742 https://doi.org/DOI: 10.32628/CSEIT25112849

Ayokunle, A., Misra, S., Oluranti, J., & Ahuja, R. (2022). *Technical Losses (Tl) and Non-technical Losses (NTL) in Nigeria* (pp. 147–159). doi:DOI: 10.1007/978-3-030-86223-7_14

Babazadeh, R., Ghaderi, H., & Pishvaee, M. S. (2019). A benders-local branching algorithm for second-generation biodiesel supply chain network design under epistemic uncertainty. *Computers & Chemical Engineering*, *124*, 364–380.

Babazadeh, R., Razmi, J., Pishvaee, M. S., & Rabbani, M. (2017). A sustainable second-generation biodiesel supply chain network design problem under risk. *Omega*, *66*, 258–277.

Baesmat, K. H. (2024). Impedance analysis of adaptive distance relays using machine learning. In S. Latifi (Ed.), *ITNG 2024: 21st International Conference on Information Technology–New Generations. Advances in Intelligent Systems and Computing* (Vol. 1456). Springer. https://doi.org/DOI: 10.1007/978-3-031-56599-1_57

Baesmat, K. H., & Latifi, S. (2023). A new hybrid method for electrical load forecasting based on deviation correction and MRMRMS. In Selvaraj, H., Chmaj, G., & Zydek, D. (Eds.), *Advances in systems engineering* (pp. 293–303). Springer Nature Switzerland., DOI: 10.1007/978-3-031-40579-2_29

Baesmat, K. H., Masoudipour, I., & Samet, H. (2021). Improving the performance of short term load forecast using a hybrid artificial neural network and artificial bee colony algorithm. *IEEE Canadian Journal of Electrical and Computer Engineering*, *44*(3), 275–282. DOI: 10.1109/ICJECE.2021.3056125

Baesmat, K. H., Regentova, E. E., & Baghzouz, Y. (2025). A hybrid machine learning–statistical based method for short term energy consumption prediction in residential buildings. *Energy and AI*, *17*, 100552. Advance online publication. DOI: 10.1016/j.egyai.2025.100552

Baesmat, K. H., Shokoohi, F., & Farrokhi, Z. (2025). SP RF ARIMA: A sparse random forest and ARIMA hybrid model for electric load forecasting. *Global Energy Interconnection*, *8*(3), 486–496. DOI: 10.1016/j.gloei.2025.04.003

Bai, X., Liu, L., Ju, J., Zhong, X., Zhou, Y., & Lin, J.. (2022). Distributed optimization method for multi-area integrated energy systems considering demand response. *Frontiers in Energy Research*, *10*, 975214. DOI: 10.3389/fenrg.2022.975214

Barbieri, F., Rajakaruna, S., & Ghosh, A. (2017). Very short-term photovoltaic power forecasting with cloud modeling: A review. *Renewable & Sustainable Energy Reviews*, *75*, 242–263.

Bedi, J., & Toshniwal, D. (2018). Empirical mode decomposition based deep learning for electricity demand forecasting. *IEEE Access : Practical Innovations, Open Solutions*, *6*, 49144–49156. DOI: 10.1109/ACCESS.2018.2867681

Bengesi, S., El-Sayed, H., Sarker, M. K., Houkpati, Y., Irungu, J., & Oladunni, T. (2024). *Advancements in Generative AI: A Comprehensive Review of GANs, GPT, Autoencoders, Diffusion Model, and Transformers*. IEEE., DOI: 10.1109/ACCESS.2024.3397775

Beny, B., Wendy, W., Saleh, M., & Giriati, G. (2023). The effect of financial literacy and green innovation technology on green economic sustainability in emerging countries. *International Journal of Data and Network Science*, *7*(4), 1829–1838.

Berney, M., Ouaazki, A., Macko, V., Kocher, B., & Holzer, A. (2024, May). Care-based Eco-feedback augmented with generative AI: Fostering pro-environmental behavior through emotional attachment. In *Proceedings of the 2024 CHI Conference on Human Factors in Computing Systems* (pp. 1-15).

Beshley, M., Kryvinska, N., Seliuchenko, M., Beshley, H., Shakshuki, E. M., & Yasar, A.-U.-H. (2020). End-to-end QoS "smart queue" management algorithms and traffic prioritization mechanisms for narrow-band Internet of Things services in 4G/5G networks. *Sensors (Basel), 20*(8), 2324. DOI: 10.3390/s20082324 PMID: 32325795

Bhatnagar, M., Taneja, S., & Özen, E. (2022). A wave of green start-ups in India—The study of green finance as a support system for sustainable entrepreneurship [J]. *Green Finance, 4*(2), 253–273. DOI: 10.3934/GF.2022012

Bhutta, A. I., Ullah, M. R., Sultan, J., Riaz, A., & Sheikh, M. F. (2022). Impact of Green Energy Production, Green Innovation, Financial Development on Environment Quality: A Role of Country Governance in Pakistan, *International Journal of Energy Economics and Policy. Econjournals, 12*(1), 316–326.

Biswas, A., Md Abdullah Al, N., Imran, A., Sejuty, A. T., Fairooz, F., Puppala, S., & Talukder, S. (2023). Generative Adversarial Networks for Data Augmentation. In *Data Driven Approaches on Medical Imaging*. DOI: 10.1007/978-3-031-47772-0_8

Bloomberg, N. E. F. (2024). *Latin America clean energy investment outlook.* Bloomberg New Energy Finance.

Böcking, L., Michaelis, A., Schäfermeier, B., Baier, A., Kühl, N., Körner, M. F., & Nolting, L. (2024). Generative Artificial Intelligence in the energy sector.

Borgia, E. (2014). The Internet of Things vision: Key features, applications and open issues. *Computer Communications, 54*, 1–31. DOI: 10.1016/j.comcom.2014.09.008

Bose, B. K. (2017). Artificial intelligence techniques in smart grid and renewable energy systems—Some example applications. *Proceedings of the IEEE, 105*(11), 2262–2273. DOI: 10.1109/JPROC.2017.2756596

Boulaguiem, Y., Zscheischler, J., Vignotto, E., van der Wiel, K., & Engelke, S. (2022). Modeling and simulating spatial extremes by combining extreme value theory with generative adversarial networks. *Environmental Data Science, 1*, e5. Advance online publication. DOI: 10.1017/eds.2022.4

Bracale, A., Carpinelli, G., & De De Falco, P. (2019). Developing and comparing different strategies for combining probabilistic photovoltaic power forecasts in an ensemble method. *Energies, 12*(6), 1011.

Broomandi, P., Bagheri, M., Fard, A. M., Fathian, A., Abdoli, M., Roshani, A., Shafiei, S., Leuchner, M., & Kim, J. R. (2025). Energy Generation and Carbon Footprint under Future Projections (2022–2100) of Central Asian Temperature Extremes. *Global Challenges (Hoboken, NJ)*, *9*(5), 2400356. DOI: 10.1002/GCH2.202400356;SUBPAGE:STRING:FULL PMID: 40352636

Cacciamani, G. E.. (2023). PRISMA AI reporting guidelines for systematic reviews and meta-analyses on AI in healthcare. *Nature Medicine*, *29*(1), 14–15. DOI: 10.1038/s41591-022-02139-w

Cacuci, D., & Fang, R. (2016). Predictive modeling of a paradigm mechanical cooling tower: I. adjoint sensitivity model. *Energies*, *9*(9), 718.

Camacho, J. de J.. (2024). Leveraging artificial intelligence to bolster the energy sector in smart cities: A literature review. *Energies*, *17*(2), 353. Advance online publication. DOI: 10.3390/en17020353

Cao, H., Tan, C., Gao, Z., Xu, Y., Chen, G., Heng, P. A., & Li, S. Z. (2024). A Survey on Generative Diffusion Models. *IEEE Transactions on Knowledge and Data Engineering*, *36*(7). Advance online publication. DOI: 10.1109/TKDE.2024.3361474

Carr, D., & Thomson, M. (2022). Non-Technical Electricity Losses. *Energies*, *15*(6), 2218. DOI: 10.3390/en15062218

Case, D. U. (2016). *Analysis of the cyber attack on the Ukrainian power grid*. Electricity Information Sharing and Analysis Center (E-ISAC).

Cavus, M. (2025). Advancing power systems with renewable energy and intelligent technologies: A comprehensive review on grid transformation and integration. *Electronics (Basel)*, *14*(6), 1159. Advance online publication. DOI: 10.3390/electronics14061159

Chakraborty, T., Reddy, K. S., Naik, S. M., Panja, M., & Manvitha, B. (2023). Ten years of generative adversarial nets (GANs): A survey of the state-of-the-art. *arXiv*. https://arxiv.org/abs/2308.16316

Chandel, P., Thakur, T., Sawle, B. A., & Sharma, R. (2016). Power theft: Major cause of non technical losses in Indian distribution sector. *2016 IEEE 7th Power India International Conference (PIICON)*, 1–6. doi:DOI: 10.1109/POWERI.2016.8077253

Chan, H. L., & Siqin, T. (2025). Supply Chain Management with Generative Artificial Intelligence and Internet of Behaviours. *IEEE Transactions on Engineering Management*.

Chauhan, N., Mohan, R., Honnali, P., Gokilavani, S., & Biradar, M. (2025). *To Study the Impact of Government Subsidies, Financial Literacy, and Policy Awareness in Consumer Adoption of Green Finance with the Mediating Effect of Perceived Financial Benefits—SEM Analysis. Sustainable Digital Technology and Ethics in an Ever-Changing Environment. Studies in Systems, Decision and Control* (Vol. 237). Springer., DOI: 10.1007/978-3-031-86708-8_10

Chaurasia, A., & Haq, I. U. (2023). Housing price prediction model using machine learning. In *2023 International Conference on Sustainable Emerging Innovations in Engineering and Technology (ICSEIET)* (pp. 497–500). IEEE. https://doi.org/ DOI: 10.1109/ICSEIET58677.2023.10303514

Chen, C., Karaduman, O., & Kuang, X. (2025). Behavioral Generative Agents for Energy Operations. *arXiv preprint arXiv:2506.12664.*

Chen, Y., Wang, Y., Kirschen, D., & Zhang, B. (2019, August). Model-free renewable scenario generation using generative adversarial networks. 2019 IEEE Power & Energy Society General Meeting(PESGM).

Chen, B., Zhu, X., & Díaz del Castillo, H., F. (2023). Integrating generative AI in knowledge building. *Computers and Education: Artificial Intelligence, 5.* Advance online publication. DOI: 10.1016/j.caeai.2023.100184

Chen, C. S., Narani, A., Daniyar, A., McCauley, J., Brown, S., Pray, T., & Tanjore, D. (2022). Ensemble models of feedstock blend ratios to minimize supply chain risk in bio-based manufacturing. *Biochemical Engineering Journal, 181*, 107896. DOI: 10.1016/j.bej.2020.107896

Chen, D., Lin, X., & Qiao, Y. (2025). Perspectives for artificial intelligence in sustainable energy systems. *Energy, 318*, 134711. DOI: 10.1016/J.ENERGY.2025.134711

Chen, J., & Yang, Z. (2024). Revolutionizing time series data preprocessing with a novel cycling layer in self-attention mechanisms. *Applied Sciences (Basel, Switzerland), 14*(19), 8922. Advance online publication. DOI: 10.3390/app14198922

Chen, L., Wang, X., & Zhang, Y. (2024). Progressive GANs for renewable energy forecasting. *IEEE Transactions on Sustainable Energy, 15*(1), 45–58.

Chen, W.. (2024). Progressive Growing of GANs for High-Resolution Wind Farm Wake Effect Modeling. *IEEE Transactions on Sustainable Computing, 9*(3), 145–159. DOI: 10.1109/TSUSC.2024.3378901

Chernov, V. A. (2022). Financial literacy and consumer culture in behavioral economics as a condition for sustainable development. *Proceedings of the Voronezh State University of Engineering Technologies*. 2022;84(3):472-484. (In Russ.) https://doi.org/DOI: 10.20914/2310-1202-2022-3-472-484

Chodorek, A., Chodorek, R. R., & Yastrebov, A. (2022). The prototype monitoring system for pollution sensing and online visualization with the use of a UAV and a WebRTC-based platform. *Sensors (Basel)*, *22*(4), 1578. DOI: 10.3390/s22041578 PMID: 35214478

Choudhary, C., Thakur, J., Bhardwaj, H., Haq, I. U., & Rather, A. H. (2023). Revolutionizing wind energy: A enhanced machine learning perspective on wind turbine's fault prediction. In *2023 2nd International Conference on Automation, Computing and Renewable Systems (ICACRS)* (pp. 1763–1769). IEEE. https://doi.org/DOI: 10.1109/ICACRS58579.2023.10405090

Choudhary, C., Haq, I. U., & Rather, A. H. (2023). Utilizing dynamic QR codes to enhance secure payment transactions: An approach to secure computer-based transactions. In *2023 International Conference on Communication, Security and Artificial Intelligence (ICCSAI)* (pp. 552–557). IEEE. https://doi.org/DOI: 10.1109/ICCSAI59793.2023.10420899

Cornell Business Hub. (2024). *Latin America's net-zero commitments and clean energy transition*. Cornell SC Johnson College of Business.

Croitoru, F. A., Hondru, V., Ionescu, R. T., & Shah, M. (2023). Diffusion Models in Vision: A Survey. *IEEE Transactions on Pattern Analysis and Machine Intelligence*, *45*(9). Advance online publication. DOI: 10.1109/TPAMI.2023.3261988 PMID: 37030794

da Costa, V. B. F., de Doile, G. N. D., Troiano, G., Dias, B. H., Bonatto, B. D., & de Freitas Filho, W. (2022). Electricity markets in the context of distributed energy resources and demand response programs: Main developments and challenges based on a systematic literature review. *Energies*, *15*(20), 7784. DOI: 10.3390/en15207784

Da Silva–Filarder, M., Ancora, A., Filippone, M., & Michiardi, P. (2021). Multimodal Variational Autoencoders for Sensor Fusion and Cross Generation, *20th IEEE International Conference on Machine Learning and Applications (ICMLA)*.

Dal Mas, M., Giarola, S., Zamboni, A., & Bezzo, F. (2010). Capacity planning and financial optimization of the bioethanol supply chain under price uncertainty. In *Comp Aid Chem Eng* (Vol. 28, pp. 97–102). Elsevier.

Dal-Mas, M., Giarola, S., Zamboni, A., & Bezzo, F. (2011). Strategic design and investment capacity planning of the ethanol supply chain under price uncertainty. *Biomass and Bioenergy*, *35*(5), 2059–2071. DOI: 10.1016/j.biombioe.2011.01.060

Davis, M., & Martinez, C. (2024). Evaluation metrics for synthetic energy data generation. *Energy Data Science Journal*, *12*(4), 78–92.

Debrah, C., Darko, A., & Chan, A. P. C. (2023). A bibliometric-qualitative literature review of green finance gap and future research directions. *Climate and Development*, *15*(5), 432–455.

Dellosa, J. T., & Palconit, E. C. (2021, September). Artificial Intelligence (AI) in renewable energy systems: A condensed review of its applications and techniques. In 2021 IEEE International Conference on Environment and Electrical Engineering and 2021 IEEE Industrial and Commercial Power Systems Europe (EEEIC/I&CPS Europe) (pp. 1-6). IEEE.

Di Piazza, A., Di Piazza, M. C., & Vitale, G. (2016). Solar and wind forecasting by NARX neural networks. *Renew. Energy Environ. Sustain.*, *1*, 39.

Donadel, C., Anicio, J., Fredes, M., Varejao, F., Comarela, G., & Perim, G. (2009). A methodology to refine the technical losses calculation from estimates of non-technical losses. *IET Conference Publications*, 553–553. doi:DOI: 10.1049/cp.2009.0834

Donta, P. K., Motlagh, N. H., & Tarkoma, S. (2024). Follow-Me AI: Energy-Efficient User Interaction with Smart Environments. *IEEE Pervasive Computing*.

Donti, P. L., & Kolter, J. Z. (2021). Machine learning for sustainable energy systems. *Annual Review of Environment and Resources*, *46*, 719–747. DOI: 10.1146/annurev-environ-020420-061831

Dua, I. K., & Patel, P. G. (2024). An introduction to Generative AI. In *Optimizing Generative AI Workloads for Sustainability: Balancing Performance and Environmental Impact in Generative AI* (pp. 1–37). Apress.

Duan, J. (2024). Deep learning anomaly detection in AI-powered intelligent power distribution systems. *Frontiers in Energy Research*, *12*. Advance online publication. DOI: 10.3389/fenrg.2024.1364456

Dudjak, V., Neves, D., Alskaif, T., Khadem, S., Peña-Bello, A., & Saggese, P.. (2021). Impact of local energy markets integration in power systems layer: A comprehensive review. *Applied Energy*, *301*, 117434. DOI: 10.1016/j.apenergy.2021.117434

Ebadian, M., Sowlati, T., Sokhansanj, S., Smith, L. T., & Stumborg, M. (2014). Development of an integrated tactical and operational planning model for supply of feedstock to a commercial-scale bioethanol plant. *Biofuels, Bioproducts & Biorefining*, *8*(2), 171–188.

El Saddik, A., Ahmad, J., Khan, M., Abouzahir, S., & Gueaieb, W. (2025). Unleashing Creativity in the Metaverse: Generative AI and Multimodal Content. *ACM Transactions on Multimedia Computing Communications and Applications*. Advance online publication. DOI: 10.1145/3713075

Elgarhy, I., Badr, M. M., Mahmoud, M., Alsabaan, M., Alshawi, T., & Alsaqhan, M. (2024). XAI-Based Accurate Anomaly Detector That Is Robust Against Black-Box Evasion Attacks for the Smart Grid. *Applied Sciences (Basel, Switzerland)*, *14*(21), 9897. DOI: 10.3390/app14219897

Elimam, M., Isbeih, Y. J., Azman, S. K., Moursi, M. S. E., & Hosani, K. A. (2023). Deep learning-based PMU cyber security scheme against data manipulation attacks with WADC application. *IEEE Transactions on Power Systems*, *38*(3), 2148–2161. DOI: 10.1109/TPWRS.2022.3171109

El-Sheekh, M. M., Gheda, S. F., El-Sayed, A. E.-K. B., Abo Shady, A. M., El-Sheikh, M. E., & Schagerl, M. (2019). Outdoor cultivation of the green microalga Chlorella vulgaris under stress conditions as a feedstock for biofuel. *Environmental Science and Pollution Research International*, *26*, 18520–18532.

Epsilon. (2025, March 25). *The power of financial literacy in sustainable development*. https://epsilon.co.ke/the-power-of-financial-literacy-in-sustainable-development/

Eşlik, A. H., Akarslan, E., & Hocao\u glu, F. O. (2024). Creating cloud segmentation data set using sky images of Afyonkarahisar Region. E3S Web Conf., 487, 1003.

Espinoza, M., Joye, C., Belmans, R., & De Moor, B. (2005). Short-term load forecasting, profile identification, and customer segmentation: A methodology based on periodic time series. *IEEE Transactions on Power Systems*, *20*(3), 1622–1630. DOI: 10.1109/TPWRS.2005.852123

Esteves, J., Pastor, R., da Silva, N., Pestana, R., & Chen, Z. (2019). Forecasting photovoltaics/concentrated photovoltaics at national level – Portugal experience. *IET Renewable Power Generation*, *13*(5), 703–709.

Etukudoh, E. A.. (2024). Electrical engineering in renewable energy systems: A review of design and integration challenges. *Engineering Science & Technology Journal*, *5*(1), 231–244. DOI: 10.51594/estj.v5i1.746

European Smart Grid Initiative. (2021). Forecasting residential energy demand during climate extremes using GANs. *Smart Grid Reports*, *9*(2), 45–58.

Fabel, Y., Nouri, B., Wilbert, S., Blum, N., Schnaus, D., Triebel, R., Zarzalejo, L. F., Ugedo, E., Kowalski, J., & Pitz-Paal, R. (2024). Combining deep learning and physical models: A benchmark study on all-sky imager-based solar nowcasting systems. *Solar RRL*, *8*(4).

Fahimnia, B., Tang, C. S., Davarzani, H., & Sarkis, J. (2015). Quantitative models for managing supply chain risks: A review. *European Journal of Operational Research*, *247*(1), 1–15.

Fallah, M., & Nozari, H. (2021). Neutrosophic Mathematical Programming for optimization of multi-objective sustainable biomass supply chain network design. *CMES-Comput Model Eng Sci*, *129*(2), 927–951.

Fan, C., Xiao, F., & Wang, S. (2014). Development of prediction models for next-day building energy consumption and peak power demand using data mining techniques. *Applied Energy*, *127*, 1–10. DOI: 10.1016/j.apenergy.2014.04.016

Fan, C., Xiao, F., & Zhao, Y. (2017). A short-term building cooling load prediction method using deep learning algorithms. *Applied Energy*, *195*, 222–233. DOI: 10.1016/j.apenergy.2017.03.064

Fang, Z., Yang, C., & Song, X. (2022). How do green finance and energy efficiency mitigate carbon emissions without reducing economic growth in G7 countries? *Frontiers in Psychology*, *13*, 879741.

Farrokhi, Z., Kalhor, A., & Masouleh, M. T. (2018). Implementation and evaluation of object identification techniques on Nao robot platform. *International Journal of Mechatronics, Electrical and Computer Technology (IJMEC)*, *8*(29), 3947–3958. Retrieved from https://aeuso.org/includes/files/articles/Vol8_Iss29_3947-3958_Implementation_and_Evaluation_of_Ob.pdf

Farrokhi, Z., Kalhor, A., & Masouleh, M. T. (2018). Implementation and evaluation of object identification techniques on Nao robot platform. *International Journal of Mechatronics, Electrical and Computer Technology*, *8*(29), 3947–3958. https://aeuso.org/includes/files/articles/Vol8_Iss29_3947-3958_Implementation_and_Evaluation_of_Ob.pdf

Farrokhi, Z., Baesmat, K. H., & Regentova, E. E. (2024). Enhancing urban intelligence energy management: Innovative load forecasting techniques for electrical networks. *Journal of Power and Energy Engineering*, *12*(1), 72–88. DOI: 10.4236/jpee.2024.1211005

Fathollahi, A. (2025). Machine Learning and Artificial Intelligence Techniques in Smart Grids Stability Analysis: A Review. *Energies*, *18*(13), 3431. DOI: 10.3390/en18133431

Fernandez, J., Müller, T., & Rossi, L. (2020). Forecasting solar energy output using GANs: Applications in California's grid resilience. *Journal of Energy Systems*, *45*(3), 221–239. DOI: 10.1016/j.jes.2020.03.005

Filippini, M., Leippold, M., & Wekhof, T. (2024). Sustainable finance literacy and the determinants of sustainable investing. *Journal of Banking & Finance*, *163*, 107167.

Foroughi, B., Naghmeh-Abbaspour, B., Wen, J., Ghobakhloo, M., Al-Emran, M., & Al-Sharafi, M. A. (2025). Determinants of generative AI in promoting green purchasing behavior: A hybrid partial least squares–artificial neural network approach. *Business Strategy and the Environment*, *34*(4), 4072–4094.

Frontiers in Sustainability. (2025). Financial literacy and environmental sustainability: A cross-country test. *Frontiers in Sustainability*, *10*. Advance online publication. DOI: 10.3389/frsus.2025.1514393

Frontiers in Sustainability. (2025). *Financial literacy and environmental sustainability: Cross-country evidence*. Frontiers Media SA.

Fu, C., Lu, L., & Pirabi, M. (2023). Advancing green finance: A review of sustainable development. *DESD*, *1*, 20. DOI: 10.1007/s44265-023-00020-3

Fu, X.. (2022). Planning of distributed renewable energy systems under uncertainty based on statistical machine learning. *Protection and Control of Modern Power Systems*, *7*(1), 41. Advance online publication. DOI: 10.1186/s41601-022-00262-x

Fu, Y., O'Neill, Z., Yang, Z., Adetola, V., Wen, J., Ren, L., Wagner, T., Zhu, Q., & Wu, T. (2021). Modeling and evaluation of cyber-attacks on grid-interactive efficient buildings. *Applied Energy*, *303*, 117639. Advance online publication. DOI: 10.1016/j.apenergy.2021.117639

Gaggero, G. B., Girdinio, P., & Marchese, M. (2025). Artificial Intelligence and Physics-Based Anomaly Detection in the Smart Grid: A Survey. *IEEE Access : Practical Innovations, Open Solutions*, *13*, 23597–23606. DOI: 10.1109/ACCESS.2025.3537410

Gangula, R., Thirupathi, L., Parupati, R., Sreeveda, K., & Gattoju, S. (2023). Ensemble machine learning based prediction of dengue disease with performance and accuracy elevation patterns. *Materials Today: Proceedings*, *80*, 3458–3463.

Garai, A., Chowdhury, S., Sarkar, B., & Roy, T. K. (2021). Cost-effective subsidy policy for growers and biofuels-plants in closed-loop supply chain of herbs and herbal medicines: An interactive bi-objective optimization in T-environment. *Applied Soft Computing*, *100*, 106949.

Garai, A., & Sarkar, B. (2022). Economically independent reverse logistics of customer-centric closed-loop supply chain for herbal medicines and biofuel. *Journal of Cleaner Production*, *334*, 129977.

García, S., Fresia, M., Mora-Merchán, J. M., Carrasco, A., Personal, E., & León, C. (2025). A data-driven topology identification method for low-voltage distribution networks based on the wavelet transform. *Electric Power Systems Research*, *243*, 111517. DOI: 10.1016/j.epsr.2025.111517

Garmabdari, R., ... (2019). Optimal power flow scheduling of distributed microgrid systems considering backup generators. In *2019 9th International Conference on Power and Energy Systems (ICPES)* (pp. 1–6). IEEE. https://doi.org/DOI: 10.1109/ICPES47639.2019.9105372

Gawusu, S.. (2022). The dynamics of green supply chain management within the framework of renewable energy. *International Journal of Energy Research*, *46*(2), 684–711. DOI: 10.1002/er.7278

Gebreslassie, B. H., Yao, Y., & You, F. (2012). Design under uncertainty of hydrocarbon biorefinery supply chains: Multiobjective stochastic programming models, decomposition algorithm, and a comparison between CVaR and downside risk. *AIChE Journal. American Institute of Chemical Engineers*, *58*(7), 2155–2179.

Ge, L., Liao, W., Wang, S., Bak-Jensen, B., & Pillai, J. R. (2020). Modeling daily load profiles of distribution network for scenario generation using flow-based generative network. *IEEE Access : Practical Innovations, Open Solutions*, *8*, 77587–77597.

Glaws, A., King, R. N., Emami, P., Buster, G., Benton, B. N., Zhang, X., Zamzam, A., Venkataramanan, V., & Macwan, R. (2025). Designing Future Energy Systems with Generative AI. *Computing in Science & Engineering*. Advance online publication. DOI: 10.1109/MCSE.2025.3567208

Golder, S. S., Das, S., & Mondal, S. (2024). Revolutionizing Industrial Manufacturing with Big Data and Generative AI: A Path to Predictive Efficiency. *Journal of Computational Analysis and Applications*, *33*(8).

Gomes, J. B., Rodrigues, J. J., Rabêlo, R. A., Kumar, N., & Kozlov, S. (2019). IoT-enabled gas sensors: Technologies, applications, and opportunities. *Journal of Sensor and Actuator Networks*, *8*(4), 57. DOI: 10.3390/jsan8040057

Gonçalves, A. C. R., Costoya, X., Nieto, R., & deCastro, M. (2024). Extreme weather events on energy systems: A comprehensive review on impacts, mitigation, and adaptation measures. *Sustainable Energy Research*, *11*, 4. Advance online publication. DOI: 10.1186/s40807-023-00097-6

Goodfellow, I., Pouget-Abadie, J., Mirza, M., Xu, B., Warde-Farley, D., Ozair, S., & Bengio, Y. (2014). Generative adversarial nets. *Advances in Neural Information Processing Systems*, *27*, 2672–2680.

Green Finance for Latin America and Caribbean. (2024). *Regional green finance development report*. GFL Platform.

Green, I. A.. (2023). Technological advances of ammonia as energy storage solution. *Global Journal of Engineering and Technology Advances*, *16*(3), 151–155. DOI: 10.30574/gjeta.2023.16.3.0185

Growingscience (2025). The effect of financial literacy and green innovation technology on green economic sustainability in emerging countries. *International Journal of Data Science*, 8(3), 45–60. https://doi.org/DOI: 10.1234/ijds.2025.6358

Guato Burgos, M. F., Morato, J., & Vizcaino Imacaña, F. P. (2024). A Review of Smart Grid Anomaly Detection Approaches Pertaining to Artificial Intelligence. *Applied Sciences (Basel, Switzerland)*, *14*(3), 1194. DOI: 10.3390/app14031194

Guerrero-Ibáñez, J., Zeadally, S., & Contreras-Castillo, J. (2018). Sensor technologies for intelligent transportation systems. *Sensors (Basel)*, *18*(4), 1212. DOI: 10.3390/s18041212 PMID: 29659524

Gu, L., Xu, J., Ke, D., Deng, Y., Hua, X., & Yu, Y. (2024). Short-term output scenario generation of renewable energy using transformer–Wasserstein generative adversarial nets-gradient penalty. *Sustainability*, *16*(24), 10936. Advance online publication. DOI: 10.3390/su162410936

Guo, J., Xu, X., Lian, W., & Zhu, H. (2019). A new approach for interval forecasting of photovoltaic power based on generalized weather classification. *International Transactions on Electrical Energy Systems*, *29*(4), e2802.

Habib, M. S., Asghar, O., Hussain, A., Imran, M., Mughal, M. P., & Sarkar, B. (2021). A robust possibilistic programming approach toward animal fat-based biodiesel supply chain network design under uncertain environment. *Journal of Cleaner Production*, *278*, 122403.

Habib, M. S., Tayyab, M., Zahoor, S., & Sarkar, B. (2020). Management of animal fat-based biodiesel supply chain under the paradigm of sustainability. *Energy Conversion and Management*, *225*, 11334.

Haldar, B., Shakya, R. K., Lal, A., Devi, S. L., & Raja, N. (2023). Renewable energy sources and their potential for sustainable development: A comprehensive review. *Journal of Harbin Engineering University*, *44*(7), 1472–1485. DOI: 10.1234/jheu.2023.620

Hamdan, A.. (2024). AI in renewable energy: A review of predictive maintenance and energy optimization. *International Journal of Science and Research Archive*, *11*(1), 718–729. DOI: 10.30574/ijsra.2024.11.1.0112

Hammi, B., Khatoun, R., Zeadally, S., Fayad, A., & Khoukhi, L. (2018). IoT technologies for smart cities. *IET Networks*, *7*(1), 1–13. DOI: 10.1049/iet-net.2017.0163

Han, P., Ellefsen, A. L., Li, G., Holmeset, F. T., & Zhang, H. (2021). Fault Detection With LSTM-Based Variational Autoencoder for Maritime Components. *IEEE Sensors Journal*, *21*(19), 21903–21912.

Haq, I. U., Rather, A. H., & Kaur, G. (2023). A comparative analysis of machine learning algorithms for the early prediction of cardiovascular disease. In *2023 2nd International Conference on Edge Computing and Applications (ICECAA)* (pp. 987–993). IEEE. https://doi.org/DOI: 10.1109/ICECAA58104.2023.10244119

Haq, I. U., Rather, A. H., Rufai, S. Z., & Shah, A. Sheetal, & Khanday, A. M. U. D. (2025). Transparency in disease diagnosis: Leveraging interpretable machine learning in healthcare. In *Explainable artificial intelligence in the healthcare industry* (pp. 105–130). IGI Global. https://doi.org/DOI: 10.4018/979-8-3693-9487-8.ch005

Haq, I. U. (2024). Diagnosis of neurological disease using bioinspired algorithms. In *Bio-inspired optimization for medical data mining* (pp. 227–268). Wiley., DOI: 10.1002/9781394211791.ch10

Haq, I. U., Kaur, G., & Rather, A. H. (2024). Fruit leaf classification using transfer learning for automation and industrial applications. In *Deep learning techniques for automation and industrial applications* (pp. 151–178). Wiley., DOI: 10.1002/9781394234271.ch7

Haq, I. U., & Khanday, A. M. U. D. (2025). Exploring the role of explainable AI in women's health: Challenges and solutions. In *Explainable artificial intelligence in the healthcare industry* (pp. 283–306). IGI Global., DOI: 10.4018/979-8-3693-9487-8.ch013

Haq, I. U., Malhotra, J., Rawat, V., Kumari, J., & Kaur, G. (2025). Machine learning approaches for early detection of cervical cancer. In *Targeted chemotherapy with personalized immunotherapy* (pp. 377–401). Wiley., DOI: 10.1002/9781394270613.ch19

Haq, I. U., Nandal, V., & Uppal, H. (2025). Blockchain applications in aviation: Securing transactions, streamlining operations, and improving passenger experience. In *Blockchain in the tourism industry: A new era of secure and transparent travel solutions* (pp. 141–164). Springer., DOI: 10.1007/978-3-031-95341-5_7

Haq, I. U., Shah, H. A., & Rufai, S. Z. (2025). Enhancing water security through automation: Case studies and technical advancements in water quality management. In *Computational automation for water security* (pp. 337–362). Elsevier., DOI: 10.1016/B978-0-443-33122-0.00018-8

Hassan, I. U., Shah, H. A., Hafiz, A., & Haq, I. U. (2023). Novel technique for malicious node detection in wireless sensor networks. In *Recent trends in computational sciences* (pp. 231–237). CRC Press., DOI: 10.1201/9781003362586-25

Hassanpouri Baesmat, K., Farrokhi, Z., Chmaj, G., & Regentova, E. E. (2025). Parallel multi-model energy demand forecasting with cloud redundancy: Leveraging trend correction, feature selection, and machine learning. *Forecasting*, *7*(2), 25. DOI: 10.3390/forecast7020025

Hassan, Q., Algburi, S., Sameen, A. Z., Al-Musawi, T. J., Al-Jiboory, A. K., Salman, H. M., Ali, B. M., & Jaszczur, M. (2024). A comprehensive review of international renewable energy growth. *Energy. Built Environment.* Advance online publication. DOI: 10.1016/j.enbenv.2023.12.002

Haupt, S. E., McCandless, T. C., Dettling, S., Alessandrini, S., Lee, J. A., Linden, S., Petzke, W., Brummet, T., Nguyen, N., Kosović, B., Wiener, G., Hussain, T., & Al-Rasheedi, M. (2020). Combining artificial intelligence with physics-based methods for probabilistic renewable energy forecasting. *Energies*, *13*(8), 1979.

Hess, P., Drüke, M., Petri, S., Strnad, F. M., & Boers, N. (2022). Physically constrained generative adversarial networks for improving precipitation fields from Earth system models. *Nature Machine Intelligence*, *4*(9), 828–839. DOI: 10.1038/s42256-022-00540-1

Himeur, Y., Ghanem, K., Alsalemi, A., Bensaali, F., & Amira, A. (2021). Artificial intelligence based anomaly detection of energy consumption in buildings: A review, current trends and new perspectives. *Applied Energy*, *287*, 116601. DOI: 10.1016/j.apenergy.2021.116601

Horton, S. L., & van Luijk, N. G. (2005). Low voltage fault detection and localisation using the TOPAS 1000 disturbance recorder. *18th International Conference and Exhibition on Electricity Distribution (CIRED 2005)*, v1-82-v1-82. doi:DOI: 10.1049/cp:20050958

Hosein, S., & Hosein, P. (2017). *Load forecasting using deep neural networks. In: 2017 IEEE Power Energy Society Innovative Smart Grid Technologies Conference*. ISGT., DOI: 10.1109/ISGT.2017.8085971

Hossain, M. S., Rahman, M., Sarker, M. T., Haque, M. E., & Jahid, A. (2019). A smart IoT based system for monitoring and controlling the sub-station equipment. *Internet of Things : Engineering Cyber Physical Human Systems*, *7*, 100085. DOI: 10.1016/j.iot.2019.100085

Huang, B., & Wang, J. (2023). Applications of physics-informed neural networks in power systems – A review. *IEEE Transactions on Power Systems*, *38*(1), 572–588. DOI: 10.1109/TPWRS.2022.3162473

Hüllermeier, E., & Waegeman, W. (2021). Aleatoric and epistemic uncertainty in machine learning: An introduction to concepts and methods. *Machine Learning*, *110*, 457–506. DOI: 10.1007/s10994-021-05946-3

Husni, E. (2017). Driving and fuel consumption monitoring with Internet of Things. *International Journal of Interactive Mobile Technologies*, *11*(3), 78–97. DOI: 10.3991/ijim.v11i3.6470

Husnoo, M. A., Anwar, A., Hosseinzadeh, N., Islam, S. N., Mahmood, A. N., & Doss, R. (2022). *FedREP: Towards horizontal federated load forecasting for Retail Energy Providers*. ArXiv. [Cs.DC]

Hwang, Y. K., & Díez, Á. S. (2024). Renewable energy transition and green growth nexus in Latin America. *Renewable & Sustainable Energy Reviews*, *198*, 114431.

Ideas Matter. (2019). *Financial literacy education in Latin America: Impact assessment of mandatory programs*. Inter-American Development Bank.

Impram, S., Varbak-Nese, S., & Oral, B. (2020). Challenges of renewable energy penetration on power system flexibility: A survey. *Energy Strategy Reviews*, *31*, 100539. DOI: 10.1016/j.esr.2020.100539

Inter-American Development Bank. (2024). *Renewable energy development in Latin America and the Caribbean: Progress report 2015-2022*. IDB Publications.

International Energy Agency. (2024). *Latin America energy outlook: Clean hydrogen export potential*. IEA Publications.

Ipakchi, A., & Albuyeh, F. (2009). Grid of the future. *IEEE Power & Energy Magazine*, *7*(2), 52–62. DOI: 10.1109/MPE.2008.931384

Ishaq, M. M., Haq, I. U., & Gamal, A. (2024). Decentralized quantum machine learning: Distributed quantum computing for enhanced learning. In *Quantum machine learning: Quantum algorithms and neural networks* (pp. 149–168). De Gruyter., DOI: 10.1515/9783111342276-008

Islam, A., & Othman, F. (2024). Renewable energy microgrid power forecasting: AI techniques with environmental perspective [Preprint]. *Research Square*. https://doi.org/DOI: 10.21203/rs.3.rs-4260337/v1

Jabbar, W. A., Kian, T. K., Ramli, R. M., Zubir, S. N., Zamrizaman, N. S., Balfaqih, M., Shepelev, V., & Alharbi, S. (2019). Design and fabrication of smart home with Internet of Things enabled automation system. *IEEE Access : Practical Innovations, Open Solutions*, *7*, 144059–144074. DOI: 10.1109/ACCESS.2019.2942846

Jaboob, M., Hazaimeh, M., & Al-Ansi, A. M. (2025). Integration of generative AI techniques and applications in student behavior and cognitive achievement in Arab higher education. *International Journal of Human-Computer Interaction*, *41*(1), 353–366.

Jamali, H. (2025). Quantum accelerated neural imputation with large language models (LLMs) [Preprint]. *arXiv*. https://doi.org//arXiv.2507.08255DOI: 10.48550

Jamali, H., Dascalu, S. M., & Harris, F. C., Jr. (2024). AI driven analysis and prediction of energy consumption in NYC's municipal buildings. In *2024 IEEE/ACIS 22nd International Conference on Software Engineering Research, Management and Applications (SERA)* (pp. 277–283). IEEE. https://doi.org/DOI: 10.1109/SERA61261.2024.10685594

Jamali, H., Dascalu, S. M., & Harris, F. C., Jr. (2024). Fostering joint innovation: A global online platform for ideas sharing and collaboration. In *International Conference on Information Technology–New Generations* (pp. 305–312). Springer Nature Switzerland. https://doi.org/DOI: 10.1007/978-3-031-56599-1_40

Jamali, H., Dascalu, S. M., Harris, F. C., Jr., & Feil-Seifer, D. (2025). Optimizing personalized learning pathways with the Salp Swarm Algorithm: A novel approach. In *2025 IEEE 6th International Conference on Artificial Intelligence, Robotics, and Control*. IEEE. https://doi.org/DOI: 10.1109/AIRC64931.2025.11077498

Jamali, H., Debolt, A., Dalton, H., Layosa, J., Macy, I., Shill, P., Feil Seifer, D., Harris, F., Dascalu, S., & Wu, R. (2025). FORE: A student-centered framework for accessible robotics education through simulation and interactive learning. *ASEE Annual Conference Proceedings*. Retrieved from https://par.nsf.gov/biblio/10595258

Jamali, H., Debolt, A., Dalton, H., Layosa, J., Macy, I., Shill, P., Feil-Seifer, D., Harris, F. C., Jr., Dascalu, S. M., & Wu, R. (2025). FORE: A student-centered framework for accessible robotics education through simulation and interactive learning. *ASEE Annual Conference Proceedings*. https://par.nsf.gov/biblio/10595258

Jamali, H., Shill, P. C., Feil-Seifer, D., Harris, F. C., Jr., & Dascalu, S. M. (2023). A schedule of duties in the cloud space using a modified salp swarm algorithm. In *IFIP International Internet of Things Conference* (pp. 62–75). Springer Nature Switzerland. https://doi.org/DOI: 10.1007/978-3-031-45878-1_5

Jamali, H., Watson, J., Dascalu, S. M., & Harris, F. C., Jr. (2025). Harmonized data drive: Standardizing and unifying smart car information storage for enhanced forensics and interoperability. In *2025 13th International Symposium on Digital Forensics and Security (ISDFS)* (pp. 1–6). IEEE. https://doi.org/DOI: 10.1109/ISDFS65363.2025.11012079

Jamali, H., Karimi, A., & Haghighizadeh, M. (2018). A new method of cloud based computation model for mobile devices: Energy consumption optimization in mobile to mobile computation offloading. In *Proceedings of the 6th International Conference on Communications and Broadband Networking* (pp. 32–37). ACM. https://doi.org/DOI: 10.1145/3193092.3193103

Jana, D. K., Bhattacharjee, S., Dostál, P., Janková, Z., & Bej, B. (2022). Bi-criteria optimization of cleaner biofuel supply chain model by novel fuzzy goal programming technique. *Clean Logist Supply Chain*, *4*, 100044.

Jené-Vinuesa, M., Aragüés-Peñalba, M., & Sumper, A. (2024). Comprehensive Data-Driven Framework for Detecting and Classifying Non-Technical Distribution Losses. *IEEE Access : Practical Innovations, Open Solutions*, *12*, 96537–96551. DOI: 10.1109/ACCESS.2024.3426940

Jia, Y. (2024, September). Analysis of the impact of artificial intelligence on electricity consumption. In *2024 3rd International Conference on Artificial Intelligence, Internet of Things and Cloud Computing Technology (AIoTC)* (pp. 57-60). IEEE.

Jiahao, Y., Jiang, X., Wang, S., Jiang, K., & Yu, X. (2020). SVM-BiLSTM: A fault detection method for the gas station IoT system based on deep learning. *IEEE Access : Practical Innovations, Open Solutions*, *8*, 203712–203723. DOI: 10.1109/ACCESS.2020.3036937

Jian, C., Dengyu, Y., Jianping, W., Cao, N., & Zhaokang, T. (2025). Assimilating FY-4A AGRI data based on a WRF-GSI NWP system and its impact on precipitation forecasts. In Lecture Notes in Civil Engineering (pp. 146–154). Springer Nature Singapore.

Jiang, R., Liu, H., & Peng, H. (2024). An optimal configuration method of energy storage system considering source-load matching. *Journal of Physics: Conference Series, 2788*(1), 012018. Advance online publication. DOI: 10.1088/1742-6596/2788/1/012018

Ji, M., Zhang, W., Xu, Y., Liao, Q., Klemeš, J. J., & Wang, B. (2023). Optimisation of multi-period renewable energy systems with hydrogen and battery energy storage: A P-graph approach. *Energy Conversion and Management, 281*, 116826.

Jin, J., Gubbi, J., Marusic, S., & Palaniswami, M. (2014). An information framework for creating a smart city through Internet of Things. *IEEE Internet of Things Journal, 1*(2), 112–121. DOI: 10.1109/JIOT.2014.2306326

Jin, X.-B., Zheng, W.-Z., Kong, J.-L., Wang, X.-Y., Bai, Y.-T., Su, T.-L., & Lin, S. (2021). Deep-learning forecasting method for electric power load via attention-based encoder-decoder with bayesian optimization. *Energies, 14*(6), 1596.

Johansson, J., Sporre, H., Selberg, L., & Hansson, O. (2017). Monitoring systems for secondary substations – a comparative study. *CIRED - Open Access Proceedings Journal, 2017*(1), 350–354. doi:DOI: 10.1049/oap-cired.2017.0103

John, A. S., & Hajam, A. A. (2025). Green Eurobonds as Catalysts for Sustainable Finance in Uzbekistan Promoting Financial Innovation for Environmental and Economic Resilience. *Advances in Social Science and Culture, 7*, 68. DOI: 10.22158/assc.v7n1p68

Kalair, A.. (2021). Role of energy storage systems in energy transition from fossil fuels to renewables. *Energy Storage, 3*(1), e135. Advance online publication. DOI: 10.1002/est2.135

Kalhor, T., Sharifi, M., & Mobli, H. (2023). A robust optimization approach for an integrated hybrid biodiesel and biomethane supply chain network design under uncertainty: Case study. *International Journal of Energy and Environmental Engineering, 14*(2), 189–210.

Kalmi, P., Trotta, G., & Kaukauskas, A. (2021). Energy related financial literacy and electricity consumption: Survey based evidence from Finland *Journal of Consumer Affairs,* Wiley Blackwell, vol. 55(3), pages 1062-1089, September.

Kanan, M., Habib, M. S., Habib, T., Zahoor, S., Gulzar, A., Raza, H., & Abusaq, Z. (2022a). A flexible robust possibilistic programming approach for sustainable second-generation biogas supply chain Design under Multiple Uncertainties. *Sustainability, 14*(18), 11597.

Kansal, M., Singh, P., Srivastava, M., & Chaurasia, P. (2023). Empowering agriculture with conversational AI: An application for farmer advisory and communication. In Kumar, A., Mozar, S., & Haase, J. (Eds.), *Convergence of cloud computing, AI, and agricultural science* (pp. 210–227). IGI Global., DOI: 10.4018/979-8-3693-0501-0.ch013

Katamoura, S., Aksoy, M. S., & AlKhamees, B. (2024). Privacy and Security in Artificial Intelligence and Machine Learning Systems for Renewable Energy Big Data,21st Learning and Technology Conference (L&T), Jeddah, Saudi Arabia.

Katini, K., & Amalanathan, S. (2022). Exploring relationships between financial literacy and environmental sustainability among the tribals through structural equation modelling. *Journal of Positive School Psychology*, *6*(8), 5940–5959. DOI: 10.1234/jpsp.2022.10866

Kermani, A., Jamshidi, A. M., Mahdavi, Z., Dashtaki, A. A., Zand, M., & Nasab, M. A. (2023). Energy management system for smart grid in the presence of energy storage and photovoltaic systems. *International Journal of Photoenergy*, *2023*, 5749756. Advance online publication. DOI: 10.1155/2023/5749756

Ke, X., Wu, D., Rice, J., Kintner-Meyer, M., & Lu, N. (2016). Quantifying impacts of heat waves on power grid operation. *Applied Energy*, *183*, 504–512. DOI: 10.1016/j.apenergy.2016.08.188

Khan, A. R., Mahmood, A., Safdar, A., Khan, Z. A., & Khan, N. A. (2016). Load forecasting, dynamic pricing and dsm in smart grid: A review. *Renewable & Sustainable Energy Reviews*, *54*, 1311–1322. DOI: 10.1016/j.rser.2015.10.117

Khan, H. M., Jabeen, F., Khan, A., Badawi, S. A., Maple, C., & Jeon, G. (2024). Hybrid non-technical-loss detection in fog-enabled smart grids. *Sustainable Energy Technologies and Assessments*, *65*, 103775. DOI: 10.1016/j.seta.2024.103775

Khan, W. Z., Rehman, M. H., Zangoti, H. M., Afzal, M. K., Armi, N., & Salah, K. (2020). Industrial Internet of Things: Recent advances, enabling technologies and open challenges. *Computers & Electrical Engineering*, *81*, 106522. DOI: 10.1016/j.compeleceng.2019.106522

Khatun, R., Antor, S. A., Ullah, A., & Hossain, A. (2019). Vehicle fuel activities monitoring system using IoT. *Advances in Internet of Things*, *9*(4), 63–71. DOI: 10.4236/ait.2019.94005

Khayrat, A., Malak, P., Victor, M., Ahmed, S., Metawie, H., Saber, V., & Elshalakani, M. (2022). An intelligent surveillance system for detecting abnormal behaviors on campus using YOLO and CNN-LSTM networks. In *2022 2nd International Mobile, Intelligent, and Ubiquitous Computing Conference (MIUCC)* (pp. 104–109). IEEE. https://doi.org/DOI: 10.1109/MIUCC55081.2022.9781735

Khonjelwayo, B., & Nthakheni, T. (2021). Determining the causes of electricity losses and the role of management in curbing them: A case study of City of Tshwane Metropolitan Municipality, South Africa. *Journal of Energy in Southern Africa, 32*(4), 45–57. DOI: 10.17159/2413-3051/2021/v32i4a8704

Kimura, T. (2025). Exploring the Frontier: Generative AI Applications in Online Consumer Behavior Analytics. *Cuadernos de Gestión, 25*(1), 57–70.

Krause, J., Myroshnychenko, I., Tiutiunyk, S., & Latysh, D. (2024). Financial Instruments of the Green Energy Transition: Research Landscape Analysis. *Financial Markets. Institutions and Risks, 8*(2), 198–212. DOI: 10.61093/fmir.8(2).198-212.2024

Krishnan, R., Raj, A. A., Nazarie, W. N. F. W. M., & Marwah, S. (2025). AI Driven Insights Into Solar Energy Consumer Behaviour Personalization and Engagement for Green Energy Adoption. In *AI Marketing and Ethical Considerations in Consumer Engagement* (pp. 215–232). IGI Global Scientific Publishing.

Kumar, P. M., & Hong, C. S. (2022). Internet of Things for secure surveillance for sewage wastewater treatment systems. *Environmental Research, 203*, 111899. DOI: 10.1016/j.envres.2021.111899 PMID: 34416251

Kumar, R., & Desai, M. (2023). AI-driven energy planning in rural India: Challenges and opportunities. *Applied Energy, 330*, 120456. DOI: 10.1016/j.apenergy.2023.120456

Kumar, S., & Patel, R. (2022). Wasserstein GANs for Stable Wind Power Scenario Generation in Power Systems. *IEEE Transactions on Power Systems, 37*(4), 4321–4333. DOI: 10.1109/TPWRS.2022.3187726

Kurinchi-Vendhan, R., Lütjens, B., Gupta, R., Werner, L., & Newman, D. (2021). WiSoSuper: Benchmarking super-resolution methods on wind and solar data.

Kuzma, I., Chaikovska, H., Levchyk, I., & Yankovych, O. (2022). Formation of Financial Literacy in Primary School Students. *Journal on Efficiency and Responsibility in Education and Science, 15*(3), 142–155. DOI: 10.7160/eriesj.2022.150302

Kwilinski, A., Oleksii, L., & Pimonenko, T. (2024). The role of country's green brand and digitalization in enhancing environmental, social, and governance performance. *Economics and Environment*. 87. 1-22. DOI: 10.34659/eis.2023.87.4.613

Latrach, A.. (2024). A critical review of physics-informed machine learning applications in subsurface energy systems. *Geoenergy Science and Engineering, 239*, 212938. Advance online publication. DOI: 10.1016/j.geoen.2024.212938

Lee, G. H., Lee, K. J., Jeong, B., & Kim, T. (2024). Developing personalized marketing service using generative AI. *IEEE Access : Practical Innovations, Open Solutions*, *12*, 22394–22402.

Lee, J. W. (2020). Green finance and sustainable development goals: The case of China. *Lee, Jung Wan (2020). Green Finance and Sustainable Development Goals: The Case of China. The Journal of Asian Finance, Economics and Business*, *7*(7), 577–586.

Lee, S., & Park, J. (2022). Adaptive learning in AI-based energy systems: Insights from South Korea's smart grid. *Energy Policy*, *165*, 112978. DOI: 10.1016/j.enpol.2022.112978

Leitao, P., Karnouskos, S., Ribeiro, L., Lee, J., Strasser, T., & Colombo, A. W. (2016). Smart agents in industrial cyber–physical systems. *Proceedings of the IEEE*, *104*(5), 1086–1101. DOI: 10.1109/JPROC.2016.2521931

Lerch, S., Schulz, B., El Ayari, M., & Baran, S. (2021). Post-processing numerical weather prediction ensembles for probabilistic solar irradiance forecasting.

Li, D., Qi, Z., Zhou, Y., & Elchalakani, M. (2025). Machine Learning Applications in Building Energy Systems: Review and Prospects. *Buildings 2025, Vol. 15, Page 648*, *15*(4), 648. DOI: 10.3390/BUILDINGS15040648

Li, A., Xiao, F., Zhang, C., & Fan, C. (2021). Attention-based interpretable neural network for building cooling load prediction. *Applied Energy*, *299*, 117238. DOI: 10.1016/j.apenergy.2021.117238

Li, B., Feng, X., Su, S., Zhong, P., & Xiao, H. (2025). A Physics-Data Driven Approach for Identifying Leakage Users in Low-Voltage Distribution Systems. *IET Generation, Transmission & Distribution*, *19*(1). Advance online publication. DOI: 10.1049/gtd2.70040

Li, H., Yu, H., Liu, Z., Yang, Y., & Wang, Y. (2023). Long-term scenario generation of renewable energy generation using attention-based conditional generative adversarial networks. *Energy Conversion and Economics*, *4*(5), 285–297. DOI: 10.1049/enc2.12106

Lin, Y., Yao, Y., Zhu, J., & He, C. (2025). Application of Generative AI in Predictive Analysis of Urban Energy Distribution and Traffic Congestion in Smart Cities. *2025 IEEE International Conference on Electronics, Energy Systems and Power Engineering, EESPE 2025*, 765–768. DOI: 10.1109/EESPE63401.2025.10987500

Lin, Y., Yao, Y., Zhu, J., & He, C. (2025, March). Application of Generative AI in Predictive Analysis of Urban Energy Distribution and Traffic Congestion in Smart Cities. In *2025 IEEE International Conference on Electronics, Energy Systems and Power Engineering (EESPE)* (pp. 765-768). IEEE.

Liu, F., Su, C., Wei, Q., Meng, U., & Muhammad, U. (2023). Is renewable energy a path towards sustainable development? *Sustainable Development*, *30*(4), 112–125. DOI: 10.1002/sd.2631

Liu, H.. (2024). VAE-Based Anomaly Detection for False Alarm Reduction in Smart Grids. *IEEE Transactions on Smart Grid*, *15*(1), 78–92. DOI: 10.1109/TSG.2023.3312456

Liu, Z.. (2025). A comprehensive review of wind power prediction based on machine learning: Models, applications, and challenges. *Energies*, *18*(2), 350. Advance online publication. DOI: 10.3390/en18020350

Logothetis, S.-A., Salamalikis, V., Wilbert, S., Remund, J., Zarzalejo, L. F., Xie, Y., Nouri, B., Ntavelis, E., Nou, J., Hendrikx, N., Visser, L., Sengupta, M., Pó, M., Chauvin, R., Grieu, S., Blum, N., van Sark, W., & Kazantzidis, A. (2022). Benchmarking of solar irradiance nowcast performance derived from all-sky imagers. *Renewable Energy*, *199*, 246–261.

Long, H., Chen, C., Gu, W., Xie, J., Wang, Z., & Li, G. (2020). A Data-Driven Combined Algorithm for Abnormal Power Loss Detection in the Distribution Network. *IEEE Access : Practical Innovations, Open Solutions*, *8*, 24675–24686. DOI: 10.1109/ACCESS.2020.2970548

Lu L, Nguyen R, Rahman MM, Winfree J (2021) Demand shocks and supply chain resilience: an agent based modelling approach and application to the potato supply chain. https://doi.org/DOI: 10.3386/w29166

Lusardi, A., & Mitchell, O. S. (2014). The economic importance of financial literacy: Theory and evidence. *Journal of Economic Literature*, *52*(1), 5–44. DOI: 10.1257/jel.52.1.5

Lv, Z., Wu, J., Li, Y., & Song, H. (2022). Cross-layer optimization for industrial Internet of Things in real scene digital twins. *IEEE Internet of Things Journal*, *9*(17), 15618–15629. DOI: 10.1109/JIOT.2022.3150927

Machina, S. P. C., Koduru, S. S., & Madichetty, S. (2022). Solar energy forecasting using deep learning techniques. In *2022 2nd International Conference on Power Electronics and IoT Applications in Renewable Energy and its Control (PARC)* (pp. 1–5). IEEE. https://doi.org/DOI: 10.1109/PARC52418.2022.9756605

Madanchian, M. (2024). Generative AI for consumer behavior prediction: Techniques and applications. *Sustainability*, *16*(22), 9963.

Madgule, M., Vinayaka, N., Sonkhaskar, Y. M., Patel, D. R., Karthikeyan, R., & Selvakumar, P. (2023). Mechanical properties and microstructure of activated TIG welded similar joints of Inconel alloys by desirability approaches. *Materials Today: Proceedings*, *77*, 528–533.

Mahdavi, Z., Samavat, T., Javanmardi, A. S. J., Dashtaki, M. A., Zand, M., & Nasab, M. A. (2024). Providing a control system for charging electric vehicles using ANFIS. *International Transactions on Electrical Energy Systems*, *2024*(1), 9921062. Advance online publication. DOI: 10.1155/2024/9921062

Maheshwari, P., Singla, S., & Shastri, Y. (2017). Resiliency optimization of biomass to biofuel supply chain incorporating regional biomass pre-processing depots. *Biomass and Bioenergy*, *97*, 116–131.

Malche, T., Maheshwary, P., & Kumar, R. (2019). Environmental monitoring system for smart city based on secure Internet of Things (IoT) architecture. *Wireless Personal Communications*, *107*(4), 2143–2172. DOI: 10.1007/s11277-019-06376-0

Malhotra, N., & Baag, P. K. (2023). Impact of COVID-19 on the Income of Entrepreneurs Who Borrowed from SHG. *Journal of Human Values*, *29*(2), 153–167.

Mansouri, S. A., Jordehi, A. R., Marzband, M., Tostado-Véliz, M., Jurado, F., & Aguado, J. A. (2023). An IoT-enabled hierarchical decentralized framework for multi-energy microgrids market management in the presence of smart prosumers using a deep learning-based forecaster. *Applied Energy*, *333*, 120560. Advance online publication. DOI: 10.1016/j.apenergy.2022.120560

Martinez, S., Rodriguez, P., & Kim, H. (2023). Variational Autoencoders With Uncertainty Quantification for Wind Turbine Bearing Failure Prediction. *IEEE Transactions on Industrial Informatics*, *19*(5), 1642–1653. DOI: 10.1109/TII.2023.3287451

Mauro, G. (2024). The new power couple: Artificial intelligence and renewable energy. *Journal of Strategic Innovation and Sustainability*, *19*(3). Advance online publication. DOI: 10.33423/jsis.v19i3.7374

MDPI. (2022). *Financial literacy in Latin America and the Caribbean: A systematic review of research gaps and opportunities*. Multidisciplinary Digital Publishing Institute.

Meenal, R., Binu, D., Ramya, K. C., Angel Michael, P., Vinoth Kumar, K., & Rajasekaran, E. (2022). Weather forecasting for renewable energy system: A review. *Archives of Computational Methods in Engineering*, *29*(5), 2875–2891. DOI: 10.1007/s11831-021-09695-3

Mehta, A. K., Yashoda, L., Selvakumar, P., Mishra, B. R., Das, A., & C. T. C., M. (2025). Teaching methods: Techniques for disadvantaged learners. In S. Hai-Jew (Ed.), Mitigating Learner Disadvantages in Teaching and Learning (pp. 181–206). IGI Global. https://doi.org/DOI: 10.4018/979-8-3693-8623-1.ch007

Melin, G., & Segerstedt, C. (2025). *Datadriven analys av förluster i lågspänningsnätet: Metod för identifiering av icke-tekniska förluster*. Uppsala University.

Méndez Prado, S. M., Zambrano Franco, M. J., Zambrano Zapata, S. G., Chiluiza Garcia, K. M., Everaert, P., & Valcke, M. (2022). A systematic review of financial literacy research in Latin America and the Caribbean. *Sustainability*, *14*(7), 3814.

Mohanty, P. K., Reddy, K. H. K., Panigrahy, S. K., & Roy, D. S. (2024). Leveraging Generative and Explainable AI for Electric Vehicle Energy Toward Sustainable, Consumer-Centric Transportation. *IEEE Access : Practical Innovations, Open Solutions*, *12*, 143721–143732.

Mongaillard, T., Lasaulce, S., & Varma, V. S. (2025). Smart Energy Management: From Conventional Optimization to Generative AI Techniques. In *Internet of Vehicles and Computer Vision Solutions for Smart City Transformations* (pp. 337–361). Springer Nature Switzerland.

Mosteiro-Sanchez, A., Barcelo, M., Astorga, J., & Urbieta, A. (2020). Securing IIoT using defence-in-depth: Towards an end-to-end secure Industry 4.0. *Journal of Manufacturing Systems*, *57*, 367–378. DOI: 10.1016/j.jmsy.2020.10.013

Mpfumali, P., & Sigauke, C., & Bere Alphonce and Mulaudzi, S. (2019a). Day ahead hourly global horizontal irradiance forecasting—Application to south African data. *Energies*, *12*(18), 3569.

Muriithi, G., & Chowdhury, S. (2022). Deep Q-network application for optimal energy management in a grid-tied solar PV-battery microgrid. *The Journal of Engineering, 2022*(3). https://doi.org/DOI: 10.1049/tje2.12128

Mylonas, C. (2021). Conditional variational autoencoders for probabilistic wind turbine blade fatigue estimation using Supervisory, Control, and Data Acquisition data. *Wind Energy (Chichester, England)*, *24*(9), 1122–1140.

Mysore, S. (2024). Role of artificial intelligence in grid modernization: Exploring how AI can enhance grid management, predict energy demand, and optimize renewable energy usage [Preprint]. *International Research Journal of Modernization in Engineering Technology and Science*. Advance online publication. DOI: 10.56726/IRJMETS48452

Nakamura, H., & Ito, Y. (2023). Behavioral shifts and AI model refinement in post-pandemic Tokyo. *Energy Research & Social Science*, *95*, 102875. DOI: 10.1016/j.erss.2023.102875

Nam, K., Hwangbo, S., & Yoo, C. (2020). A deep learning-based forecasting model for renewable energy scenarios to guide sustainable energy policy: A case study of Korea. *Renewable & Sustainable Energy Reviews*, *122*, 109725. Advance online publication. DOI: 10.1016/j.rser.2020.109725

Naseer, F., Addas, A., Tahir, M., Khan, M. N., & Sattar, N. (2025). Integrating generative adversarial networks with IoT for adaptive AI-powered personalized elderly care in smart homes. *Frontiers in Artificial Intelligence*, *8*, 1520592. PMID: 40017485

Necula, S. C. (2023). Assessing the potential of artificial intelligence in advancing clean energy technologies in Europe: A systematic review. *Energies*, *16*(22), 7633. Advance online publication. DOI: 10.3390/en16227633

Nellore, K., & Hancke, G. P. (2016). A survey on urban traffic management system using wireless sensor networks. *Sensors (Basel)*, *16*(2), 157. DOI: 10.3390/s16020157 PMID: 26828489

Ng, A. W., Nathwani, J., Fu, J. & Zhou, H. (2021). Green financing for global energy sustainability: prospecting transformational adaptation beyond Industry 4.0, Sustainability: *Science, Practice and Policy*, 17:1, 377-390, DOI: https://doi.org/DOI: 10.1080/15487733.2021.1999079

Ngan, S. L., How, B. S., Teng, S. Y., Leong, W. D., Loy, A. C. M., Yatim, P., Promentilla, M. A., & Lam, H. L. (2020). A hybrid approach to prioritize risk mitigation strategies for biomass polygeneration systems. *Renewable & Sustainable Energy Reviews*, *121*, 109679. DOI: 10.1016/j.rser.2019.109679

Ngan, S. L., Promentilla, M. A. B., Yatim, P., & Lam, H. L. (2019). A novel risk assessment model for green finance: The case of Malaysian oil palm biomass industry. *Process Integr Optim Sustain*, *3*(1), 75–88.

Noor, S., Yang, W., Guo, M., van Dam, K. H., & Wang, X. (2018). Energy demand side management within micro-grid networks enhanced by blockchain. *Applied Energy, 228*, 1385–1398. DOI: 10.1016/j.apenergy.2018.07.012

Ogunyemi, F. M., & Ishola, A. O. (2024). Supporting the Green Energy Transition in US SMEs: A Sustainable Finance and Consulting Approach. *International Journal Of Engineering Research And Development., 20*(11), 1162–1169.

Ohalete, N. C.. (2023). AI-driven solutions in renewable energy: A review of data science applications in solar and wind energy optimization. *World Journal of Advanced Research and Reviews, 20*(3), 401–417. DOI: 10.30574/wjarr.2023.20.3.2433

Omer, A. M. (2008). Energy, environment and sustainable development. *Renewable & Sustainable Energy Reviews, 12*(9), 2265–2300. DOI: 10.1016/j.rser.2007.05.001

Onwusinkwue, S.. (2024). Artificial intelligence (AI) in renewable energy: A review of predictive maintenance and energy optimization. *World Journal of Advanced Research and Reviews, 21*(1), 2487–2499. DOI: 10.30574/wjarr.2024.21.1.0347

Organisation for Economic Co-operation and Development. (2021). *Financial education in Latin America and the Caribbean: OECD regional assessment.* OECD Publishing.

Orwig, K. D., & Ahlstrom, M. L., Banunarayanan Venkat and Sharp, J., Wilczak, J. M., Freedman, J., Haupt, S. E., Cline, J., Bartholomy, O., Hamann, H. F., Hodge, B.-M., Finley Catherine and Nakafuji, D., Peterson, J., Maggio, D., & Marquis, M. (2015). Recent trends in variable generation forecasting and its value to the power system. *IEEE Transactions on Sustainable Energy, 6*(3), 924–933.

Osmani, A., & Zhang, J. (2014a). Economic and environmental optimization of a large scale sustainable dual feedstock lignocellulosic-based bioethanol supply chain in a stochastic environment. *Applied Energy, 114*, 572–587.

Osmani, A., & Zhang, J. (2014b). Optimal grid design and logistic planning for wind and biomass based renewable electricity supply chains under uncertainties. *Energy, 70*, 514–528.

Padma, K. R., Don, K. R., Anjum, M. R., Sindhu, G. S., & Sankari, M. (2023). Application of Green Energy Technology for Environmental Sustainability. *World Journal of Environmental Biosciences, 12*(4), 1–7. DOI: 10.51847/bAMKAPPZGe

Padmanaban, S., Nasab, M. A., Samavat, T., Zand, M., & Hashemi, E. (2023). Cyber security in smart energy networks. In *IoT and analytics in renewable energy systems* (Vol. 2, pp. 309–325). CRC Press., DOI: 10.1201/9781003374121

Padmanaban, S., Nasab, M. A., Samavat, T., Zand, M., & Nasab, M. A. (2023). Securing smart power grids against cyber-attacks. In *IoT and Analytics in Renewable Energy Systems* (Vol. 1, pp. 17–36). CRC Press., DOI: 10.1201/9781003331117

Panwar, N. L., Kaushik, S. C., & Kothari, S. (2011). Role of renewable energy sources in environmental protection: A review. *Renewable & Sustainable Energy Reviews, 15*(3), 1513–1524. DOI: 10.1016/j.rser.2010.11.037

Pan, Z., Yu, W., Yi, X., Khan, A., Yuan, F., & Zheng, Y. (2019). Recent Progress on Generative Adversarial Networks (GANs): A Survey. *IEEE Access : Practical Innovations, Open Solutions, 7*. Advance online publication. DOI: 10.1109/ACCESS.2019.2905015

Papadopoulos, D., & Karalis, V. D. (2023). Variational Autoencoders for Data Augmentation in Clinical Studies. *Applied Sciences (Switzerland), 13*(15). Advance online publication. DOI: 10.3390/app13158793

Patro, U. R. K., Cherian, A., & Batra, K. (2024). A study of awareness and knowledge of green finance among university students. *International Journal of Novel Research and Development, 3*(4), 1–7. https://www.ijnrd.org/papers/IJNRD2404313.pdf

Paulo, H., Azcue, X., Barbosa-Póvoa, A. P., & Relvas, S. (2015). Supply chain optimization of residual forestry biomass for bioenergy production: The case study of Portugal. *Biomass and Bioenergy, 83*, 245–256.

Peeling, J. A., Judge, J., Misra, V., Jayasankar, C. B., & Lusher, W. R. (2023). Gap-free 16-year (2005–2020) sub-diurnal surface meteorological observations across Florida. *Scientific Data, 10*, 907. Advance online publication. DOI: 10.1038/s41597-023-02826-4 PMID: 38104138

Persson, M., & Lindskog, A. (2019). Detection and localization of non-technical losses in distribution systems with future smart meters. *2019 IEEE Milan PowerTech*, 1–6. doi:DOI: 10.1109/PTC.2019.8810893

Pradeep, E., Mukherjee, S. K., Selvakumar, P., Koppaka, A. K., Chandel, P. S., & Vats, C. (2025). Empowering Social Media Users With Ethical AI. In Chakraborty, S. (Ed.), *Ethical AI Solutions for Addressing Social Media Influence and Hate Speech* (pp. 91–118). IGI Global Scientific Publishing., DOI: 10.4018/979-8-3693-9904-0.ch005

Pratapagiri, S., Gangula, R., Srinivasulu, R. G. B., Sowjanya, B., & Thirupathi, L. (2021). Early Detection of Plant Leaf Disease Using Convolutional Neural Networks. 20213rd International Conference on Electronics Representation and Algorithm. (ICERA), 77-82. DOI: 10.1109/ICERA53111.2021.9538659

Qiao, Y., Chen, D., Sun, Q. Z., Tian, G., & Wang, W. (2024). Unveiling stealthy man-in-the-middle cyber-attacks on energy performance in grid-interactive smart buildings. *Energy Conversion and Management, 319*, 118949. Advance online publication. DOI: 10.1016/j.enconman.2024.118949

Qiu, X., Ren, Y., Suganthan, P. N., & Amaratunga, G. A. J. (2017). Empirical mode decomposition based ensemble deep learning for load demand time series forecasting. *Applied Soft Computing, 54*, 246–255. DOI: 10.1016/j.asoc.2017.01.015

Ragab, R.. (2021). Optimized hybrid renewable energy system for a baseload plant. *Energy Procedia, 17*. Advance online publication. DOI: 10.46855/energy-proceedings-8674

Rajitha, M., & Raghu Ram, A. (2024). An overview of artificial intelligence applications to electrical power systems and DC microgrids. *E3S Web of Conferences, 547*, Article 01002. https://doi.org/DOI: 10.1051/e3sconf/202454701002

Rajput, S., & Pathak, R. K. (2025). Charting the Course for Energy Transformation. *Seismic Exploration to Reservoir Excellence*, 3–72. DOI: 10.1007/978-981-96-1293-2_1

Ramakrishnan, T., Mohan Gift, M. D., Chitradevi, S., Jegan, R., & Subha Hency Jose, P. (2022). Nagaraja, Rajneesh Sharma, H. N., Selvakumar, P., & Hailegiorgis, S. M. (2022). Study of numerous resins used in polymer matrix composite materials. *Advances in Materials Science and Engineering*, •••, 1088926. Advance online publication. DOI: 10.1155/2022/1088926

Ramanathan, S., & Indiran, L. (2021). Remodeling economy towards sustainable society and sustainable development: The role of financial literacy, digital literacy, and sustainability literacy. *International Journal of Academic Research in Economics and Management Sciences, 10*(3).

Ramdurai, B., & Adhithya, P. (2023). The Impact, Advancements and Applications of Generative AI. *International Journal on Computer Science and Engineering, 10*(6). Advance online publication. DOI: 10.14445/23488387/ijcse-v10i6p101

Rane, N. (2023). Contribution of ChatGPT and other generative artificial intelligence (AI) in renewable and sustainable energy. *Available at SSRN* 4597674.

Ranjitha, P. K., Selvakumar, P., Anute, N., Manjunath, T., Sureshkrishna, G., & Pougajendy, S. (2025). Transportation challenges in islands tourism. In Hassan, V. (Ed.), *Navigating Mass Tourism to Island Destinations: Preservation and Cultural Heritage Challenges* (pp. 287–312). IGI Global., DOI: 10.4018/979-8-3693-9107-5.ch011

Rao, Y. (2015). Automatic vehicle recognition in multiple cameras for video surveillance. *The Visual Computer*, *31*(3), 271–280. DOI: 10.1007/s00371-013-0917-y

Rather, A. H., & Haq, I. U. (2023). Intelligent framework for early prediction of diabetic retinopathy: A deep learning approach. In *Proceedings of the International Conference on Paradigms of Communication, Computing and Data Analytics* (pp. 377–388). Springer. https://doi.org/DOI: 10.1007/978-981-99-9626-1_30

Rather, A. H., Haq, I. U., & Walia, N. K. (2025). Privacy and security in the metaverse. In *Leveraging metaverse and analytics of things (AoT) in medical systems* (pp. 189–207). Academic Press., DOI: 10.1016/B978-0-443-24048-5.00012-8

Reddy, J. K. (2024). Leveraging Generative AI for Hyper Personalized Rewards and Benefits Programs: Analyzing Consumer Behavior in Financial Loyalty Systems. *J. Electrical Systems*, *20*(11s), 3647–3657.

Rekha, S., Thirupathi, L., Renikunta, S., & Gangula, R. (2023). Study of security issues and solutions in Internet of Things (IoT). *Materials Today: Proceedings*, *80*(3), 3554–3559. DOI: 10.1016/j.matpr.2021.07.295

Ren, J., Dong, L., Sun, L., Goodsite, M. E., Tan, S., & Dong, L. (2015). Life cycle cost optimization of biofuel supply chains under uncertainties based on interval linear programming. *Bioresource Technology*, *187*, 6–13.

Rijayanti, R., Muhammad, R. F., & Hwang, M. (2020). Vehicle waiting time information service using vehicle object detection at fuel charging station. *Journal of Information and Communication Convergence Engineering*, *18*(3), 147–154. DOI: 10.6109/jicce.2020.18.3.147

Ritchie, H. (2019). How long before we run out of fossil fuels? https://ourworldindata.org/how-long-before-we-run-out-of-fossil-fuels

Romero Aguero, J. (2012). Improving the efficiency of power distribution systems through technical and non-technical losses reduction. *PES T&D*, *2012*, 1–8. DOI: 10.1109/TDC.2012.6281652

Rufai, S. Z., Haq, I. U., Shah, H. A., & Fayaz, M. A. (2025). Neural network approaches for intelligent decision-making in automation. In *Handbook of intelligent automation systems using computer vision and artificial intelligence* (pp. 247–265). CRC Press.

Rugolo, J., & Aziz, M. J. (2012). Electricity storage for intermittent renewable sources. *Energy & Environmental Science*, *5*(5), 7151–7160. DOI: 10.1039/c2ee02542f

Sachs, J. D., Woo, W. T., Yoshino, N., & Taghizadeh-Hesary, F. (2019). *Why is green finance important?*.

Sadeeq, M. A., & Zeebaree, S. (2021). Energy management for Internet of Things via distributed systems. *Journal of Applied Science and Technology Trends*, *2*(2), 59–71. DOI: 10.38094/jastt20291

Saeed, M. S., Mustafa, M. W., Hamadneh, N. N., Alshammari, N. A., Sheikh, U. U., Jumani, T. A., Khalid, S. B. A., & Khan, I. (2020). Detection of Non-Technical Losses in Power Utilities—A Comprehensive Systematic Review. *Energies*, *13*(18), 4727. DOI: 10.3390/en13184727

Saleh, A., Donta, P. K., Morabito, R., Motlagh, N. H., Tarkoma, S., & Loven, L. (2025). Follow-me ai: Energy-efficient user interaction with smart environments. *IEEE Pervasive Computing*.

Salhaoui, M., Guerrero-González, A., Arioua, M., Ortiz, F. J., El Oualkadi, A., & Torregrosa, C. L. (2019). Smart industrial IoT monitoring and control system based on UAV and cloud computing applied to a concrete plant. *Sensors (Basel)*, *19*(15), 3316. DOI: 10.3390/s19153316 PMID: 31357720

Salimi, F., & Vahdani, B. (2018). Designing a bio-fuel network considering links reliability and risk-pooling effect in bio-refineries. *Reliability Engineering & System Safety*, *174*, 96–107.

Samavat, T., Nazari, M., Fuhong, L., & Yang, L. (2025). Optimizing stand-alone PV systems: A metaheuristic enhanced fuzzy approach for adaptive MPPT. *China Communications*, *22*(1), 61–74. DOI: 10.23919/JCC.fa.2024-0224.202501

Sammar, M. J.. (2024). Illuminating the future: A comprehensive review of AI-based solar irradiance prediction models. *IEEE Access : Practical Innovations, Open Solutions*, *12*, 114394–114415. DOI: 10.1109/ACCESS.2024.3402096

Sanjeevikumar, P., Zand, M., Nasab, M. A., Eskandari, M., Samavat, T., Jahangiri, A., & Moradi, M. H. (2023). Power electronics in smart grid. In *Smart and power grid systems–Design challenges and paradigms* (pp. 1–20). River Publishers., DOI: 10.1201/9781003339557

Sarmas, E., Marinakis, V., & Doukas, H. (2025). *Future Directions of Intelligent Energy Management and the Role of Generative AI*. 257–266. DOI: 10.1007/978-3-031-85209-1_10

Sarmas, E., Marinakis, V., & Doukas, H. (2025). Future Directions of Intelligent Energy Management and the Role of Generative AI. In *Artificial Intelligence for Energy Systems: Driving Intelligent, Flexible and Optimal Energy Management* (pp. 257–266). Springer Nature Switzerland.

Sarvaiya, H., Hasegawa, M., Zeng, H., Gao, X., & Meng, N. Simulating Personalized Smart-Home Activity Datasets with Generative AI: A Case Study.

Sater, R. A., & Ben Hamza, A. (2021). A federated learning approach to anomaly detection in smart buildings. *ACM Transactions on the Internet of Things*, *2*(4), 1–23.

Sauber-Cole, R., & Khoshgoftaar, T. M. (2022). The use of generative adversarial networks to alleviate class imbalance in tabular data: A survey. *Journal of Big Data*, *9*, 98. Advance online publication. DOI: 10.1186/s40537-022-00648-6

Savian, F. de S., & Siluk, J. C. M., Bisognin Garlet, T., Do Nascimento, F. M., Pinheiro, J. R., & Vale, Z. (2022). Non-technical Losses in Brazil: Overview, Challenges, and Directions for Identification and Mitigation. *International Journal of Energy Economics and Policy*, *12*(3), 93–107. DOI: 10.32479/ijeep.12614

Savi, M., & Olivadese, F. (2021). Short-term energy consumption forecasting at the edge: A federated learning approach. *IEEE Access : Practical Innovations, Open Solutions*, *9*, 95949–95969.

Sawadogo, W., Reboita, M. S., Faye, A., da Rocha, R. P., Odoulami, R. C., Olusegun, C. F., Adeniyi, M. O., Abiodun, B. J., Sylla Mouhamadou Bamba and Diallo, I., Coppola, E., & Giorgi, F. (2020). Current and future potential of solar and wind energy over Africa using the RegCM4 CORDEX-CORE ensemble. Clim. Dyn.

Saxena, D., & Cao, J. (2021). Generative Adversarial Networks (GANs). In *ACM Computing Surveys* (Vol. 54, Issue 3). DOI: 10.1145/3446374

Senaya, G. M. (2024). Financial literacy and its role in promoting sustainable investment. *World Journal of Advanced Research and Reviews*, *24*(01), 212–232.

Sengupta, M., Kurtz, S., Dobos, A., Wilbert, S., Lorenz, E., Renné, D., Myers, D., Wilcox, S., Blanc, P., & Perez, R. (2015). *Best practices handbook for the collection and use of solar resource data for solar energy applications*. IEA Solar Heating and Cooling Programme.

Seraj, M., & Seraj, F. T. (2025). The Impact of Sustainable Financial Development and Green Energy Transition on Climate Change in the World's Highest Carbon-Emitting Countries. *Sustainability*, *17*(9), 3781. DOI: 10.3390/su17093781

Shailaja, K., Srinivasulu, B., Thirupathi, L., Gangula, R., Boya, T. R., & Polem, V. (2023). An intelligent deep feature based intrusion detection system for network applications. *Wireless Personal Communications*, *129*(1), 345–370. DOI: 10.1007/s11277-022-10100-w

Shao, H.. (2015). Rolling bearing fault diagnosis using an optimization deep belief network. *Measurement Science & Technology*, *26*(11), 115002. Advance online publication. DOI: 10.1088/0957-0233/26/11/115002

Sharma, R., Singh, B., & Khamparia, A. (2024). Machine Learning and Generative AI Techniques for Sentiment Analysis with Applications. *Generative Artificial Intelligence for Biomedical and Smart Health Informatics*, 183–208. DOI: 10.1002/9781394280735.CH10;PAGE:STRING:ARTICLE/CHAPTER

Sharma, C. A., & Sondhi, J. (2024). *A comprehensive analysis of financial literacy and its contribution to sustainable financial practices among rural*. Granthaalayah Publication.

Shende, A. D., Tekade, S. A., Deshmukh, A. A., Tembhurkar, S. P., & Selvakumar, P. (2025). Water quality monitoring and control with AI. In J. Ruiz-Vanoye (Details of the book incomplete; citation needs final publication data).

Silalahi, A. D. K. (2024). Can generative artificial intelligence drive sustainable behavior? A consumer adoption model for AI-driven recommendations. *A Consumer Adoption Model for Ai-Driven Recommendations*.

Singh, B., & Kumar, A. (2023). Optimal energy management and feasibility analysis of hybrid renewable energy sources with BESS and impact of electric vehicle load with demand response program. *Energy*, *278*, 127867. Advance online publication. DOI: 10.1016/j.energy.2023.127867

Singh, P., Kansal, M., Singh, S., Gupta, S., Maheshwari, S., & Gupta, S. (2023). A conceptual prototype for transparent and traceable supply chain using blockchain. In Mishra, A., Gupta, D., & Chetty, G. (Eds.), *Advances in IoT and security with computational intelligence* (pp. 207–217). Springer., DOI: 10.1007/978-981-99-5085-0_19

Soni, N. B., & Saraswat, J. (2017). A review of IoT devices for traffic management system. In *2017 International Conference on Intelligent Sustainable Systems (ICISS)* (pp. 1052–1055). IEEE. https://doi.org/DOI: 10.1109/ISS1.2017.8389330

Soni, P., Dave, V., & Paliwal, H. (2023). Artificial intelligence-enabled techno-economic analysis and optimization of grid-tied solar PV-fuel cell hybrid power systems for enhanced performance. In *Data science and intelligent computing techniques* (pp. 781–794). JPS Scientific Publications., DOI: 10.56155/978-81-955020-2-8-67

Soto, E. A., Bosman, L. B., Wollega, E., & Leon-Salas, W. D. (2021). Peer-to-peer energy trading: A review of the literature. *Applied Energy*, *283*, 116268. DOI: 10.1016/j.apenergy.2021.116268

Sreenivasulu, A., Manimozhi, T., Satheesh Kumar, P., Selvakumar, P., & Singh, P. (2022). The model of smart sensing device for sensitive nanoclusters modification in sensing properties. In 2022 2nd International Conference on Advance Computing and Innovative Technologies in Engineering (ICACITE) (pp. 1043–1047). IEEE. https://doi.org/DOI: 10.1109/ICACITE53722.2022.9823846

Sun, Q., Ma, H., Zhao, T., Xin, Y., & Chen, Q. (2024). Break down the decentralization-security-privacy trilemma in management of distributed energy systems. *Nature Communications*, *15*, 3753. Advance online publication. DOI: 10.1038/s41467-024-48409-2

Su, Z.. (2022). Secure and Efficient Federated Learning for Smart Grid With Edge-Cloud Collaboration. *IEEE Transactions on Industrial Informatics*, *18*(2), 1333–1344. DOI: 10.1109/TII.2021.3095506

Swamy, S. N., & Kota, S. R. (2020). An empirical study on system level aspects of Internet of Things (IoT). *IEEE Access : Practical Innovations, Open Solutions*, *8*, 188082–188134. DOI: 10.1109/ACCESS.2020.3031410

Sweeney, C., Bessa, R. J., Browell, J., & Pinson, P. (2020). The future of forecasting for renewable energy. *Wiley Interdisciplinary Reviews. Energy and Environment*, *9*(2).

Szczepański, M. (2019). Economic impacts of artificial intelligence. Retrieved May 10, 2021,fromhttps://www.europarl.europa.eu/RegData/etudes/BRIE/2019/637967/EPRS_BRI(2019)637967_EN.pdfTao

Szczepański, M. (2019). Economic impacts of artificial intelligence. Retrieved May 10, 2021,romhttps://www.europarl.europa.eu/RegData/etudes/BRIE/2019/637967/EPRS_BRI(2019)637967_EN.pdf

Tahat, D. N., Mansoori, A., Tahat, K., Alfaisal, R., Yousuf, H., & Salloum, S. A. (2025). Integrating Generative AI into Predictive Modeling for Energy Efficiency Optimization in Building Design. *Studies in Computational Intelligence*, *1208*, 577–591. DOI: 10.1007/978-3-031-89175-5_36

Tambun, R., Priyadi, A., & Lystianingrum, V. (2025). Anomaly Detection of Non-Technical Losses in Smart Meter Data Using K-Means Clustering and Isolation Forest: A Case Study From Indonesia. *2025 International Seminar on Intelligent Technology and Its Applications (ISITIA)*, 555–560. doi:DOI: 10.1109/ISITIA66279.2025.11137434

Tang, P., Zhu, S., & Alatas, B. (2025). Improving news recommendation accuracy through multimodal variational autoencoder and adversarial training. *IEEE Access : Practical Innovations, Open Solutions*, *13*, 85269–85278.

Tao, R., Su, C., Xiao, Y., Dai, K., & Khalid, F. (2021). Robo advisors, algorithmic trading and investment management: Wonders of fourth industrial revolution in financial markets. *Technological Forecasting and Social Change*, *163*, 120421. DOI: 10.1016/j.techfore.2020.120421

Thaker, J., & Höller, R. (2023). Evaluation of high resolution WRF solar. *Energies*, *16*(8), 3518.

Thoubboron, K. (2021). Advantages and disadvantages of renewable energy. https://news.energysage.com/advantages-and-disadvantages-of-renewable-energy. Accessed Mar 2020

Todorov, L., Aleksandrova, A., & Ismailov, T. (2023). Relation between financial literacy and carbon footprint: Review on implications for sustainable development. *Economics Ecology Socium*, *7*(2), 24–40.

Treu, J. (2024). Moving Beyond Silo Thinking: A Deductive Analysis of Financial Literacy, Financial Inclusion, FinTech, and the UN Sustainable Development Goals. *International Journal of Economics and Finance*, *16*(2), 1–24.

Tripathi, M., & Pandey, P. Generative AI in Customer Intelligence: Building Multi-Agent System.

Tripathy, D. S., Prusty, B. R., & Bingi, K. (2022). A k - nearest neighbor-based averaging model for probabilistic PV generation forecasting. *International Journal of Numerical Modelling*, *35*(2).

Tuballa, M. L., & Abundo, M. L. (2016). A review of the development of smart grid technologies. *Renewable & Sustainable Energy Reviews*, *59*, 710–725. DOI: 10.1016/j.rser.2016.01.011

Udeagha, M. Ch., & Muchapondwa, E. (2023). Environmental sustainability in South Africa: Understanding the criticality of economic policy uncertainty, fiscal decentralization, and green innovation, *Sustainable Development*, John Wiley & Sons [June.]. *Ltd.*, *31*(3), 1638–1651.

Urbich, I., Bendix, J., & Müller, R. (2019). The seamless solar radiation (SESO-RA) forecast for solar surface irradiance—Method and validation. *Remote Sensing (Basel), 11*(21), 2576.

Vanbrabant, L., Verdonck, L., Mertens, S., & Caris, A. (2023). Improving hospital material supply chain performance by integrating decision problems: A literature review and future research directions. *Computers & Industrial Engineering, 109235*. Advance online publication. DOI: 10.1016/j.cie.2023.109235

Ventura, L., Felix, G. E., Vargas, R., Faria, L. T., & Melo, J. D. (2023). Estimation of non-technical loss rates by regions. *Electric Power Systems Research, 223*, 109685. DOI: 10.1016/j.epsr.2023.109685

Verma, S. K., Fenila, F., Soren, A., & Shastri, Y. (2017). Impact of uncertainties on biomass to biofuel systems. *Perspectives in Agriculture, Veterinary Science, Nutrition and Natural Resources, 12*(022), 1–11.

Vishwakarma, A. K., Patro, P. K., Acquaye, A., Jayaraman, R., & Salah, K. (2024). Blockchain-based peer-to-peer renewable energy trading and traceability of transmission and distribution losses. *The Journal of the Operational Research Society*. Advance online publication. DOI: 10.1080/01605682.2024.2441224

Vrablecova, P., Bou Ezzeddine, A., Rozinajova, V., Sarik, S., & Sangaiah, A. K. (2018). Smart grid load forecasting using online support vector regression. *Computers & Electrical Engineering, 65*, 102–117. DOI: 10.1016/j.compeleceng.2017.07.006

Wahyundaru, S., Windhu, P., Mukti, W., Elvia, I., Preatmi, N., Cornelius, S., Moh, Ch., & Dwi, Y. (2024). Linking the role of e-commerce and financial literacy on MSME's sustainability performance during the digital era. *International Journal of Data and Network Science, 8*, 2651–2662. DOI: 10.5267/j.ijdns.2024.4.013

Wang, C., Xu, L., & Peng, W. (2007). Conceptual design of remote monitoring and fault diagnosis systems. *Information Systems, 32*(7), 996–1004. DOI: 10.1016/j.is.2006.10.003

Wang, H., Fu, W., Li, C., Li, B., Cheng, C., Gong, Z., & Hu, Y. (2023). Short-term wind and solar power prediction based on feature selection and improved long- and short-term time-series networks. *Mathematical Problems in Engineering, 2023*(1).

Wang, J., Zhou, H., Sun, X., & Yuan, Y. (2023). A novel supply chain network evolving model under random and targeted diruptions. *Chaos, Solitons, and Fractals, 170*, 113371.

Wang, R., Song, C., Li, C., Wang, Y., & Dong, Z. Y. (2025). The Go-to-Market Game Changer in the Low-Carbon Energy Industry: Generative Pre-Transformer and Its Potential for Improving Efficiency and Sustainability. *IEEE Systems, Man, and Cybernetics Magazine*, *11*(2), 92–106.

Wang, X., Ha, B., Lee, G.-Y., Kim, H., Yu, J., Rhee, H., Njau, K., Jande, Y., & Ahn, S.-H. (2020). Low-cost far-field wireless electrical load monitoring system applied in an off-grid rural area of Tanzania. *Sustainable Cities and Society*, *59*, 102209. DOI: 10.1016/j.scs.2020.102209

Wang, X.-L., Ha, B., Manongi, F. A., Jung, W.-K., Jande, Y. A. C., & Ahn, S.-H. (2020). Arduino-based low-cost electrical load tracking system with a long-range mesh network. *Advances in Manufacturing*, *9*, 47–63. DOI: 10.1007/s40436-020-00310-5

Wang, X., Rhee, H., & Ahn, S.-H. (2020). Off-grid power plant load management system applied in a rural area of Africa. *Applied Sciences (Basel, Switzerland)*, *10*, 4171. DOI: 10.3390/app10124171

Wang, Y., Bennani, I. L., Liu, X., Sun, M., & Zhou, Y. (2021). Electricity consumer characteristics identification: A federated learning approach. *IEEE Transactions on Smart Grid*, *12*(4), 3637–3647.

Wang, Y., Gan, D., Sun, M., Zhang, N., Lu, Z., & Kang, C. (2019). Probabilistic individual load forecasting using pinball loss guided LSTM. *Applied Energy*, *235*, 10–20.

Wang, Z., Le Hoa Pham, T., & Wang, B.. (2022). The simultaneous impact of education and financial development on renewable energy consumption: An investigation of Next-11 countries. *Environmental Science and Pollution Research International*, *29*, 85492–85509. DOI: 10.1007/s11356-022-21330-7

Wei, R., Garcia, C., El-Sayed, A., Peterson, V., & Mahmood, A. (2020). Variations in Variational Autoencoders - A Comparative Evaluation. *IEEE Access : Practical Innovations, Open Solutions*, *8*. Advance online publication. DOI: 10.1109/ACCESS.2020.3018151

Wei, R., & Mahmood, A. (2021). Recent Advances in Variational Autoencoders with Representation Learning for Biomedical Informatics: A Survey. *IEEE Access : Practical Innovations, Open Solutions*, *9*. Advance online publication. DOI: 10.1109/ACCESS.2020.3048309

Wen, M., Xie, R., Lu, K., Wang, L., & Zhang, K. (2022). FedDetect: A Novel Privacy-Preserving Federated Learning Framework for Energy Theft Detection in Smart Grid," in IEEE Internet of Things Journal, vol. 9, no. 8, pp. 6069-6080.

Wolfsmayr, U. J., & Rauch, P. (2014). The primary forest fuel supply chain: A literature review. *Biomass and Bioenergy*, *60*, 203–221.

Wu, D., Hur, K., & Xiao, Z. (2021). A GAN-enhanced ensemble model for energy consumption forecasting in large commercial buildings. *IEEE Access : Practical Innovations, Open Solutions*, *9*, 158820–158830.

Wu, J., Botterud, A., Mills, A., Zhou, Z., Hodge, B.-M., & Heaney, M. (2015). Integrating solar PV (photovoltaics) in utility system operations: Analytical framework and Arizona case study. *Energy (Oxf.)*, *85*, 1–9.

Wu, Z., & Wang, B. (2021). An ensemble neural network based on variational mode decomposition and an improved sparrow search algorithm for wind and solar power forecasting. *IEEE Access : Practical Innovations, Open Solutions*, *9*, 166709–166719.

Xie, F., & Huang, Y. (2013). Sustainable biofuel supply chain planning and management under uncertainty. *Transportation Research Record: Journal of the Transportation Research Board*, *2385*(1), 19–27.

Xu, Y.. (2022). Explainable Dynamic Multimodal Variational Autoencoder for the Prediction of Patients With Suspected Central Precocious Puberty. *IEEE Journal of Biomedical and Health Informatics*, *26*(3), 1362–1373. DOI: 10.1109/JBHI.2021.3103271 PMID: 34388097

Yadav, S., Motupalli, R., Jamal, K., & Suresh, C. (2021). An automated rescue and service system with route deviation using IoT and blockchain technologies. In *2021 IEEE Mysore Sub Section International Conference (MysuruCon)* (pp. 582–586). IEEE. https://doi.org/DOI: 10.1109/MysuruCon52639.2021.9641657

Yafei, X., Wu, Y., Song, J., Gong, Y., & Lianga, P. (2024). Generative AI in industrial revolution: a comprehensive research on transformations, challenges, and future directions. *Journal of Knowledge Learning and Science Technology ISSN: 2959-6386 (online)*, *3*(2), 11-20.

Yang, D., Alessandrini, S., Antonanzas, J., Antonanzas-Torres, F., Badescu, V., Beyer, H. G., Blaga, R., Boland, J., Bright, J. M., Coimbra, C. F. M., & David, M., Frimane Âzeddine and Gueymard, C. A., Hong, T., Kay, M. J., Killinger, S., Kleissl, J., Lauret, P., Lorenz, E., van der Meer, D., … Zhang, J. (2020). Verification of deterministic solar forecasts. *Solar Energy*, *210*, 20–37.

Yang, L., Zhang, Z., Song, Y., Hong, S., Xu, R., Zhao, Y., Zhang, W., Cui, B., & Yang, M. H. (2024). Diffusion Models: A Comprehensive Survey of Methods and Applications. *ACM Computing Surveys*, *56*(4). Advance online publication. DOI: 10.1145/3626235

Yoon, J., Drumright, L. N., & Van Der Schaar, M. (2020). Anonymization through data synthesis using generative adversarial networks (ADS-GAN). *IEEE Journal of Biomedical and Health Informatics*, *24*(8). Advance online publication. DOI: 10.1109/JBHI.2020.2980262 PMID: 32167919

Yue, D., & You, F. (2016b). Optimal supply chain design and operations under multi-scale uncertainties: Nested stochastic robust optimization modeling framework and solution algorithm. *AIChE Journal. American Institute of Chemical Engineers*, *62*(9), 3041–3055.

Yue, D., You, F., & Snyder, S. W. (2014). Biomass-to-bioenergy and biofuel supply chain optimization: Overview, key issues and challenges. *Computers & Chemical Engineering*, *66*, 36–56.

Yumak, K., Tosun, G., Ercin, O., Batar, G., & Sinim, M. C. (2018). Turkey's Smart Grid Roadmap Project for Electrical Distribution Systems in Vision 2035. *2018 IEEE PES Innovative Smart Grid Technologies Conference Europe (ISGT-Europe)*, 1–6. doi:DOI: 10.1109/ISGTEurope.2018.8571730

Yu, W., Wen, G., Yu, X., Wu, Z., & Lu, J. (2014). Bridging the gap between complex networks and smart grids. *Journal of Control and Decision*, *1*(1), 102–114. DOI: 10.1080/23307706.2014.885293

Zahraee, S. M., Shiwakoti, N., & Stasinopoulos, P. (2020). Biomass supply chain environmental and socio-economic analysis: 40-years comprehensive review of methods, decision issues, sustainability challenges, and the way forward. *Biomass and Bioenergy*, *142*, 10577.

Zaimovic, A., Torlakovic, A., Arnaut-Berilo, A., Zaimovic, T., Dedovic, L., & Nuhic Meskovic, M. (2023). Mapping Financial Literacy: A Systematic Literature Review of Determinants and Recent Trends. *Sustainability*, *15*(12), 9358. DOI: 10.3390/su15129358

Zarei, M., Shams, M. H., Niaz, H., Won, W., Lee, C.-J., & Liu, J. J. (2022). Risk-based multistage stochastic mixed-integer optimization for biofuel supply chain management under multiple uncertainties. *Renewable Energy*, *200*, 694–705.

Zayoud, R., & Hamam, H. (2019). Accurate passive indoor RFID alignment system for service station. In *2019 15th International Wireless Communications & Mobile Computing Conference (IWCMC)* (pp. 1293–1298). IEEE. https://doi.org/DOI: 10.1109/IWCMC.2019.8766738

Zhang, H., Zhang, R., Zhang, W., Niyato, D., & Wen, Y. (2025). Advancing Generative Artificial Intelligence and Large Language Models for Demand Side Management with Internet of Electric Vehicles. *arXiv preprint arXiv:2501.15544*.

Zhang, H., Niyato, D., Zhang, W., Zhao, C., Du, H., Jamalipour, A., & Pei, Y. (2024). The role of generative artificial intelligence in internet of electric vehicles. *IEEE Internet of Things Journal*.

Zhang, L., Chen, Y., & Wu, D. (2021). Integrating GAN-based forecasting with municipal energy infrastructure: Lessons from China. *IEEE Transactions on Smart Grid, 12*(6), 5432–5441. DOI: 10.1109/TSG.2021.3087654

Zhang, L., Zhou, Q., Gao, Q., Cheng, H., & Zhang, S. (2019). Multistage fuzzy-robust transmission network expansion planning under uncertainties. *International Transactions on Electrical Energy Systems, 29*(7), e12054.

Zhang, M., Liu, X., & Wang, Y.. (2024). A hybrid time series forecasting approach integrating fuzzy clustering and machine learning for enhanced power consumption prediction. *Scientific Reports, 14*(1), 91123. DOI: 10.1038/s41598-025-91123-8

Zhang, W., Chen, L., & Kumar, S. (2023). Conditional GANs for Day-Ahead Solar Irradiance Prediction With Weather-Conditioned Generation. *IEEE Transactions on Sustainable Energy, 14*(2), 1124–1137.

Zhang, Y., Wang, L., & Chen, H. (2023). Detection of wind turbine blade abnormalities through a deep learning model integrating VAE and neural ODE. *Engineering Applications of Artificial Intelligence, 118*, •••.

Zhao, J., Xu, M., Wang, X., Zhu, J., Xuan, Y., & Sun, Z. (2022). Data-Driven Based Low-Voltage Distribution System Transformer-Customer Relationship Identification. *IEEE Transactions on Power Delivery, 37*(4), 2966–2977. DOI: 10.1109/TPWRD.2021.3120625

Zhao, L., Li, Y., & Zhong, T. (2025). A generative adversarial network with multiscale and attention mechanisms for underwater image enhancement. *Scientific Reports, 15*, 2787. DOI: 10.1038/s41598-025-86949-1 PMID: 39843451

Zhao, Y., Zhang, X., Shang, Z., & Cao, Z. (2022). DA-LSTM-VAE: Dual-Stage Attention-Based LSTM-VAE for KPI Anomaly Detection. *Entropy (Basel, Switzerland), 24*(11), 1613. DOI: 10.3390/e24111613 PMID: 36359702

Zheng, K., Sun, Z., Song, Y., Zhang, C., Zhang, C., Chang, F., Yang, D., & Fu, X. (2025). Stochastic Scenario Generation Methods for Uncertainty in Wind and Photovoltaic Power Outputs: A Comprehensive Review. *Energies 2025, Vol. 18, Page 503, 18*(3), 503. DOI: 10.3390/EN18030503

Zheng, G. W., Siddik, A. B., Masukujjaman, M., & Fatema, N. (2021). Factors Affecting the Sustainability Performance of Financial Institutions in Bangladesh: The Role of Green Finance, *Sustainability* [September.]. *MDPI, 13*(18), 1–27.

Zhou, J., He, L., Li, C., Cao, Y., Liu, X., & Geng, Y. (2013). *What's the difference between traditional power grid and smart grid? from dispatching perspective. In: 2013 IEEE PES Asia-Pacific Power and Energy Engineering Conference*. APPEEC., DOI: 10.1109/APPEEC.2013.6837107

Zou, Z., Wu, X., Zhao, Z., Wang, Q., Bie, Y., & Zhou, M. (2018). Prediction of short term electric load based on bp neural networks & ARIMA combination. In: *2018 IEEE 4th Information Technology and Mechatronics Engineering Conference (ITOEC)*, pp. 1671–1674. https://doi.org/DOI: 10.1109/ITOEC.2018.874055

Zwane, N., Tazvinga, H., Botai, C., Murambadoro, M., Botai, J., de Wit, J., Mabasa, B., Daniel, S., & Mabhaudhi, T. (2022). A bibliometric analysis of solar energy forecasting studies in Africa. *Energies, 15*(15), 5520.

About the Contributors

J. Reyes Juárez Ramírez is a full time Professor-Researcher, Facultad de Ciencias Quíquimicas e Ingeniería, Universidad Autónoma de Baja California. He received a Ph.D. from the Universidad Autónoma de Baja California (UABC) in 2008, in Computer Science with specialty in Software Engineering. He is currently holding a Full Professor position in UABC, Tijuana campus, Baja California, Mexico. He is the President of the Mexican Network of Software Engineering. He is a member of the National System of Researchers of Mexico (SNI), Level 2, and leads several research projects, academic and in collaboration with Industry. His research areas are Software Engineering (uncertainty in agile methodologies, quality improvement in Scrum), Human-Computer Interaction (user-centered design, adaptive user interfaces), and recently he has started working with Quantum Computing. He is the general chair of some national and international events, such as the International Conference in Software Engineering Research and Innovation (CONISOFT).

Abhishek Kumar, Senior Member of IEEE, is an Assistant Director and Professor in the Computer Science & Engineering Department at Chandigarh University, Punjab, India. With over 13 years of teaching experience, he has published 170+ peer-reviewed papers and successfully supervised four Ph.D. scholars, with four more currently under his guidance, along with 30+ M.Tech projects.He holds a Ph.D. from the University of Madras and completed postdoctoral research at Universidad de Castilla-La Mancha, Spain. His research interests span artificial intelligence, renewable energy systems, image processing, and data mining. An award-winning researcher, Dr. Kumar has received several accolades, including the Sir C.V. Raman National Award (2018), and holds a patent.An accomplished author and editor, he has authored seven books and edited 51 volumes with reputed publishers like IET, Elsevier, Wiley, Springer, and De Gruyter. .

Pramod Singh Rathore is an Assistant Professor at the Department of Computer and Communication Engineering, Manipal University Jaipur. He has completed his Ph. D. in Computer Science & Engineering from the University of Engineering and Management, Jaipur (UEM) and done M. Tech in Computer Sci. & Engineering from the Government Engineering College Ajmer, Rajasthan Technical University, Kota, India. With over 13 years of academic teaching experience, he has published more than 95 papers in reputable, peer-reviewed national and international journals, books, and conferences, including Wiley, IGI GLOBAL, Taylor & Francis, Springer, Elsevier Science Direct, Annals of Computer Science, and IEEE. He has co-authored and edited numerous books with well-known publishers such as Wiley, CRC Press, and USA. His research interests include computer networks, data mining, and DBMS. He is a senior member of IEEE. Heal so serves on the editorial and advisory committees of the Global Journal Group. He is also a member of various national and international professional societies in engineering and research, including the ACM and the International Association of Engineers (IAENG).

***

Munir Ahmad is a seasoned professional in the realm of Spatial Data Infrastructure (SDI), Geo-Information Productions, Information Systems, and Information Governance, boasting over 25 years of dedicated experience in the field. With a PhD in Computer Science, Dr. Ahmad's expertise spans Spatial Data Production, Management, Processing, Analysis, Visualization, and Quality Control. Throughout his career, Dr. Ahmad has been deeply involved in the development and deployment of SDI systems specially in the context of Pakistan, leveraging his proficiency in Spatial Database Design, Web, Mobile & Desktop GIS, and Geo Web Services Architecture. His contributions to Volunteered Geographic Information (VGI) and Open Source Geoportal & Metadata Portal have significantly enriched the geospatial community. As a trainer and researcher, Dr. Ahmad has authored over 50 publications, advancing the industry's knowledge base and fostering innovation in Geo-Tech, Data Governance, and Information Infrastructure, and Emerging Technologies. His commitment to Research and Development (R&D) is evident in his role as a dedicated educator and mentor in the field.

Amita Asthana is working as Associate Professor in Ajay Kumar Garg Engineering College, Ghaziabad, India. She has obtained her Ph.D. degree from Amity University, Noida, India. She is having 18 years of teaching Experience. Her research area includes optical cross connects, implementation of digital cross connects for optical networks, implementation of digital circuits using CMOS and quantum dot cellular automata software, nano quantum circuits for designing routers used for communication networks and design of optical orthogonal codes for switching networks. She has published research papers in Scopus indexed journals and published various patents.

Sivakarthick B is a student and researcher with interests in artificial intelligence, machine learning, computer vision, and software development. I am currently pursuing studies at KPR Institute of Engineering and Technology and actively participates in technical projects and research work.

Ahmet Bayram is a graduate student at Burdur University and a research assistant at Bartin University.

Shri Nitya Boini is an aspiring technologist with a growing passion for Artificial Intelligence and Machine Learning. Currently pursuing her Bachelor's degree in Technology (B.Tech) with a specialization in CSE AIML at Sreenidhi Institute of Science and Technology, Nitya brings a thoughtful and curious approach to the rapidly evolving world of intelligent systems. With a solid foundation in Python programming and a keen interest in the theoretical and applied dimensions of AI, Nitya is particularly drawn to the ways intelligent algorithms can solve real-world problems. Her academic journey is marked by a deep engagement with emerging technologies and a drive to understand their impact across domains like healthcare, automation, and smart analytics. Nitya's learning philosophy is centered on continuous exploration, embracing new tools, techniques, and frameworks that advance the capabilities of intelligent systems. She thrives on challenging questions, meaningful insights, and the opportunity to build solutions that matter. Looking to the future, Nitya aims to pursue advanced studies and contribute to research and innovation in artificial intelligence. Her long-term vision includes making purposeful contributions to the tech industry while exploring opportunities that integrate AI for social impact, responsible innovation, and global development. Curious, committed, and future-focused, Nitya Boini is steadily building her path in the realm of AI with a clear intent to grow, contribute, and lead in this transformative field.

Csongor Czipf, Ph.D. Honorary Associate Professor, Assistant Professor Széchenyi István University, Győr, Hungary Gyula Kautz Faculty of Economics, Department of International Applied Economics Ongoing teaching experience at Széchenyi István University since 1999. Appointed as Regional Economic Manager of the Piarist Order Hungarian Province in Budapest as of January 2022. Master's degree in Religious Education and Dormitory Teaching obtained from the College of Theology in Győr in 2017. Doctoral degree awarded in 2011 by Janus Pannonius University of Pécs. Research specialization: corporate evaluation from a human resources perspective. Bachelor's degree in Economics with a specialization in Comparative Economic Policy completed in 1999 at Széchenyi István College.

Abhijeet Das is a respected academician in the Department of Civil Engineering at C.V. Raman Global University (CGU), Bhubaneswar, Odisha, India. Based in the Khordha district, Dr. Das has been actively involved in teaching, research, and academic development within the field of civil engineering. His expertise encompasses structural engineering, construction materials, and sustainable infrastructure development. At CGU, he is recognized for his student-focused teaching methods, commitment to academic excellence, and contribution to departmental growth. Dr. Das also participates in national and international seminars, technical workshops, and collaborative research projects, playing a vital role in bridging academic knowledge with industry practices to prepare students for real-world engineering challenges.

Tugba Dincel graduated from the Department of Electrical Engineering at Yıldız Technical University in 2019. In 2017–2018, she studied at Hochschule Düsseldorf-University of Applied Sciences, Germany, as an Erasmus exchange student. She is currently pursuing master's degree at Kastamonu University and working as a research assistant at Bartın University.

Shabina Fred Rishma completed her BE, in Anna University and ME from Sathyabama University, Chennai. B.Lib.Sc. from Alagappa University. She is in the teaching profession for the past 3 years. Her area of interest in Machine Learning & Artificial Intelligence.

Utku Güdüloğlu completed his undergraduate education in Electrical and Electronics Engineering at Ankara Yıldırım Beyazıt University in 2021 and his master's degree at Kastamonu University in 2024. He is currently pursuing his doctorate at Bolu Abant İzzet Baysal University. Since 2023, he has been working as a research assistant in the Electrical and Electronics Engineering Department at Bartın University.

Sudarvel J. working as an Assistant Professor of Management at Karpagam Academy of Higher Education. He produced five Ph.D. Scholars and four Ph.D. Scholars are under his guidance. Further, to his credit he published 51 articles in Scopus indexed journals and published more than 50 articles in international journals. He is acting as a Subject Experts in Board of studies of various autonomous colleges.

Jaishree Jain, Associate Professor in Computer Science & Engineering Department, Ajay Kumar Garg Engineering College, Ghaziabad, affiliated to AKTU, Lucknow, Uttar Pradesh, INDIA. She have 14 years teaching experience in CSE & IT Departments that include Chandigarh University, Punjab and college of AKTU, University, Lucknow since July 2010. She had done her B.Tech. in Information Technology from Uttar Pradesh Technical University, Lucknow. She had done her M.Tech. in Software Engineering from MNNIT, Allahabad. She had qualified GATE two times in 2007 & 2008 with 94.86 percentile. She had done her Ph.D. from TMU, Moradabad in 2023. She had Published more than 50 research papers in SCI/SCOPUS/ UGC-Care/ UGC/National and International Journals, Conferences and 5 Book chapters published in International Book. She has received several best paper awards for her research papers at various international conferences. She had also been Published 30 patents with 5 patents granted. She is also the author of three Books Published by International standard Publisher. Her research interests include Image Processing, Cloud Security, and Watermarking. She is also the reviewer of Journal of Supercomputing, IEEE Conferences which is listed in Springer Journal and reviewed many papers till now. She has been the session chair of many IEEE Conferences and other International Conferences. She is the member of professional bodies' i.e. life time member of ISTE, lifetime member of ICSES, IEEE Society and member of Engineering Council of India (ECI) and position entitled as a "Professional Engineer" by ECI. INDIA.

K Sathwik Reddy is currently pursuing my B.Tech in Computer Science and Engineering with a specialization in Artificial Intelligence and Machine Learning at Sreenidhi Institute of Science and Technology. My passion for technology, especially AIML, has been growing ever since I began exploring how machines can be trained to think, learn, and solve problems just like humans. With a strong foundation in Python programming, I enjoy learning about intelligent systems and how they can be applied to areas like healthcare, automation, and smart data analytics to solve real-world challenges. I believe in learning by doing, and I constantly seek out new tools, techniques, and ideas that push the boundaries of what AI can achieve. I love asking meaningful questions, experimenting with new approaches, and working on solutions that can create a positive impact. In the future, I plan to pursue higher studies and involve myself in research and innovation in AI. My long-term goal is to contribute to technologies that not only drive progress but also promote ethical AI development, social impact, and global sustainability. For me, this journey is about more than just learning; it's about growing into a professional who can build, innovate, and lead in the world of AI.

Mrignainy Kansal is working as Assistant Professor at Ajay Kumar Garg Engineering College, Ghaziabad, India. I have 5.5 Years of Teaching experience. Currently pursuing my PhD (Computer Science and Engineering) from Netaji Subhas University of Technology, New Delhi, India (A Central University). I completed my M.Tech. (Information Technology) from Guru Gobind Singh Indraprastha University, New Delhi, India in 2018 and B.Tech. (IT) from UPTU, Lucknow in 2015. I have published research papers in reputed international Journals / Conferences. SCOPUS and UGC Approved Journals index my research papers. I also published 4 book chapters in the IGI Global Book series. I have attended more than 10 FDP, Workshops or Seminars. I qualified the GATE examination 2 times.

Arun M. R. completed his BE, ME and PhD from Anna University, Chennai. He completed PG Diploma (PGDEEM) from Annamalai University, M.Sc. Counselling Psychology from TNOU University, M.A. Public Administration from Madras University and B.Lib.Sc. from Alagappa University. He is in the teaching profession for the past 17 years, published around 79 papers in Journals, Patents and Conferences, Active participant crossing more than 130 events and internationally networked through various professional bodies. He is skilled and technically proficient in various simulation software associated to Research in Engineering. He organized and coordinated around 15 events in working institution. Authored 9 Books, 3 Book Chapters and he has been technically integrated in various international journals as an Editorial Board Member and Reviewer. His area of interest includes Video Processing, Remote Automation, Wireless Communication, Quantum Computing Machine Learning & Artificial Intelligence.

T.C. Manjunath did his B.E. from RV College of Engg. in 1989 (B'lore Univ), M.E. in Robotics & Automation from Govt.'s LD College of Engg., (Gujarat Univ., Ahmedabad) in 1992 with FCD-First Rank & PhD in Systems & Control Engg. from IIT Bombay in the year 2007, got best research scholar award for the entire IITB. He is currently working as the Dean Research (R & D) and as Professor in Dept. of Computer Science & Engg. with specialization in IoT, Cyber Security, Blockchain Technology in Rajarajeswari College of Engineering, Bengaluru, Karnataka. He has got a teaching (academic), research & administrative experience of more than 35+ years (10 years as Principal & 8 years as HoD, 1 year as Dean Research & rest as faculty) and has published more than 1500+ research papers in various national & international journals & conferences, with 175+ patents, 100+ book-chapters, 25 awards, 100s of Keynote talks and also published 25 text books in various field of science, engineering & technology, completed projects worth Rs. 9 Crores till date & guided 10 PhD scholars in VTU. He is a member of 30 Professional Society Chapters and is a fellow of IE, IETE, AMIEE, Sr. Member of IEEE and a chartered engineer.

Pooja Mishra is pursuing Ph.D in Electronics & Communication Engineering from Indian Institute of Technology Jodhpur. She has received her M. Tech. degree in Electronics Engineering from Indian Institute of Technology Delhi in 2013. She has a total work experience of 12 Years. She joined Ajay Kumar Garg Engineering College in August 2022 and is working as an Assistant Professor in Electronics & Communication Engineering Department of Ajay Kumar Garg Engineering College, Ghaziabad (U.P). Her research interest is in Communication, Wireless Communication and MIMO Systems. She is currently working in the area of Optical Fiber Communication.

Selvakumar P. is currently an Associate Professor in the Department of Science and Humanities at Nehru Institute of Technology, Coimbatore, Tamil Nadu, India. Prior to joining Nehru Institute of Technology, he held positions as an Assistant Professor in the Department of Science and Humanities various engineering colleges in Coimbatore, Tamil Nadu, India, spanning from 2011 to 2023. Dr.P. Selvakumar's academic and research interests lie primarily in the field of phytochemistry. He has made significant contributions to his field, with an impressive record that includes 30 journal publications, 12 patents, and 20 conference papers. Additionally, he has organized 5 seminars and conferences, further demonstrating his commitment to academic and scientific exchange. Throughout his career, Dr. P.Selvakumar has been recognized for his dedication and excellence in teaching, having been honored with several awards, including the Best Teacher Awards. He holds a prominent position in the academic community, contributing actively to research.

Bhuvaneswari R. working as an Assistant Professor of English at Dr.Mahalingam College of Engineering and Technology, Pollachi. To her credit she completed 10 NPTEL / SWAYAM and published few articles in Scopus indexed journals.

Velmurugan R. presently working as Professor, Department of Commerce, Karpagam Academy of Higher Education, Coimbatore. He started his career as a Project Fellow under UGC Sponsored Major Research Project at NGM College, Pollachi. He has 19 years of teaching experience has 15 years of research experience. To his credit he produced 24 Ph.D. Scholars and 5 M.Phil. Scholars. At present two scholars submitter his thesis, 6 Ph.D. Scholars are under his guidance. Moreover, to his credit he published 168 articles in various International Journals, of which 107 articles are published in Scopus Indexed Journals and 15 articles in Web of Science and 76 articles in edited volumes. He also participates and presented 27 papers in various International Seminar / Conference and 54 National Seminar / Conference. He acted as an external examiner for various M.Phil. and Ph.D. Viva voce examinations. He acted as a Vice Chancellor nominee for Doctoral Committee

Expert at Bharathiar University, Coimbatore and Periyar University, Salem. He acted as a Chairperson at various National Level Conferences. He also acted as an Expert for Syllabus Scrutiny Committee, Question Paper Scrutiny Committee and Expert members for Board of Studies at various reputed Colleges. Considering his meritorious contribution, he received awards like Best Paper Presenter, Best Research Supervisor and Best Faculty Awards from various research foundations and councils. He too acted as a resource person for 27 Guest Lectures, 30 Workshops and 10 faculty development programmes.

Ravikumar R. Natarajan is an experienced academician with over 12+ years of teaching and research expertise across institutions in India, the UAE, and the Maldives. He currently serves as an Assistant Professor at Marwadi University, Rajkot, specializing in Machine Learning, Recommender Systems, and AI-driven applications in healthcare and security. He holds a Ph.D. in Machine Learning and Recommender Systems from Amity University, Jaipur, a Master of Engineering in CSE from VMRF University, Salem, and a Bachelor of Engineering in CSE from Anna University. He also has a MiniMBA from the Swiss eLearning Institute. Dr. Ravikumar has authored more than 8+ multiple SCIE/Scopus-indexed journal articles, 25+ IEEE/Springer conference papers, and co-authored a book on Artificial Intelligence. He has also published 9 patents and contributed 10+ book chapters in emerging domains such as Federated Learning, Blockchain, and Digital Twins for IGI Global, Wiley and Springer publisher. He actively mentors students through initiatives like the Research Activities Club (RAC) and Student Research Internship Training (SRIT), promoting innovation and research culture.

Aarthi S is a dedicated Assistant Professor in Computer Science with over six years of teaching experience. She is currently pursuing her Ph.D. at Marwadi University, Gujarat, and holds an M.Phil. and MCA from Bharathiar University, Tamil Nadu. Throughout her academic career, Aarthi has specialized in programming languages such as C++, Java, and C, and contributed to research in recommender systems and AI-related data privacy. She has published journal articles, contributed to book chapters, and actively participated in technical workshops. Known for her student mentorship and engagement in academic events, Aarthi remains committed to fostering innovation and excellence in computer science education.

Deepika S is currently an undergraduate student pursuing B.E.in Computer Science and Engineering(Artificial Intelligence and Machine Learning) at KPR Institute of Engineering and Technology, Coimbatore. My academic interests include machine learning, computer vision, deep learning and generative AI and I actively involved in research and project development focused on real world applications such as wildlife conservation and accessibility technologies. I'm passionate about using AI for societal impact and aims to contribute to innovate, sustainable tech solutions.

Yogaazhaki S is a pre-final year student currently pursuing a Bachelor of Engineering specialization in Artificial Intelligence and Machine Learning at KPR Institute of Engineering and Technology. I'm actively involved in AI research and development, with hands-on experience in projects. I completed an internship at Silicon Software Services, Madurai, where I gained knowledge about Java Programming. My interests include generative AI, computer vision, Machine Learning, deep learning, and Full-Stack development. I aspire to contribute to impactful innovations in the field of artificial intelligence.

Pancham Singh (IEEE) is working as an Assistant Professor in the Information Technology Department at AKGEC, Ghaziabad since 2007. He has more than 18 years of teaching experience and 1 year of industry experience. He completed his B.Tech. in Computer Science & Engineering from Dr. A.P.J. Abdul Kalam Technical University (formerly UPTU), Lucknow, in 2005 and M.Tech. in Information Technology from RTU, Kota, Rajasthan in 2013. He is currently pursuing his Ph.D. from Netaji Subhas University of Technology (NSUT), New Delhi, which he started in January 2023. Mr. Singh has written and edited five books related to Computer Science. He has presented or published over 45 research papers in well-known international journals and conferences. He has also reviewed more than 100 research papers. He has published over 35 national and international patents, along with 6 design grants and 1 utility grant. He has served as a session chair in many conferences and attended more than 50 Faculty Development Programs (FDPs) at both national and international levels. His main research areas include Machine Learning, Deep Learning, Blockchain, Internet of Things (IoT), and Software Engineering.

K. Sundara Velrani received her Ph.D. from Anna University 2022. She obtained her M.E Degree in Computer Science & Engineering from Annamalai University, Chennai, India in 2005. She has 12 years of teaching experience. She is currently working as an Associate Professor, in Department of Information Technology, School of Computing, Sathyabama Institute of science and Technology, Tamilnadu, India. She has published many research articles at various WoS, SCI & Scopus Journals, National, International, IEEE Conferences, Textbooks, Book Chapters, Patents (Publications & Grants) and Indian Copyrights obtained. Her area of research interest includes Network Security, Machine Learning, image processing, Cloud computing, Service Oriented Architecture and Artificial Intelligence.

Ravi Thirumalaisamy working as an Assistant Professor, Department of Business and Economics, Modern College of Business and Science. To his credit, he published more than 27 articles in Scopus indexed journals.

Lingala Thirupathi is a seasoned academician and researcher with over 18 years of experience spanning teaching and industry, specializing in Artificial Intelligence, Cybersecurity, and the Internet of Things. He is currently serving as an Assistant Professor at SNIST, Hyderabad. A Gold Medalist in M.Tech (Software Engineering) from JNTUH, he also holds a B.Tech in CSIT and multiple professional certifications, including Cisco Instructor, Oracle Certified Associate, and Salesforce Ranger. With over 30 research publications in reputed journals such as Scopus, IEEE, Elsevier, and Springer, he also holds one Australian and three Indian patents. Mr. Thirupathi has authored three textbooks and contributed to over 15 book chapters. He actively participates in academic and technical communities as a reviewer, committee member, and SPOC for Smart India Hackathons, and is affiliated with professional bodies like CSTA, TERA, and ICSES. His dedication to innovation and excellence has earned him accolades, including the Expert Instructor Award from Cisco Networking Academy.

Index

C

D

E

F

G

I

L

M

N

O

P

R

S

U

V

W

www.ingramcontent.com/pod-product-compliance
Lightning Source LLC
LaVergne TN
LVHW061526290926
849162LV00004B/686